Underground Escape

by Masanobu Tsuji

David G. Thomas, Editor

<u>and</u>

The God of Strategy, A Biography of Masanobu Tsuji

By David G. Thomas

World War II History Series, Vol 2

Doc45 Publishing, Las Cruces, N. M.
books@doc45.com

To obtain books, visit:
doc45.com

YouTube Channel
youtube.com/c/Doc45Publications

Cover artwork by Dusan Arsenic

ISBN 978-1-952580-22-2

000h

DOC45 PUBLISHING

Mesilla Valley History Series

La Posta – From the Founding of Mesilla, to Corn Exchange Hotel, to Billy the Kid Museum, to Famous Landmark – by David G. Thomas

Giovanni Maria de Agostini, Wonder of The Century – The Astonishing World Traveler Who Was A Hermit – by David G. Thomas

Screen with a Voice – A History of Moving Pictures in Las Cruces, New Mexico – by David G. Thomas

Billy the Kid's Grave – A History of the Wild West's Most Famous Death Marker – by David G. Thomas

Killing Garrett, The Wild West's Most Famous Lawman – Murder or Self-Defense? – by David G. Thomas

The Stolen Pinkerton Reports of Colonel Albert J. Fountain Investigation – David G. Thomas, Editor

The Trial of Billy the Kid – by David G. Thomas

The Frank W. Angel Report on the Death of John H. Tunstall – by David G. Thomas

Water in a Thirsty Land – by Ruth R. Ealy, David G. Thomas, Editor

"Dirty Dave" Rudabaugh, Billy the Kid's Most Feared Companion – by David G. Thomas

World War II History Series

Torpedo Squadron Four – A Cockpit View of World War II – by Gerald W. Thomas

Torpedo Squadron Four – Photo Supplement – by Gerald W. Thomas

Incident at Ple Tonan – An Imperial Japanese War Crime and the Fate of U.S. Navy Airmen in French Indochina – by David G. Thomas

Carrier Warriors, The Men of Air Group 4 in World War II – by David G. Thomas

Underground Escape and *The God of Strategy,A Biography of Masanobu Tsuji* – by Masanobu Tsuji and David G. Thomas

New Mexico History Series

The Academic Ecosystem: Issues Emerging in a University Environment – by Gerald W. Thomas

A Winding Road To The Land of Enchantment – by Gerald W. Thomas

Mesilla Valley Reprints

When New Mexico Was Young – by Harry H. Bailey

Dave Rudabaugh, Border Ruffian – by F. Stanley, David G. Thomas, Editor

Table of Contents

List of Images

Underground Escape

by Masanobu Tsuji

David G. Thomas, Editor

Masanobu Tsuji, 1943 photo, Records of Bob Wurth, John Curtin Prime Ministerial Library, Perth, Australia.

Introduction to 2025 Edition by Editor

When *Underground Escape* was first published in 1950, it became an instant bestseller in Japan – a mixture of confession, manifesto, and survival narrative by one of the most controversial military warriors of World War II. Long out of print, it is now available again, inviting today's readers to peer into the mind of a man who witnessed and shaped some of the most decisive events of World War II.

On August 15, 1945, Japan surrendered unconditionally to the Allies. In Bangkok, Thailand, Lieutenant Colonel Masanobu Tsuji – then stationed with the 39th Japanese Army – listened as Emperor Hirohito's voice declared the end of the war. Tsuji, long associated with Imperial Japanese militarist policies and implicated in wartime atrocities, immediately grasped what this meant. "The fact that I had gone underground, without committing suicide after the war," he wrote later, "was an action sufficient to spur my wife to tell me how to die." The very next day, disguised in the robes of a Shinshu Buddhist priest and using the alias Norinobu Aoki, Tsuji vanished.

What followed was an extraordinary three-and-a-half-year odyssey through Southeast Asia and China, during which Tsuji not only avoided arrest as a suspected war criminal but also managed to document the collapse of Japanese power and the rise of Communist China with remarkable prescience and narrative force.

Tsuji's observations are raw, vivid, and at times disturbing. He offers stomach-turning details about his physical suffering – "Soon I was seized with spasms of vomiting and my pulse beat exceeded 120 a minute. It was another attack of acute malaria." With no medical care, he recorded how "my urine turned red as blood," and, recalling an earlier diagnosis, added grimly: "I was told at the Army Medical School I would die."

But illness was only one challenge. Hounded by fear of arrest, unsure whom to trust, and deeply shamed by the consequences of his disappearance, Tsuji wrestled with doubt and disgrace. "It was painful to think that people were cursing me as a coward," he admitted. "I wondered how long I would have to bear with this."

He passed through ruined cities and burned villages, lived among Chinese peasants, wretched laborers, and displaced Japanese, and bore witness to atrocities committed not only by Imperial forces but by the Soviets and Nationalists alike. His account of Soviet occupation in Manchuria is unflinching: "Such acts, at which even beasts would turn aside their faces, were carried out in broad daylight… The true characteristics of the Soviet soldier were to be seen in their view that it was no crime to rape women in broad daylight…." [1]

Underground Escape is also a prophetic geopolitical document. Tsuji saw early on the rotting decay of Chiang Kai-shek's regime and the inexorable rise of the Chinese Communist Party. "I realized that there was no way left of saving the Chinese Nationalist Government," he wrote. "There was no meaning left in my continuing longer to be fed by the Defense Department." He described the Nationalists as "weak in military efficiency, but very good at amusing themselves and in making money," and was struck by the fluidity with which Nationalist troops defected: "The reason why… they should ally themselves so easily to be disarmed arises from just this sort of feeling."

He was equally critical of Chiang himself: "The Generalissimo became a man on clouds, far removed from this earth. Below him… hundreds of devils ran rampage… the chief leaders of this evil band were the men closest to the Generalissimo."

The book's most astonishing moment may be Tsuji's handwritten letter to Chiang Kai-shek, composed not with ink, but with his own blood: "I washed the lid of a tea-cup, prepared a new brush and then with a straight-back razor cut deep into my left thumb…. Holding back the surging blood... I wrote to the Generalissimo." It was a plea for understanding, not only for himself, but for Emperor Hirohito, whose true intentions, he claimed, were twisted by reckless officers during the war. He urged Chiang to look past wartime bitterness and embrace a vision of Japanese-Chinese cooperation as the foundation for Asia's future.

In the same breath, Tsuji excoriates Communist ideology: "Marxism views men as machines and is not surprised at the possession of women as common property." Yet he noted the discipline and effectiveness of the Communist forces and foresaw what would come. His prediction was chilling: "World War Three will occur sometime between 1950 and 1955… The Soviet Union will also succeed in making the atom bomb itself."

But there is also the mundane and grotesque. He described a water seller drawing drinking water from a bomb crater with a dead dog floating in it. He wrote of sleepless nights plagued by "succeeding waves of blood thirsty bed bugs," and recalled the most humiliating hardship of his stay in his quarters in Nanking: "There was not a single toilet…. Chamber pots were used indiscriminately…. In my life… this was the hardest ordeal I had to bear."

In quieter passages, Tsuji wrote with longing for his family. "One of my daily worries was how my fatherless children were faring…. I did not even know where they were…. Or perhaps, the whole family had ended their lives in mass suicide." These confessions reveal a man tormented not just by external persecution, but by the collapse of his world, his ideals, and perhaps even his soul.

Tsuji is a problematic figure. He was unapologetic about his role in the war, argued that the guilt for war crimes should be borne by all officers – "It cannot be claimed that company and platoon commanders… were absolutely without sin" – and returned to Japan determined to shape post-war Japan in his image, becoming a member of the Diet. Yet *Underground Escape* is not an exercise in self-exoneration. It is, instead, a rare, candid, often brutal look at war's aftermath, filtered through the eyes of a man who had reasons to stay hidden.

This edition brings back into print a remarkable and controversial Japanese wartime memoir. It is an extraordinary personal narrative, a meditation on political collapse, and a first-hand account of the birth of a new China. Whatever one's judgment of Tsuji's wartime conduct, *Underground Escape* is crucial reading for anyone seeking to understand the chaotic years between the fall of the Japanese Empire and the Communist victory in China.

In 1961, during a "private" trip to Southeast Asia, Tsuji mysteriously disappeared. Nothing certain has been learned of his fate since.

For a biography of Tsuji's life and what details are known of his wartime service, see the second volume in this book, *The God of Strategy, A Biography of Masanobu Tsuji.*

Foreward by Masanobu Tsuji

It was the summer before the defeat of Japan when I was chased out of the Supreme Headquarters of the Japanese Expeditionary Forces in China at Nanking. Then, as a staff member of the 33rd Army, I found myself on the battlefields of North Burma immediately after the failure of the Imperial Campaign.

Following a year of bitter and unsuccessful fighting, I nurtured my wounds in the steaming jungles of Burma, and wracked my brains for some means of saving friendly forces left behind on the west banks of the Sittang River when the center of gravity of the war in Southeast Asia seemed to shift to within the boundaries of Thailand.

In the first week of June 1944, I was transferred without previous warning to a staff position with the Japanese garrison forces in Thailand. With my right arm in a sling and with dragging heels, I bid farewell to my colleagues and soon found myself in Bangkok, capital of Thailand.

In contrast to the fury of battle in Burma, Thailand, lulled into a sense of security and quiet, I, with the smell of blood and battle still clinging to me, was not a welcome guest.

The Japanese Army anticipated a revolt by the Thailand police and army forces. To nip the threat of revolt in the bud, the Japanese Command had decided on a rough surgical operation–the disarming of the Thai forces. And, it seemed that I had been chosen, and extracted from my post as the most suitable doctor to undertake the task. Such a plot against all morality and in contravention of the principles of the Japan-Thailand alliance could not be allowed.

With only a force of less than 10,000 demoralized and lax soldiers, the Japanese garrison headquarters dreamed that it could disarm some 150,000 Thai soldiers and police. It awoke in time, in face of a threatening crisis, tightened its discipline and began an extensive drive to train troops and prepare for battle. At the same time as achieving preparedness against any eventuality, the garrison headquarters placed greater confidence in the people of Thailand. In less than two months, the Japanese garrison forces were reborn into a fine army and regained confidence as an army.

Then came August 15–the tragedy of the Japanese Nation.

Shouldered with the responsibility of the crime of defeat, worthy of ten thousand deaths, I changed my uniform for the garb of a priest and let myself be swallowed into the great Asiatic continent. This I did in the hope of making amends for even a part of my sin.

This volume is the record of the following 7,500 miles of travel in disguise–a trip that look roughly three years, as I made an underground escape.

I liquidated my past as a career solider–I , who had enjoyed through the power of my uniform an authority and prestige far beyond my intrinsic worth within the organization and the tradition of a great army that had once boasted of invincibility, was a member of a defeated nation, shorn of all personal guarantees. I managed to escape from one corner to another eluding foes and swift pursuers. My very life hung on a strand of rosaries as I continued my stealthy travels.

This is a log of that journey of escape. The sights I saw and the things I heard in an Asia writhing and struggling in a storm of Communism, form the contents of this book.

The fate of Japan whether Japan so desires or not, is influenced most sensitively and most deeply by the repercussions of events on the neighboring continent. The aim of this work has been to depict in naked reality the picture of Asia as I saw it, not with the eyes of a victor national, but with the sympathetic suffering of one that tasted bitter defeat. I would deem it an unexpected happiness if from these pages the throbbing feelings and aspiration of the peoples of Asia can be caught.

This work was published last spring by the *Mainichi Newspaper*. However, due to certain circumstances, the book was allowed to go out of print midway. Since then, I have been urged from all sides to reissue the volume. As a result of changes in the objective world around, I now am able to include various facts which I could not in last spring's edition, and happily present this complete volume to the world as a final work.

Several of my colleagues and seniors have died on the gallows as war criminals, and many still live a life of suffering behind prison walls. I who have sinned in no less degree than these colleagues and seniors, escaped and lived in disguise for three years. I feel that I have done great wrong in escaping my responsibilities. Now, when the grim shadow of an approaching atomic war, I have no other wish than to present this work as reference material to those who worry about the future of Asia.

Looking back I realize that we Japanese brought no small measure of suffering to our Asian brothers during the long years of the war. But, to another degree, were also sown seeds of good. Today, these seeds have sprouted to provide Asia with a consciousness that forms the prime force behind her awakening. None can deny this in face of the realities of Asia's problem today. The friendships proffered me by the youths of Asia during my 7,500-mile travel in disguise is a concrete example of this fact. At the same time as being grateful for this, we Japanese's should liquidate our past mistaken sense of racial superiority–a superiority which we felt only toward our brothers of Asia. Then, with a feeling of mutual pity for those afflicted with the same malady, we should through mutual trust and mutual assistance join forces with our one billion brothers of Asia in preventing the world from being dragged into an atomic war. Surely, this is the destiny of we Asians.

What struck me most forcibly during my long travels on the continent was the sight of internecine bloodshed in all parts of Asia. China, which emerged victorious from the war, is torn against the wishes of her people, by a bloody struggle several times the magnitude of her eight-year war against Japan–the struggle between the Chinese Nationalists and the Chinese Communists. Likewise, the Korean race, which for 38 years was of one nationality with us, is crushed again by a red and white juggernaut that rolls up and down the peninsula, tearing and smashing all in its path. In sharp contrast, Germany, though the defeated in war and divided into east and west, has been held back through the sensibleness of her people from engaging in internecine war.

In defeated Japan, not a gunshot is heard. Is this the result of the wakening of the Japanese people? Perhaps, I am not alone in my presentiments of threatening clouds that might split Japan into two camps and bring down torrents of blood. Which should be all preceding–"isms" ideologies or the common feelings of a common race?

Japanese should not shoot Japanese. Asians should never shoot Asians. This is the first step toward ultimate peace for the whole human race.

January 25, 1951, upon the eve of Japan's greeting Ambassador John Foster Dulles, ***the author***.

Chapter 1 | Man To Man–1945

It was a lonely send-off which the jungle headquarters gave me as I left the dusk-wrapped hamlet of Bilin by truck. When I reached the Moulmein Area Army Headquarters, it was well past midnight. Staff Officer Aoki welcomed me warmly and led me to his quarters. There I flung my tired body into bed and snatched a few short hours of sleep on my last night in Burma [Myanmar]. Early next morning, we sped down the squall-drenched Moulmein Highway to the airfield. My wounds pained to the bones each time the car hit an uneven spot on the road. Staff Officers Aoki, and Lance Corporal Kubo, who had looked after me the last six months as my orderly, were the only ones to see me off.

When I reached the airstrip which has previously been cleared from the jungles tenacious clutches, I found a small two-seater plane awaiting my arrival, its engines roaring impatiently.

My bone-weary body with half-useless hands and feet was lifted bodily from the ground and deposited in the narrow cockpit. The pilot, obviously a veteran flyer, paid scant attention to the steaming squall and brought the plane down the runway. In a haze of spray from mud puddles on the airstrip, the pane lifted itself form the soul of Burma.

I had only been in Burma short of a year, but I felt a pang of nostalgia at leaving the mountains and rivers where I had fought and suffered. The plane knitted its way eastward through openings in the dirt-grey rain clouds.

Time and time again, air pockets sought to suck us down close to the treacherous terrain.

Far below, the silver thread of the Salween River thinned into the distance. Rising in the mountain fastness of Yunnan Province, it flows past the La-Meng fortifications, where an isolated army fought for 120 days steaming an enemy several times its size, only to fall to the last man [Salween River Offensive, May 10 1944]. It seemed the yellow-brown turbulence of the Salween was colored by the blood of countless men.

The bloody battlefields of South Burma, the site of a hundred days of hard fighting, spread its panorama before us. Try as I did to wipe its spread from my eyes, it refused to fade.

We crossed the mountains forming the Thai-Burma boundary. Anxieties over a new task took the place of memories to suggest a similar course by the entire Thai Army. Should n Southeast Asia would lose its supply base.

Two weeks ago, I had been summoned to the Supreme Headquarters at Saigon. The drift of the conversation of staff officers indicated that they were thinking of surprise disarmament of the Thai police and army forces.

I pondered to myself what new and heavy task heaven intended to place on this derelict soldier, a defeated warrior with arm in sling and leaning heavily on a green bamboo staff for support.

As long as there was logic, the heavier the duty the more welcome it would be. But, I hated to chose to be the perpetrator of a dastardly deed–to stab a weak Thailand in the back in disregard of the principles of our alliance. Dark intrigue was absolutely against my nature.

With dark forebodings filling a wounded body, I landed at Don Maung Airfield in the northern suburbs of Bangkok.

Staff Officer Yano was at the airfield to greet me, he had brought a stretcher. It was only four o'clock in the afternoon, but already all the officers at headquarters had gone home. I was advised that I was to report to headquarters the next day. For the time being, I had to content myself with reporting for my new duties at the quarters of Chief of Staff Hamada.

I was treated to ice-cold beer and salted rice cake pretzels, brought all the way to Thailand from Tokyo.

Unpretentious, blunt and trustful as Chief of Staff Hamada was, the military leaders of that time did not recognize his qualities. But he was trusted with absolute confidence by the leaders of Thailand.

I was taken to my quarters. It had been the home of a staff officer who had recently been transferred to the Kwantung Army in far-away Manchuria. It turned out to be a gleaming palace with marble stairways and toilets. Numberless times, I slipped and stumbled on the polished floors. The double bed in the spacious bedroom swallowed half my body into its folds. I found I could not sleep well in such luxury. The jungles had been my home, torn tents my bedding and the wild grass my food. I could not get over the sudden switch from my life hardship on jungle battlefields to this unlooked for comfort. I wanted for a jungle animal like myself a home fit for a jungle ape, a home where I would not be slipping constantly.

The next morning I rode to work in an automobile. I felt I was taking unfair advantage in using an automobile to ride a distance of only five or six blocks.

"It's only until your wounds heal," they said, and so I complied and got into the car. However, wonder of wonders I found that this habit of using a car for even the shortest distance was not limited to staff officers alone. There were hardly any officers walking along the streets. This was especially so in the case of officers above the rank of major. It seemed that every officer had a car at his disposal. Sedans, flying small red and yellow flags, glided through the streets of Bangkok. And seated within, in majestic indolence were officers setting off to the headquarters or returning to their quarters. Official cars were used for shopping and for geisha parties, without discrimination or compunction.

I reported my arrival to Commander [Aketo] Nakamura. I had met the general several times before when he was Commander of the Gendarmerie. He had also been my instructor at the Military Academy. He was thus an old and respected senior.

He rose slowly from his chair, and a merry twinkle filled his eyes; small in proportion to his size like the eyes of an elephant.

"Welcome. I was waiting for you. Now that you're here, everything is going to be fine. I'm counting on you… and, how about your wounds?"

I had heard by chance that before I was chosen for my new job, Supreme Headquarters had questioned the choice of me. The answer had been a clear, "Not welcome." I had steeled myself to a cold greeting. Contrary to expectations, however, I was being told:

"We waited for you. We are counting on you!"

To this, I had nothing to answer.

"It's the Emperor's orders. Yes, I'll do my best–for my land and for my country," I thought. It was no time, with the war at such a critical stage, to allow myself to indulge in private feelings, and criticisms.

I resolved to help the Commander overcome the crisis and never let the atmosphere of the headquarters, no matter how hostile it might be, bother me in the pursuit of my duties. My classmate at the Military Academy, Staff Officer Konishi, welcomed me warmly. I have never forgotten to this day his friendship in mollifying the hostile feelings surrounding men of equal rank.

Ambassador Kumaichi Yamamoto was an old acquaintance of mine from my days in the Kwantung Army. Despite the differences in our positions and professions, I went to pay him a call as one whom I respected. The sparse hair on his elegantly balded head was beautifully combed. I found in his eyes and in his mustache a deep nostalgic attachment. He shook my hand firmly and spoke in a quiet voice:

"We appreciate all your hardships in Burma. The situation here in Thailand has become extremely critical. All I hope to do is to abide my principles and die at my post. I remember words which you once spoke in China, 'The only way to approach the Chinese people is to trust them implicitly. Even if you are fooled twice, even if you are cheated thrice, continue trusting them without showing anger. You will eventually meet up with Chinese who will never fool or cheat you.'"

"Even if we are thrown overboard by the people of Thailand, I would not like to play a foul trick on Thailand. I would like to leave as a final gift in Thailand, the remembrance of the trustworthiness of the Japanese people."

What an admirable attitude, I thought. He, surely, is the true Japanese diplomat. I could feel in his words his strong determination never to shame the Emperor and to die for his principles. If this Ambassador were only the commander, how easy my task would be. Even if he had no military experience or knowledge, I felt he was a man who would fill the post of commander even in the darkest hours.

There was a small Shinto shrine behind the official residence of the commander. A wooden plaque, inscribed with clear black calligraphy gave its name as ""Daigi Shrine" (Loyalty Shrine).

On June 8, I attended my first monthly rites there. Every single member of the headquarters form the commander to privates, as well as representatives of Japanese residents in Bangkok, used to attend these ceremonies held early in the morning.

Within the shrine compound, the officers were lined along the right side facing the shrine and the Japanese residents along the left side. Outside the "tori" were lined the non-commissioned officers and the rank and file.

Among the reservist officers called to colors were men who were Shinto priests in civilian life. These men presided at the rites. As they solemnly chanted the Shinto prayers, dressed in the age-old white robes and tall headgear, I noticed a group of young women lined opposite the officers signaling the officers with sensual eyes. It was a sight unbecoming to the solemnity of the sacred compound.

Upon my return from the shrine, I inquired as to the nature of the girls. I learned they were waitresses in the officer's club and girls from the army-operated officer comfort stations (brothels).

These women in a sense might have been performing certain duties of their own. They were, I saw no reason why they should extend their sweet dreams of the night before to the sacred compounds of the shrine the next morning and speak of love with their eyes to the officers. I asked the reason for their presence at the ceremonies form a senior aide-de-camp. He replied that it was the orders of the commander. At the next regular monthly rites, I quietly changed the position of the young women.

But my arbitrary decision immediately stirred up unexpected repercussions. In front of the entire gathering at the shrine, the senior aide-de-camp was being mercilessly scolded.

"No. This is something that the aide-de-camp does not know anything about. The arrangement was changed arbitrarily by the senior staff officer," I explained.

However, the perversity was not easily allayed. While desperate battles raged on the banks of the Irawaddy River, the Japanese Army stationed in Bangkok, the capital city of Thailand, were roused to passion over such a trivial matter. I pondered over the truth of the adage, "A change of place and a change of heart."

Night after night, practically the whole headquarters staff was invited to one party after another. They were given no time to study or rest.

I realized that these parties could not be completely abolished. But it was no easy task to try and limit the attending officers to the men directly connected with the guest of honor.

I felt that at this rate we could never engage in real battle. It might take only one day to lift a person from a low to a high level of loving, but it was not easy to bring down the scale of living.

I felt within me flames of righteous indignation–an indignation fearless of recrimination or of resentment.

The flames of war raging in Burma stretched long tongues of fire toward the Burma-Thai border. Showers of flaming sparks threatened at any moment to wrap Thailand in a mighty conflagration. All the while, the Japanese forces in Bangkok slept on in indolent lethargy. Enemy Fifth Columnists ran rampant underground and had joined hands with the Free Thai Movement under General Aung Sun, waiting for the moment to strike at the Japanese. It was a clear substantiation of old Japanese adage: "Only the husband is blind to his wife's infidelities."

I started the day after my arrival on the task of sweeping aside the torpidity that had seized the Japanese forces in Thailand. I called together the unit commanders and threw my first bomb.

I explained in detail the actualities of the death struggle of Japanese forces in Burma. I stressed that if 150,000 Thai troops and police revolted, it would be the less-than-10,000 Japanese forces that would be disarmed. I pointed out bluntly that American and British airborne troops could land with ease in the suburbs of Bangkok, and I gave directions for immediate and serious preparations for battle, preparations that were to begin that very day and not the morrow.

The next day, we began a study of troop disposition. On the basis of this study an order to begin the construction of fortifications was given out. It came as a bolt form the blue to the Japanese garrison. Geisha houses were closed down and all leaves for private purposes from garrisons and billets were cancelled. Soldiers and officers alike launched into the construction of fortifications. The work was pushed day and night. The barracks of each garrison unit became strong points surrounded by rows of barbed wire. The routes linking these scattered strong points with the nuclear headquarters became a maze of strongly fortified defense lines. The garrison could now sleep in peace despite the possibilities of revolt by the Thai forces.

Hospitals were not excepted. Doctors took spades in hand. Nurses dug trenches and carried earth. Even the patients were set to the work of making sandbags.

The Geisha houses and the restaurants that had been packed the day before were deserted. Only the red lights on their doors shone forlornly at night. Not a single Japanese uniform was to be seen in the streets.

In order to set an example, the fortifications around the Garrison Headquarters building were constructed thoroughly and formidably. For a garrison force that had no experience in fortifications, field supervision was more important than the most ingenious of planning.

In one month's time, the Buddhist capital of Bangkok had been transformed in appearance. At the Don Maung Airfield, concrete artillery emplacements were erected to take care of possible landings of enemy airborne troops. Each munitions depot and each supply go down was surrounded by rows and rows of barbed wire entanglements and numerous concrete pill-boxes. The handful of troops stationed at each point could now hold out for several months against Thai troops and police forces several times greater in number.

Such preparations were not restricted to Bangkok alone. From the smallest–gendarmerie outposts of a few men–to the largest–divisional encampments–every place where Japanese troops were stationed, at key costal points or at communications hubs, all were strengthened by concrete fire points and barbed wire entanglements, erected by the soldiers themselves.

The first week of July was crammed with days of inspection of these new fortifications. The commander, wiping great beads of perspiration from his brow, carefully inspected in detail the completed works. During these days, he lost all his excess weight. Staff officers and soldiers had thrown themselves into the work with heart and soul. Now we were ready to meet any surprise incident.

News of our feverish preparations were undoubtedly reported daily and in detail by Fifth Columnists to the enemy command. But, we did not fear in particular this information reaching the enemy. Our purpose was thorough preparedness to make the enemy give up his plans to land airborne troops, to prevent an uprising of Thai forces and to save Thailand from being dragged into the vortices of war. The only way to avoid fighting is to complete preparations that fear no fighting. This truth gradually dawned upon the Japanese garrison forces, privates and officers, who had been lulled into a false sense of security.

Almost a month of day and night digging of fortifications wrought a change in the appearance of the Japanese forces. Eyes sunk with fatigue and wan cheeks replaced looks of well-fed indolence. But, strangely, the number of sick fell rather than increased.

These preparations, aimed at enabling less than 10,000 troops to defend themselves against 150,000 Thai forces, were camouflaged on the surface as measures to cope with enemy airborne attacks, and every effort was made to eliminate any uneasiness on the part of the Thai peoples.

Two nights after my arrival at my new post in Bangkok, I received a visit from Major Att Chalenshilba.

During the days when I was a company commander at the Military Academy in 1934, he was my sole Thai student. He was a beloved and intelligent pupil, and spoke Japanese with exceptional fluency.

Eleven year's later, I found him a nuclear figure in the staff headquarters of the Thai Army engaged in important negotiations and liaison between the Thailand and Japanese forces. We shook hands firmly, both teacher and pupil deeply moved by this meeting after so many years.

He addressed me as his company commander as in days gone by and said:

"Sir! I am ready to do anything for you. Please let me know without any reservations."

On his face was written a genuine sincerity, transcending national distinctions, and a sincere desire to repay the little kindness that I had once tendered him.

A group in Headquarters sought to utilize this man as a spy to obtain secret information from the Thai Army. And, if I had asked him for old time's sake to engage in such acts of espionage, he undoubtedly would have acted on my request. However, my conscience would not allow me to stoop to such a level.

It was not that I had chosen him for special favors during the days at Ichigaya; I had taught him as any other Japanese student, without making any distinctions because of his race.

"Thank you, I appreciate your good wishes. Italy has been beaten and Germany has lost. It should be more than clear to you what fate lies in store for Japan, now pitted alone against the entire world. The only thing I desire is not to let relations between Japan and Thailand be broken up," I told him.

"I'd like you to believe me, if only on this one point. The relations of two countries at times cannot be influenced by the puny weight of two people. I firmly believe that it is to the benefited both Japan and Thailand not to allow Bangkok to become a battlefield. However, if a time should come when my country and your country will have to fight each other, remember, you will have to fight for Thailand against the most courageous of company commanders. That which I taught you in the lecture rooms at Ichigaya can be summed up in a nutshell–'Be loyal to your country.' This is true today as it was at that time. The friendship between your company commander and his pupil will never change, but this friendship should never take precedence over the welfare of you fatherland. I know that you believe that I am a person that cannot tell a lie. I would like to see that war does not break out between Thailand and Japan."

I saw the color rise to his face. We gripped our hands together in silence and tears flowed down our cheeks.

Several days later, the Chief of Staff of the Thai Army invited Chief of Staff Hamada and staff officers to a banquet. I must have been an unwelcomed guest, with the smell of gore still clinging about me, my heels dragging from battle wounds and my arm still hanging in a sling around my neck.

After I was introduced to the gathering by Chief of Staff Hamada, I gave a simple address:

"My task is to prevent Bangkok from becoming a battlefield. This is not only for the sake of Thailand alone, but also for the sake of the Japanese Army."

The Chief of Staff of the Thai Army then rose and said:

"I have heard from Major Chalenshilba of Staff Officer Tsuji. I know you well and I believe in you. No matter how the world situation may change, I pray from my heart that the friendly relations between Japan and Thailand, both tied together by Buddhism, will not change to the last."

It was a sincere welcome.

Half way through the banquet, this imposing general took out from an inner pocket of his tunic a bag containing a cherished talisman. From this bag, he pulled a gilt statute of Buddha, the size of a thumb joint, and gave it to me as a memento.

I heard that it was a rare gesture for Thailanders to give another person a talisman handed down from generation to generation.

Attending the banquet was a vice-admiral of the Thai Navy, who claimed to be an exceptional phrenologist. He came to me and asked to be allowed to read my features. For a few minutes, he glared ominously at my head and felt the bumps. Then he pronounced my fortune:

"No matter how great the dangers that you may meet, you will never die but will live to a ripe old age and carry out a great work."

Perhaps it was a courtesy speech at a social gathering, but I felt deeply the warm friendship manifested toward me by the Thai military leaders.

Probably I owed all this too Att. However, the prediction made by the fortune-telling vice-admiral that I would not die no matter what the dangers flashed back to mind time and time again later whenever I was brought face to face with death.

After the banquets, we were shown a famous Thai drama.

There was dancing, there was a display of military arts; I sat enthralled by the dazzling brilliance and the inspired skill of the participants, when suddenly a beautiful dancing girl placed a beautiful wreath of flowers around my neck. What an incongruous sight it must have been. Gorgeous flowers around the neck of a wounded and somber soldier. I never felt as abashed as this before.

Prime Minister Apaiyon, who took over the reins of Government after the mass resignation of the Phibul Cabinet, enjoyed the overwhelming confidence of the Thai Parliament. When he spoke, even the Opposition listened to him with close attention and applause.

I was the middle of July when even the trees and the grass seemed to seat with the oppressive heat. The Commander was in the middles of an inspection of the fornications of the garrison headquarters. Suddenly, an urgent messenger from Prime Minister Apaiyon arrived on the scene. The emissary brought the following message:

"An urgent matter has arisen. The Prime Minister requests the Commander to come as soon as possible together with Ambassador Yamamoto."

Just at this time, the Thai Parliament was in the midst of considering the demand of the Japanese Garrison forces for a budget ten times that of the previous year to cover occupation expenditures. Both the Commander and the Ambassador thought that they were called in some connection with this matter. However, they were surprised at the totally unexpected business that awaited them.

The Prime Minister greeted them with a pale face, with blue veins standing out on his forehead and lips bitten tightly together. In a rough tone he said:

"The Japanese Gendarmerie has a man following me. Both of your excellencies are fully aware that I have faithfully adhered to the Japan-Thai Alliance, despite being seriously misunderstood by a portion of the Thai people. But now what do I find? If you have no confidence in me, I shall resign immediately. Please convey my intentions to the Japanese Government."

The Prime Minister's attitude was more than just strong. He was furious with anger.

Stunned by this unexpected blow, the Commander wiped the fine beads of perspiration off his brow and strove to calm the Prime Minister without success. He promised that he would take steps after a careful investigation and returned home. However, there was nothing that he could do. Ever since the surrender of Germany, the gendarmerie had reported almost daily that a group of Thai police and armed forces had secretly succeeded in contacting the British and Americans and were waiting for the opportune moment to stab the Japanese forces in the back. These reports said that the central figure in this movement was Lieutenant-General Adon, Chief of the Thai police. In addition, there were puzzling reports and rumors that ex-Prime Minister Phibul Songgram was the person behind all this plotting. Unrest and anxiety filled the peoples' hearts.

Of course, the Commander himself was well aware that the gendarmerie was keeping a strict watch on the movements of all important Thai leaders in anticipation of any untoward event.

Completely at a loss as what to do, the Commander called together all his staff officers to discuss the possibly of a way out of this impasse. But, of course, there was no quick solution. In face of the critical war situation, the resignation of Prime Minister Apaiyon would stir up Thailand's political world into a hornets nest. The slightest mistake could easily touch off a political fuse leading to the revolt of the Thai army and police forces. It was no wonder the Commander's face seemed twisted with anxieties.

"Isn't there some way out this difficulty?" He said addressing me.

"There's nothing. The only thing is for me to go and apologize. Please leave it up to me," and I took over the touchy task. I added, "There is no other way than to apologize from one's heart. You can't move a man's heart by make-shift explanations or by utilizing third persons."

Thirty minutes after the Commander returned from the Prime Ministers officials' residence, four persons, Staff Officer Yano in charge of Thai affairs, the deputy commander of the Gendarmerie, Interpreter Hatano and myself visited the Prime Minister [Khuang Aphaiwong].

The Prime Minister's home was a small and unimposing structure. I could not bring myself to believe that the Prime Minister of an independent state could live in such a place. There was not even a single police standing guard in front of the residence. I felt that in this respect he was on a higher level than the Japanese prime minister. Entering at the gate, I presented my calling card. We told you that surprise, this youth in a simple sport shirt said;

"I am the Prime Minister."

I could not believe that this unpretentious and rather thin youth was the prime minister who had thrown the Commander into such a fit of anxiety.

We were led into a simply-furnished living room. On the wall opposite the door hung a full-length oil painting. Father and son were like two twins. The father was appointed in the full uniform of a general. I bowed respectfully toward the painting and then took the seat offered me.

"I have been acquainted with you Excellency's name and reputation for some time. I had hoped that someday I would be honored to meet you. However, I did not expect to meet you accidentally over such an unfortunate incident. Up to a month ago, I was engaged in fighting in Burma. Without being fully acquainted with the customs and the situations in your country, I took over the position of high-ranking staff officer only to encounter almost immediately such an unfortunate mistake. In reality, Commander Nakamura has told me that I must guard you unnoticed against all enemy elements because your Excellency is the most respected and most trusted of Thai's leaders. This was to be done form the standpoint of protecting the welfare of both Japan and Thai. I had transmitted the Commander's wishes directly to the deputy-Gendarmerie commander, here with me today. However, I can only apologize for having caused Your Excellency unpleasantness because of the lack of care in looking after the details of this order. This is the work of complete newcomers, unused to the situation here, and I alone am responsible. Both the Commander and the Ambassador knew nothing about it. This is the situation, and I can only hope that you will forgive us this once only."

The Prime Minister who had listened with care to my words, nodding here and there answered in a soft voice:

"This is the first time I have ever met such a blunt person as you. I know everything that is involved, but shall forget about this incident in deference to you sincerity. Incidentally, pardon me for asking, but how old are you?"

"Thank you very much. I am 44 years old."

"Why, you're the same age as I am... From now on, I'd like you to visit me often as a friend...."

"I feel deeply honored. However, Your Excellency is the head of a state, while I am nothing more than a colonel that fought a losing war in Burma. I accept your words as honoring me beyond my station. I would be extremely happy to have you teach me and guide me in the future. The Commander is very worried about this problem and so, with your leave, I would like to return and tell him as soon as possible of your magnanimous forgiveness."

The Prime Minister insisted that I stay longer, but I left and reported on what had taken place to the Commander for the first time showed relief on his face.

Several days later 209-odd American bombers suddenly raided Bangkok.

The planes dropped a hail of bombs in the vicinity of the railway station, but the Japanese anti-aircraft batteries could not reach raiders and not a single plane was downed.

Fire spread in all directions. It was daytime, but the streets were completely deserted. I sped from one anti-aircraft battery to another, encouraging the soldiers. I also went to the battery headquarters for the same purpose. But, when I got there, I found the commander, a major general, and his entire officers having a drinking party in the cafeteria in commemoration of the completion of their reorganization. I forgot my rank and roared at the gathering, asking them what in the world they thought themselves to be, drinking at noon in the middle of an air raid.

The neighborhood of the Prime Minister's residence was wrapped in roaring flames. I felt that now was the time to repay him for his kindness in forgiving my grave mistake. Through flaming streets and dropping bombs, I hurried toward his home. Happily his residence was still untouched. I entered the compound and paid my respects to the Prime Minister's wife who had sought refuge to in an air-raid shelter.

I searched for the Prime Minister. I found the youthful 44-year-old leader of Thai busily brushing away the sparks falling on is clothes and supervising the firefighters and the fleeing people. He seemed oblivious to the falling bombs.

We gripped our hands together in a warm handshake while all about us the fire roared and incendiaries burst.

The subversive activities of British and American agents in Thai were stepped up with increasing fervor with the gradual deterioration of Japan's position in Burma. From intercepted radio messages and other sources of information, we were now certain that secret enemy airfields had been prepared in northern Thai and on the north-eastern sector.

The enemies had learned the taste of successful airborne operations in the Hukong Area in North Burma and were preparing the same sort of operations for Thai itself.

Pointing out that it was an obligation under the Japan-Thai Alliance to strike at the enemy while he was still unprepared, we warned the Thai Army leaders of the enemy's preparations. But, they refused to listen or to do anything. It was now for us to find out by ourselves where these secret airfields were located. On a small plane, piloted by Air Force Staff Officer Koyanagi, I scouted every likely strategic point in Thai and succeeded in finding several secret air strips.

Without disclosing our discovery, we completed preparations to wipe out these air strips when need arose. However, we knew that any rash action would only lead to all-out war and took every precaution in our preparations.

Perhaps, the Thai leaders were unaware of the fact, but there was no room left to doubt that there were elements amount the young Thai air force officers who were in close touch with the enemy.

On the other hand, the extensive construction of fortifications by the Japanese forces in Thai since June had excited the Thai people. This was interpreted as preparations for immediate action against the Thai race and rumors and counter-rumors spread to create uneasiness among the population. We realized that we must do something to allay this popular unrest.

It was July 20. We opened a part of our fortifications for a month's inspections to Thai military leaders, Government chiefs and Japanese and Thai newspapermen. The inspections were limited to the railway regiment, whose fortifications were built most strongly.

To a group of over 200 people, I explained through Major Att:

"The reason why the Japanese forces in neighboring Burma lost was because geisha houses prospered while military preparations were neglected. The only way to prevent Bangkok from being thrown into a crucible of war is to complete thorough preparations and to discourage any enemy attacks. We are throwing open our secret fortification to you on the basis of the principles of the Japan Thai Alliance. Please feel free to look around as you see fit. It is our hope that the Thai forces also will prepare fortifications still stronger than these."

Concrete barracks had been strengthened by loopholes and sandbags and surrounded by several rows of barbed-wire entanglements. In the several holds of concrete strong-points were stored-ammunition and food sufficient to last several months. The defending soldiers lived and slept in mosquito netting in these strong-points.

The eyes of the visitors shone with astonishments. We had placed interpreters among the visitors. The interpreters, dressed in ordinary civilian clothes, kept their ears open but acted like the rest of the party Among the comments of the younger Thai officers were: "The fortifications are really well-built. We wouldn't be able to do a thing against these strong-points." It seemed as if our aims had been achieved. After the inspection, we gave a demonstration of anti-tank mines. The explosions seemed to rock the hearts of the victors even more than they did the surrounding window panes.

As a final gesture of appreciation of their attendance, we invited the entire group to a sukiyaki party in the name of the Commander in the dining hall of the Officer's Club. The group undoubtedly must have felt they were being trusted and at the same time as if they were being threatened. It was natural to expect the reports from the investigation to be transmitted to the British and American forces.

Following this, the Thai forces began to push in earnest to construction of their own fortifications. Barricades were built along all avenues offering tank passage and loop-holes were built into all army barracks.

We had no way of knowing whether these preparations were made in anticipation of British and American airborne attacks or against the Japanese forces. It seemed as if it were a contest of duplicity between the Japanese and the Thai forces. We placed strict restrictions on the Japanese soldiers and demanded uttermost discipline in their contacts with the Thai forces. Through our ever-watchful attitude and constant training and preparations, we won at least the fear and the respect of the Thai people.

Hand in hand with the gradual disintegration of the Japanese forces in Burma, the Japanese Occupation troops in Thai were raised to the status of an "area" army. The step was taken in view of the dangers of Thai becoming the next battleground.

Major-General Hamada was raised to the rank of Lieutenant-General. However, with the raise in the status of the occupation forces to that of an "area army," Lieutenant-General Hamada's position was lowered to that of a deputy-chief of staff. Despite the fact that he was a man of character, trusted by the Thai people and loved and respected by the officers and men under him, his devotion must have been made on the grounds of lack of experience in the past. It was not rare in the personal choices made by the military leaders, to see ignored the true value of an individual and see overestimated the boastful type of solider as the hero of battlefields. Thus it was in this case. The military leaders failed to recognize the quiet determination and the deep sense of faithfulness to duty of General Hamada.

When we had prepared to fight to the last solider in Bangkok and to stand siege in the headquarters and its auxiliary network to fortifications, it was the commander, Lieutenant-General Hamada, the ordinarily quiet and unimposing man, who had made the decision to die together with his men.

His new job as Vice-Chief of Staff was limited to liaison work with the Thai Government. He also acted concurrently as Military Attache to the Japanese Embassy in Thai. The man appointed commander of the Thai Area Army was Lieutenant-General Hanatani, who had fought in Burma as divisional commander. He was the exact opposite of General Hamada in character. The 15th Army Corps which had retreated from Burma was incorporated in the new Thai Area Army. The Northern Thai defenses were strengthened and the Kirmura Division sent us to as reinforcements from French Indo-

China and the Sato Division from China. These divisions were added to the Thai Area Army.

It was not an easy task to add life both spiritually and materially, and to build up a new will to fight in a conglomerate force made up of a defeated army corps, fresh divisions and the original garrison forces.

After we received news of the fight to the last man of Lieutenant-General Ushijima and his entire force on Okinawa with the General leaving the following verse to posterity before his suicide:

Through with the last arrow gone,
My blood dyes heaven and earth,
My spirit shall return, shall return,
To defend the motherland.

The war situation deteriorated daily with accelerating speed. I felt that the only way left was for the 5,000,000-strong Japanese land forces left abroad to fight a long-range war, though cut off completely from the homeland, and to wear down the enemy for another three or five years.

I determined I would be another Nagamasa Yamada in this country of rice, and made preparations to meet the worst. [Yamada was born in Japan in 1590. Travelled to Thailand in 1612. Appointed a military commander by King Songtham. Died in 1630.] It was imperative that we keep Thai an ally as long as possible, if failing that, to keep her neutral. With this in mind, Commander Nakamura thought up the idea of holding a joint memorial service in honor of the war dead of both countries. This plan to hold a memorial service for both Japanese and Thai forces who fell during the brief fight between Thai and Japan in the opening curtain-raiser to the Malay campaign and the numbers who had died outright or from wounds during enemy air raids, was motivated by a deep sincerity and was to have far-reaching repercussions in Thai.

The services were held in all solemnity at the Vat Mahatat Temple. The sense of friendship and love through a common religion helped greatly in relieving the tense atmosphere of Thai, faced as it was with a sense of impending doom.

From the beginning of August, the unrest within Thai became more and more evident. Tension between the Japanese and the Thai armed and police increased daily.

In the midst of the atmosphere, the annual Thai Army Day ceremonies were held in the plaza in front of the royal palace. A whole division of Thai forces was to participate in the event and our Government and military leaders were invited to the festivities scheduled to last through the night in drinking and dancing.

It was an ominous invitation. We had information that the plans for a revolt by the Thai forces were in an advanced stage. There was ground to believe that the nights partying might not end simply in festivities. The commander was worried and pondered whether or not he should accept the invitation and attend.

I proposed:

"Your Excellency, please attend and taken with you all the staff officers. Please attend completely unarmed and unescorted, and please drink and eat and sing the night through with such carefree abandon as to surprise the Thai leaders. I alone shall stay in the operations rooms and shall have everything ready for any crisis."

I made preparations to place all the forces on alert, ready to turn out and take actions at the ring of the bell. Then, with my phone ready in front of me, I spent several tense hours, scanning with careful eyes all incoming reports.

In this way, the Commander and his officers attended this odd party, without a single worry on their faces, while I remained prepared to meet any emergency.

When the Commander and his friend group returned late at night, I could not help but release a sigh of relief.

According to the Commander, the Thai Chief of Staff asked in the middles of the festivities what had happened to the officer with his arm slung around his neck. The Commander answered with a smiling face, as if he knew nothing:

"Oh, that man. He's the chief operations officer. He hates parties. Tonight, he's staying as usual all by himself in the Operations Room. As long as he's there, even if the British and Americans land airborne troops, there's nothing to worry about."

Irregardless of whether the Commander actually places that much trust in me or not, I felt that his answer was a masterpiece.

In this way, preparations for action were pushed day and night, with the aim of making full use of our numerically inferior forces. On the other hand, we took every care to abide by the principles of the Japan-Thai Alliance and to maintain protocol with the Thai forces.

On August 5, the first atomic bomb was dropped on Hiroshima. Although we could not obtain details, we were told that the military center of Hiroshima was destroyed by one single bomb. We felt the final crisis had finally come.

On August 8, a conference of all unit commanders was held at the General Headquarters. At this meeting, called for the purpose of transmitting final orders and instructions to the troop to meet the crisis, the unexpected news of the declaration of war against Japan by the Soviet Union was announced.

We were told that in line with the War Minister's instructions, "Smash the onrushing enemy legions and defend the Imperial line," the Kwantung Army planned to withdraw its forces southward and defend Korea.

If Thai were to turn coat, now was the time. Orders were given at the conference for troops to be placed on an emergency basis. That night, all army corps and army units took their fortifications. Scattered billets were quietly vacated and both officers and men prepared for siege in the strongly fortified headquarters. There was a certain gaiety even in the midst of this atmosphere of tragic determination.

On the night of August 12, the Bangkok Chief of the Domei News Agency Bureau rushed to the Staff Headquarters with sad news. The Domei office had monitored a foreign news broadcast to the effect that the Japanese Government had accepted the terms of the Potsdam Declaration.

In view of such a situation, the conclusion was to be reached that someone should go to Tokyo and obtain orders on future steps to be taken. As a representative of the Japanese forces in Thai, I headed for the Supreme Headquarters of the Japanese forces in Southeast Asia at Saigon, oblivious to the fact that I was in a capacity far above my rank.

I had planned to find out the attitude of the Supreme Headquarters at Saigon and then to fly immediately to Tokyo. However, at Saigon, I was told that this was a matter within the jurisdiction of Southeast Asia Army Headquarters and that high-ranking Staff Officer Kushida was flying the next day to Tokyo.

In this, I was lightly brushed off. However, I knew that it was not for me to take action. As a low-ranking officer, there was no justification for my stepping out of bounds. Though forced to watch the Imperial country sink like a crippled ship, I had to bow to logic.

I wondered whether there was anything that I could do, even if I had to commit the crime of insubordination or of desertion. When I remembered that in my thirties I had divorces my wife and left my children to participate in the movement for national reformation, I could not but feel that my passion for my country had grown colder.

However, it was fate. If I had gone against orders from Tokyo at that time, perhaps I would have pushed my way into the palace with young officers and committed "hara-kiri" or perhaps I would have been arrested and ended my life in the hangman's noose. Fate is indeed ironic.

That night, I was lectured to at length by my respected teacher, Hidesumi Hayashi. I re-examined in cold sanity and with sincerity the frightful power of the atom bomb and the future of abiding by my principles in spite of orders from above. I finally resolve to drop from the ranks, go underground on the continent and work for the reconstruction of Japan.

I told myself that the course of the Japanese Army should not be influenced by "face" or the emotions of an individual. The Army must, without any criticism, bow to the wishes of His Imperial Majesty and carry out his orders. When I considered the everlasting history of a race, I felt that surrender was a trial that we would have to accept. I told myself that misfortune must be utilized as a basis for future expansion.

When I returned to Bangkok on August 14, I heard of the suicide of War Minister Anami and that the entire Headquarters was waiting the morrow, when the Emperor himself would broadcast a message at noon.

And, thus, everything ended. In the basements of the headquarters, soldiers and officers listened with bowed heads.

From the radio came the halting voice of the Emperor.

"To bear the unbearable, to suffer the insufferable....

"Lose not again your trust both at home and abroad....

"We are with you, our subjects...."

Suppressed cries rose from the ranks of men standing at attention.

As I listened, I told myself:

"I have no excuse. I have committed a sin for which ten thousand deaths cannot atone. It would be Bushido to apologize by committing 'hara-kiri.' However, the Imperial wish is that we disarm unconditionally and abide by the orders of the Allied Powers. To obey is the duty of a subject. It is not the wish of His Majesty that we, by our actions, should lose again our trust both at home and abroad."

I felt as if my very bowels were being torn into shreds. In the midst of this intense suffering. I decided to go underground on the continent and through Buddhism to become and everlasting bond between Japan and Thai. If this action should run counter to the wishes of His Majesty, I felt that someday....

With cheeks still wet with tears, I attended the final conference in the Commanders room. At this meeting, a certain high-ranking general said:

"How can we bear to hand over our swords now hanging at our waists? We should return to Japan armed as we are and should be disarmed there. Unless we can do that, I'll cut the enemy to pieces or cut open my stomach."

There was none to refute the rantings of this officer. Unable to stand it, I said quietly:

"The Imperial Rescript asks us to bear the unbearable. If there are any who do not agree with His Majesty's wish, I see nothing wrong in his cutting up the enemy or committing "hara-kiri" on his own initiative. But, it would be a mistake to have the attitude of the entire army decided by the emotions of one individual. We must strive to send every last youth back safely home. Please go ahead and cut open your stomach in front of Loyalty Shrine."

It was a struggle for me to say what I did before a large gathering of others especially to a superior officer and a senior who had looked after me and cared for me for so long. However, there was no other way out of the difficulty.

Deputy Chief of Staff Hamada, who had listened with quiet composure to the arguments from the beginning to the end, opened his mouth and said detachedly:

"Tsuji, I have one thing to ask of you. Japan must suffer for the next ten or twenty years. If possible, I'd like you to go underground in China and open up a new way for the future of Asia. I know a Taiwanese who is head of a group of pirates. If you want to go to Taiwan, I'll introduce you to him and help you out anytime.

"Taiwan's not too bad. But, in Thailand there's Prime Minister Apaiyon and Major Att. I'll try my hardest to go underground in Thailand by becoming a priest."

And my resolution was complete.

Chapter 2 Into the Jaws of Death

There had been distributed to the Southeast Asia Army a number of "special volunteers." These "special volunteers" had been hand-picked and educated to become "special attack corps" members. Now, without any planes to fly, they had been parceled out among various ground detachments as so-called "strategic forces," to be used as spies and subversive agents. There were around 50 of these youths in Bangkok.

I spoke to these youngsters; none older than 22 or 23 that I intended to go underground in Thailand by becoming a Buddhist priest. When they heard of this, eight of these young men, lieutenants and probationary sub-lieutenants, came and asked to go with me. All these men had been real priests in their civilian life. I decided to choose two or three of the best, but they all insisted on joining me. On the youthful faces of these men was written the firm resolve to work for the reconstruction of their country with the same spirit underlying the "body-crashing" tactics for which they had been trained. They were set on joining me in going underground. I had been with them for only a few hours of lectures, yet for some reason I felt toward them a deep confidence and immeasurable love. When the majority of the older men had only one thought in mind—to return to Japan as soon as possible, here were these youngsters competing for the right to stay in Thailand.

"Alright. I'm asking you all to stay and die with me. We might have to stay underground for ten years, maybe twenty, but I want all of you to give your loves as foundation stones in the establishment of true Thai-Japan friendship."

I, too, resolved that I would be ready to die together with these youths, who seemed like my own sons, my own young brothers.

The memory of the Deputy War Minister came to my mind. He had been a commander of the Army Air Force in the Philippines. There he had thrown into the teeth of death hundreds of youthful "special attack corps" fliers. But, once the situation developed in favor of the enemy, he defied the orders of Commander Yamashita, abandoned his men and fled before anyone to Taiwan.

I pledged to myself, "No matter what happens, I'll throw away this old body of mine with these youngsters."

I immediately investigated the family backgrounds of these youths. I found that Probationary Sub-Lieutenant Tada was the only son of a widowed mother and already the head of a temple. I felt that I would have to send Tada home. I spoke to him of Karma and said:

"I want you to go home and get in touch with the families of the other seven men. Then, I want you to start on the reformation of the religious world of Japan".

He did not utter a word of dissatisfaction replied:

"I deeply regret that I cannot be accepted. But, I feel that what you tell me is also my task. I shall return as you order".

He could say no more, for the tears filled his eyes. It would have been easier for me if had said that he wanted to remain and idle with us. But, his silent acceptance of the task that had been given him, despite the depths of his youthful feelings, touched me greatly as I watched his mouth set hard with determination.

How deeply I was touched. Here these youths thought not of fame or of life. The task given them was their all.

The next thing to do was to persuade Ambassador [Kumaichi] Yamamoto to obtain the understanding of Thai leaders for our venture. This was a prime requisite.

I immediately visited the Embassy. I found the Ambassador bowed down with the swift changes of the few days. Suffering was written all over his face, but he greeted me in a quiet voice"

"I'm very sorry. The future Japan must win back the trust of the world through her moral principles. Let us work toward the re-birth of a nation that will not contain even a single criminal."

I was surprised that already he was deep in thought over the future course of Japan.

"Today's situation is the result of the sins of us military men. In the hope of atoning for even a part of this great sin, I want to stay in Thailand together with seven young men and through religion would like to die as foundation stones in the erection of a lasting friendship between Thailand and Japan. Could your excellency not ask for the understanding of the Thai leaders for this venture?"

Ambassador Yamamoto immediately acceded to my request.

"Very good. As my last service to my country, I shall immediately ask for permission for your venture form the Thai Minister of Education and the Prime Minister. I know you will do your best."

The British forces had not as yet arrived in Thailand. Although, I had not heard of my being designated as a war crimes suspect, I was fully ready to meet such a fate. If the British Army were to start a search for me, I knew that I would be causing undue trouble to the Ambassador. Although he was a civil servant, the Ambassador did not pause to this of his own position of my request.

If this man had becomes an army leader, he would surely have become like General Nogi. I could not but feel admiration for Ambassador Yamamoto, when I compared his soft-spoken determination to the proud and overbearing high-ranking officers, who thought to take full advantage of the might of their military uniforms. Ambassador Yamamoto had showed he ready to give up everything out his deep concern for his country.

I felt again that Ambassador Yamamoto was a true diplomat and true patriot.

I shook hands firmly with him and soon left. In all probability, this was an eternal farewell. I could not halt the streams of tears that flowed down my cheeks.

Priestly garbs for eight men were prepared. The costumer was Kamito Maruyama of the Hikari Mission, a priest of the Nichiren Sect and the dressing room was an inner suite in the Thailand Hotel.

We had our pictures taken to present for identification purposes to the Thai authorities. I'd not realize that it was so difficult to imitate the careless way of the Thai priests, who walked the streets of Bangkok in their yellow robes, oblivious to all around them. The yellow robe was a five by six-foot single piece of cloth. This covered the entire body. Then there was a three by six-foot cloth, which was worn like a skirt and held in place by a thin cord. In my hand I carried a Dhuta bag.

The picture taken of me in this garb was the picture of myself in the swaddling clothes of a man newly born into a new life.

I again changed to uniform and on August 16, I went to say goodbye to the commander.

"I want to thank you for your many kindnesses to me over a long period. You can rest assured that should anything happen I shall die bravely. Here's wishing you a safe return to Japan."

My words were blurred by tears. From the eyes of the easily-moved commander, tears also were flowing. I shall never forget his parting look as he watched me depart, with deep anxiety written over his face.

That night, I had "sukiyaki" with Deputy Chief of Staff Hamada at a farewell dinner in the Officers' Club. The faces of even the pretty waitresses were filled with determination and each and every one had vials of potassium cyanide in their pockets ready for use should the worst happen.

This was my farewell to 30 years of life as a solider. "With the best of luck," I dragged my feet away from a hateful parting. When I reached my quarters in the headquarters, Lieutenant-General Kenryo Sato came over to see me.

"We're banking everything on you," he said, and then could say no more. We parted after a firm handshake. And, I thought that this was my last parting from seniors and colleagues with whom I had worked together for so long.

A little later, Major Att, with crestfallen looks, appeared.

"Att, during the last two months, I've caused you only a lot of trouble just because Japan lost. Please be strong for the sake of Thailand and for the sake of Asia. If Buddha wills, let's meet again. Don't be sad and don't be anything rash."

Att replied:

"I have to thank you for all you kindness over these long years. Please take good care of yourself."

In this way, we parted in tears. I saw him to the gates–beloved pupil and I watched his figure recede through the falling rain.

I realized that just because he had been taught by Japanese and just because he had felt friendship for Japan, he would have to undergo much suffering in the future. With the bonds between teacher and pupil stronger than the barriers of race, our parting was indeed bitter. This, I knew, was fate, but I could find no words to apologize to him.

Then there was Sub-Lieutenant Shirakura. We had tasted together the bitterness of death and defeat since our Burma days. And now I had to say a final farewell to him. Under the dark shadow of the trees in the Headquarters courtyard, we embraced each other.

The canteen that I had carried with me from the days of the Nomonhan Incident [Battles of Khalkhin Gol, 11 May–16 September 1939, between Japan and USSR, at the Khalkha River, in Mongolia], several years previous, I gave as a moment to this Sub-Lieutenant, my uniform with its countless bullet holes I handed to Captain Koo, and my wooden sword of ebony to Captain Tabuchi. I threw my pistol into the pond in the grounds of the Headquarters and I entrusted my military sword to be hidden in the private home of the interpreter and not his billet. Having disposed of all my private effects, I waited until the coming of night.

The Headquarters that had literally boiled over with excitement during the day slept as if dead, the sentinels the only ones let awake. I heard the sound of the clock strike three in the morning. I bowed toward the north-east and deeply apologized to the Emperor. I then took off my uniform which I had worn with pride for 30 years and changed to my yellow robe.

Taking care not to disturb my colleagues who were deep in slumber, I tip-toed out of my room. Just as I reached the car that I had readied in secret, I noticed Sub-Lieutenants Akutsu and Shirtakura waiting in the shade of the car, shoulders fallen in despondency.

"Thanks very much for everything. Please take good care of yourself."

And we exchanged our final farewells. 30 years of military life came to an end. The lonely sentinel, perhaps ignorant of who was in the car, presented arms as I slipped out of the headquarters gate. The street lights reflected on the polished bayonet affixed to the sentinel's rifle.

Good-by army life, good-bye comrades.

And I parted with the first 44 years if my life. I felt that the great ship, "The Army," had sunk, and I, a solitary survivor, was throwing myself into the giant waves of a bottomless sea. My past life, during which I had acted–thanks to my uniform–over and above my natural abilities, was gone.

I looked back again and again at the familiar headquarters gate. The darks streets of Bangkok were thickly stationed with Thai police. It seemed as if Thailand was afraid of something. But the populace was deep in dreams. Somehow the sight of a small dog attracted my attention and burned into my mind. It seemed to suggest my own future.

It was! This was the beginning of an underground travel that spanned three countries of death barriers stretched along the way–a path that traversed a jungle of thorns.

The first mile-post was the small ossuary for the remains of Japanese erected in the compounds of Ryab Temple, mercilessly destroyed by bombings. I got out of the car at the rear gate and peeped through the hedge, but saw none.

I wondered for a moment whether I made a mistake and was looking around the neighborhood, when one of the disciples, who had completed moving the night before, woke up and opened the half-door for me. He must have been tired and deep asleep. Rubbing his yes, he said:

"Good morning, master."

Nonetheless, his greeting was cheerful.

The disciples, suddenly raised by their master, were frantically folding up their bedding. I met the guardian of the ossuary, the old priest Chino.

"I am Aoki. I hope you will be kind enough to teach me."

I learned that this old priest was a native of Wakayama Prefecture. During his youth, he had left his homeland and had gone as a pearl-buyer to Northern Australia. Since then he had worked on rubber plantations in Malay and shipped as cabin boy on steamers. He had settled down in Rangoon and had taken a wife. There he had opened a laundry and prospered. However, he had come under the influence of His Holiness Fuji of the Nihon Sanmyohoji Temple and had gone through a religious awakening. As a result, he threw away his business and dedicated his life to beating the drum. This was the strange story of this old priest.

Although he was close to 60 years of age, he daily pounded the drum with an energy surpassing that of younger men. Chanting, "Homage to the Sutra of the Lotus of the Good Law," at the top of his voice, he would stride through the streets of Bangkok, oblivious to the laughter and the stones thrown at him. He had already become a character in the city. Although he had only graduated from elementary school, he had learned difficult sutras since his conversion to Buddhism and was always engrossed, whenever he had time, in the writings of Nichiren. From two years ago, he had become the Buddhist priest looking after the funerals of Japanese in Bangkok and at the same time acted as guardian of the Japanese ossuary. He lived with an old Chinese, who was handy man around the temple and he undoubtedly must have been surprised to have his home invaded by eight unexpected itinerant priests.

The young disciples, their vibrant and energetic bodies wrapped in a single yellow robe, were busily cooking rice in earthen pots and boiling miso soup, as happy as children playing housekeeping. The food that we had brought from the Army warehouse–miso, rice, dehydrated vegetables, canned goods and dried fish–was sufficient to last us one whole year. Our quarters were the roughly remodeled resting place for visiting worshippers. Only composed of a roof and four posts, we had nailed planning on all four sides and had made eight beds within. It was like a wooden barracks for troops.

The ossuary was composed of two towers, 36 feet high and 16 foot square at the base. It contained the remains of Japanese who had died in Bangkok from about 50 years ago. It also contained the votive tablets of 120 way heroes that had died in the Pacific War in Thai and French Indo-China.

Before taking up my chopsticks for my first meal in the new latter half of my life, I had to chant three times:

"Homage to the Sutra of the Lotus of Good Law."

I felt awkward. I told myself that this would not do, I would have to learn to become completely a priest. After the meal, the old priest Chino poured hot water into his bowl, stuck his fingers into the bowl and washed the clinging bits of rice from the sides. Then he places the bowl to his lips and drank the water. Somehow, I could not bring myself to following his example and picked lose the remnant bits of rice with my chopsticks and then drank down the water.

As soon as the morning meal was over, we lined up in front of the ossuary and with young Kubo leading us, began chanting sutras. While we repeated the sutras together, with wooden and metal gongs keeping time, I found myself captivated by the rhythm and learning the sutras much quicker than I expected.

I realized the truth of the old saying, "Children near the temple chant unlearned sutras." In ones week's times, I had learned the entire Prajna Paramita Sutra.

The principal image here was a sitting statue about a man's size sent 10 years ago from the Nissen Temple in Nagoya. I did not know whether it was an image of Amitabha or Shakyamuni, but the image, with its look of happy prosperity and kindness looked down with a benign smile in a transcendent manner and seemed to view with pity the confusion and chaos of mundane life.

I felt that I would find salvation here for my wounded soul. To chant sutras before this image both morning and night seemed to me more of a comfort than a regimen of strict training.

The men who were teaching this old interact colonel-priest were the lieutenants and the probationary sub-lieutenants of yesterday. The old priest Chino did not like to hear chanted any sutra other than the Sutra of the Lotus. He would derogate the Prajna Paramita Sutra and the ritual of the Shinshu Sect. Perhaps, it is the tradition of the followers of the Nichiren to condemn other sects.

There was Bonze Kygro Sakaki, an intellectual priest of the Shinshu Sect, who arrived two year previous as a formal student of Buddhism. He had occupied himself heart and soul in a study Buddhism in Thailand. He was a fine priest, quiet and understanding, and spoke and understood the Thai language. He was so quiet and humble that at a glance he would have been mistaken for a Buddhist nun. He watched over us eight men in all ways and worried over matters of liaison with the Thai people.

He did not have the fanaticism of a priest of the Nichiren Sect, but he had the self-composure not to lose himself in the moment of national disintegration.

There was no light or water in the ossuary. Our only drinking water was rain caught from the roof through a trough into a large vat. In the back yard was a 12-foot square pond. It had not been cleaned out for years, it seemed and was covered with greenish growth and smelled muddy. However, after a day's training, it was our custom to strip to the skin and bathe in this pond.

Our greatest pleasure was to surround a lone candle after supper and talk together for two or three hours until bed-time on the future of Japan.

In our employ was the old Chinese who acted as handy-man. He was a native of Canton. We called him "Lao-Tai" (old man). He had been for 15 odd years at this ossuary. Wearing only a pair of shorts, he worked all day long, shopping, sweeping the compound and plucking the weeds. At night, he acted as watchman, and during the day he worked as if he were a machine. His sun-burnt skin shone like polished mahogany.

The novice priest, myself, who did not understand either Chinese or Thai, sent him in a shopping tours to buy bananas and fish, explaining what I wanted by gestures or by drawings pictures. He seemed to have a common-law wife somewhere in town. At the end of each month, he was paid 20 Bahts. Smiling away, he would trim his whiskers with a pair of scissors, put on his half-sleeve shirt, bleached white by many washings, and set off for home. On these occasions, he was as happy as a child. What was surprising was the fact that overtime he passed in front of the ossuary he would stop, bow reverently and clasp his hands in prayer. It seemed that his chief pleasure in old age was to serve Buddha. Unconscious of the defeat of Japan, he seemed happy working for Japanese.

Although our household had suddenly increased from two men to 10, our daily expense was limited to 20 Bahts. This was because we had brought our staple food and auxiliary food with us from the Army, but at the same time, we had to be careful with the funds we had. We could not waste a single yen, when we thought of the future and the years of uncertainty ahead.

For breakfast, we had miso soup and pickles. For lunch and supper, we had cooked dehydrated vegetables, with perhaps a piece of salt fish. The fresh fish which we brought in town was a rare luxury that made our table festive. Five yen worth of bananas were first offered before Buddha and then became dessert for the ten people.

Large and small, our laundry was done by each man. I realized fully how extravagant and spoiled my past 20 years had been–20 years during which orderlies had waited hand and foot on me. After a week's time, I was able to put on by myself the priestly robes of Thailand. There were many nights I could not sleep because of the hot and steamy weather.

Often, I noticed with distaste the dye from my yellow robes clinging to my under clothing and my bedding because of profuse perspiration.

As the monsoon season gradually drew to a close, we begin suffering from lack of water, for which we had relied upon the frequent tropical squalls. Mosquito larvae began hatching in our drinking water. The time had come for us to buy our water.

A friend of the old Chinese came daily to sell us water at one yen for two pails of water carried on a yoke. The water smelt odd. I followed the water seller and to my surprise found that he was getting the water from a puddle formed in a bomb crater. A dead dog was floating in the pond.

In face of the approach of the British forces, it became a problem for resident Japanese to obtain their identification certificates. Thai Government and police officers must have felt that their turn to boast had now arrived. Long queues of Japanese lined in front of Thai Government offices waiting for their turn. When their turns came I saw these Japanese, benefits of the backing of their country, rubbing their hands in pitiful pleas before Thai officials.

I felt that we would have to be spared the ordeal of being examined for our identification cards. I made a try through interpreter Hatano. But no, this was not allowed.

Priests were not to be excepted. We received the notification:

"The eight of you will appear all together at the Religious Education Bureau on August 20." I had not succeeded in becoming in such a short time a full-fledged priest. The robe itself would not stick to me and often fell down from my shoulders.

I took off my glasses in the hope that this would change my appearance. But, I could not reduce my muscles. For the first time in my life, I regretted my muscular build. I told myself: I must look a worse sight that Kiyomori Taira, who in haste threw over himself, wearing armor and all, a priest's robe. The wound on my right arm looked still fresh. Because of the long days in a plaster cast, the arm appended shriveled. If I was inspected with suspicion, there was plenty about me to cause further suspicion.

Led by the interpreter, we made our way at the appointed time to the Religious Education Bureau, situated close to the Japanese Army headquarters.

We found close to 100 Japanese people lined up awaiting their turn under the burning sun. I even felt that when eight hastily produced priests, with pates blue from recent shaving, appeared, we were eyed with suspicion by even the waiting Japanese.

Finally our turn came. We were asked our birthplace, age and occupation and were finally asked to sign our names to our affidavits. I took out from my Dhuta bag the ink-horn and brush that I had carried with me always on battlefields and in quieter life. However, I was handed a pen and told to sign my name in Romanized Japanese. I did not know what to do, when Kubo said he would sign in my stead.

"A bonze does not know English," he explained and somehow I was able to pass the first barrier.

I noticed Japanese, wearing dark glasses, standing by looking at me. He came close to me and prodded me in my ribs with his finger. I looked at him and realized that he was Ishida, a class-mate of mine in the Military Academy. He had come to Thai several years previously as a member of a certain Japanese firm and had engaged in underground activity. He was a specialist in gathering information.

We spoke to each other with eyes, putting heart and soul into our looks. I realized that he also for some reason was seeking to remain in Thailand and work underground.

We were then taken to the room of the Chief of the Religious Education Bureau.

Now it was our turn to be interviewed personally. I felt beads of perspiration form in my armpits and wet my robe. I was supposed to be the head of a group of student priests recently arrived from Burma and had been picked as the first one to be questioned.

In the center of the room sat the Bureau Chief, like a judge at a trial. On both sides were seated members of the bureau, as if they were jurors. All eyes were turned upon me as I entered the room, worrying whether my robe had not fallen from my shoulder. I felt that they could see through my disguise. Realizing that if I were caught, there was nothing I could do, I calmed myself.

"What sect do you belong to?"

"The Shinshu Sect."

"How many years have you been a priest?"

"Twenty years in Japan and roughly two years in Burma I have, only recently arrived in Thailand and am not as yet used to the Thai robes."

"Do you also beat drums?

"No, it is only the Nichiren Sect that beats drums."

"What is the significance behind the beating of drums?"

"Oh (and I felt cold sweat suddenly flow down my sides).... That is to chase away enemies of Buddha."

"For what purpose and for how long do you desire to study in Thailand?"

"I would like to stay for about ten years and become a link in bringing about eternal friendship between Japan and Thailand through Buddhism. If you will only allow me to stay in your country...."

"Good. Please study hard and strive for the friendship of Thailand and Japan."

This one sentence was like gospel from heaven. I had finally passed my first test. They had watched with suspicious eyes the clumsy way had worn my robe. Yet....in this way, I passed safely.

The remaining seven pupils who stood next on the inquisition, stand succeeded in passing their test, thanks to the impression left by my answers.

When I returned to the temple, I found my underarms wringing wet. Perhaps, the lenient attitude of our questioners was due to a good word put in by Prime Minister Apaiyon. In addition, the lavish gifts offered to Buddha when the joint Japan-Thai memorial services were held must also have had some effect. But, in all probability, the humble attitude taken by Interpreter Hatano and his clever answers, for exceeding the actual replies, must have been the real reason for our success.

We were thus formally certified as Buddhist students residing in Thailand. I felt more fatigued than I had ever felt on any battlefield. I was able to sleep soundly for the first time that night. It was the first sound sleep that I had had since I took off my army uniform. The snoring of my seven disciples was also loud. How great must have been the anxiety of these young men over today's examination. Undoubtedly, they surely could not have encountered in all the examinations in the past one that worried them so much.

As long as my disciples were with this aged itinerant priest, they would never be of any use no matter how long they stayed. In accordance with the old Japanese proverb, "Send a beloved child out on travels," I decided to send them as trainees to the Mahatat Temple. This Temple, the site of the joint Japan-Thai memorial services in honor of the war dead, was the second most influential temple in Thailand. In the past, the Commander had given numerous presents to the priests of this temple, and they were pleased to take over my disciples.

I first had decided on sending all the young men. However, the men felt that I would be lonely without someone beside me, and finally we agreed on sending the five oldest, leaving two of the youngest with me at the ossuary.

It was the last week in August that I dispatched the five to their new home. We had been together only ten days but I felt as if I were parting with my own younger brothers. We loaded rice, miso, dehydrated vegetables and dried fish on a three-wheeled rickshaw and went with them as far as the gate. There we waved good-bye.

The ossuary became suddenly empty after their departure. Four or five days later, the five brother disciples came to visit us. They had transformed into true Thai priests. They wore their robes with abandon and had even shaved off their eyebrows as do the native priests. They looked completely changed in appearance. One of the men who had remained with me listened with interest to the account of the new life in Mahatat Temple. Finally, this man also expressed his desire to join them and that night I let him leave with the rest.

Now only one youth, Fukuzawa, remained. His desire was to stay till the very end with me in life and death. The boy's name was Takashi. His grandfather had been a high priest of the Soda Sect, and had been a chief abbot also. Soon after his graduation from the preparatory department of Komazawa University, he volunteered for the special attack corps. He was a swimming champion and though small in body was well built.

Our life together began. There was no concept of rank or of age. We were bound together by one passionate fervor. It was Takashi's job to cook the rice and make the miso soup in the morning. It was also Takashi who taught me the Prajna Paramita Sutra and the ritual of the Shinshu Sect.

The monsoon season ended and we entered a dry period. We, who had relied on tropical squalls for our drinking water, felt the pinch of our water famine. In anticipation of the day when we would have to use the water from the pond as drinking water, the seven youths, just before the departure of the five, stripped to their waist and cleaned, out the pond. Years of accumulated filth and mud were taken out and the sight of clear water filling the pond to its brim was unforgettable.

Ignoring the opposition Lao-Tai, the Chinese handyman, Chino, Takashi and myself dug a well. Three meters below the surface, we struck water. We placed a tub, with a hole in the bottom, in the well and allowed the water to accumulate. However, the water smelt and could not be used for any purpose let alone as drinking water. Mosquito larvae appeared and soon it became the habitat of frogs and tadpoles.

Once a week, a Thai lad came from Mahatat Temple to obtain rice and miso. Our liaison with the other group was carried out through this youth, with our messages hidden in the rice that we sent them.

The training of a Thai priest was based on strict discipline. Early in the morning, the priests were sent out with begging bowls. The homes of believers would be visited and these bowls would be filled with newly-cooked rice and soup. This food would be eaten in the morning and at noon. The third meal of the day was dispensed with. Priests were not allowed to take any food after the noon hour. They were only allowed to drink tea.

Often the other six would come to visit us. In the afternoons, they would look longingly and hungrily at the bananas placed on the dining table. And it was touching to see them suppressing their desire.

Not to allow myself to be beaten by the other six, I started studying the Thai language in earnest. Bonze Sasaki introduced me to a Thailander, who had once been his own teacher. This teacher was a sub-lieutenant of the Thai Police Force, and it was felt that he might turn out be of some use to us in case of an emergency. Takashi and I studied at the same desk. I made Takashi study the very fundamentals reading, writing and speaking the language.

However, in my case, I limited my objectives to learning to understand and speak the language in one month's time. I limited my resources to my ears and mouth alone and did not bother to learn the Thai alphabet.

This was in preparation for an emergency. If ever we were pursued, Takashi and the rest would have nothing to worry about. However, I would have to be prepared to escape alone at a moment's notice. I felt that as long as I could understand and speak the language, it would be quite sufficient. There are two ways to learn a new language; to begin from the very ABC's and build up slowly, or to limit one's study to absolute essentials, cramming without pause and with ultra-speed.

At the news of the approach of the British occupation forces, unrest and tension filled the hearts of Japanese residents and the military forces alike. I knew that I must not allow my beloved Takashi to accompany me. I felt that I must in some way or the other make Takashi join the other six, and I racked my brains in search of the best way to do this. I realized that he would never agree to leave me if I spoke to him honestly of my intentions. Therefore, I said, "As soon as procedures are complete, I intend to enter the Mahatat Temple also. I want you to move there ahead of me. It would arouse suspicion if we were to leave this place together." In this way I convinced this pure-hearted youth.

Once he left, I was completely by myself. Together with the ensuing sense of aloneness, I felt a deep relief to be alone. True, there were no blood ties between us, but I felt a deep attachment grow in me for the seven men, an attachment no less strong than that between brothers. My feeling for Takashi was particularly deep. I suppose such you would call destiny.

One of the first surprises upon arrival in Thailand is the large number of priests encountered. You wonder whether a conference of priests is being held. In addition, you are surprised to note that the majority of priests are all strong and young men.

You wonder how all these priests can make a living. But, soon your doubts are cleared up. Thai youths enter temples for three month's training and undergo the life of a priest. After that period is over, some continue with their studies, while others return to their own civilian life. Of course, the latter category is predominantly larger.

When the Thailander wears priestly robes, he submits to a rigorous regimen. He abides by the strict discipline of priest life. The austerity of this life, both with respect to food and clothing and the suppression of physical desires form no comparison with that of the priestly school of Japan. But, once a lad returns to his civilian life, he feels no pangs of conscience in stealing or whoring from the very day of his return. The strict discipline that calls for a priest to turn his eyes away from any woman he meets on the street lasts only for three months. Many, it seems, think that three-months of self-denial is efficacious in obliterating all the sins that he may commit the rest of his life.

The Japanese ossuary where I had sought hiding was separated by a hedge from the adjoining Ryab Temple. Day and night, I could see watchful eyes looking at me from the holes in the hedge.

Some would be men, others would be children. During the short minutes that I would be chanting a sutra, children would slip through the hedge, run into my room and snatch everything in sight. Blankets, sheets, mosquito netting or any handy object would be thrown over the hedge to a waiting adult on the other side. In a twinkling of an eye, these objects were whisked away by a relay of waiting men.

When the seven young men were with me, these thieves did not make their appearance. But when finally all left, the old priest Chino and myself began losing things quickly. First it was my watch, next my robes, next my shoes until in the end I had only a ragged blanket left.

These thefts were not carried out with any particular disdain toward defeated Japanese or with hostile intent. It was a purely animal instinct to steal. I could only explain it as being the same instinct of a cur snatching a piece of food lying in the roadway. There were cases where parents supervised thefts by their children. I chased them away only to find them back a minute later. I complained to the police, only to be told that it was my fault for letting things be stolen.

As a final measure, we decided to take turns at chanting sutras, Chino in the mornings and I at night, with the other standing guard during these periods. But still that did no good.

One morning, while Chino was copying a sutra in his own room, à thief crept into the same room, opened a suitcase under his bed and stole 6,000 and several hundred yen which the old man had saved from his Burma days, as well as all his better clothing. The thief did not make a sound but came and went like a summer breeze.

It was only after he had chanted his sutra that Chino noticed the theft. His face showed that he was on the verge of crying. I could not bear to look at him. It seemed that all the beating of his drums had no efficacy against thieves of Thailand.

The next day, during the short period while I was in the toilet, I found several precious mementoes that I had carried with me literally through a number of valleys of the shadow of death were stolen. The young child, who had done the deed and had passed his booty to his parents, was already back at the hedge, peering intently through the openings, looking for his next haul.

The shaggy mongrels roaming the streets of Bangkok seemed as numerous as the priests and thieves. I realized that Thailand was after all a country of rice and a country of Buddhists. Left- over food was abundant and there were none to break Buddha's precepts by killing stray dogs. One day, I noticed in a ruined temple near our gate, a shepherd dog that resembled one I had cared for with attachment years ago. It lay shaggy and thin with illness and had not the energy to seek food. It looked with sorrowful eyes at passers-by as it waited for death. I took him some left over rice and water. The dog ate a mouthful or so as it lay on its side, too weak to stand; it lapped up a little water and imperceptibly wagged its tail with joy. The next morning, it lay cold in death.

I felt if I had lost my own child and I dug a hole near the gate and buried the dog. I erected a small stone over its grave, sprinkled it with water and chanted a sutra.

As the days passed, an increasing number of Thai priests came to visit us. I practiced my Thai conversation, haltingly and with the help of gestures. Somehow, I was able to make myself understood.

I would treat them to some tea and the next day they would be back, bringing others with them. Poor as I was, I did not fail to serve tea. It might have been this tea but more probably, it must have been Chino's drum that attracted the visitors.

In my interrogation by the Chief of the Religious Education Bureau, I had answered that the beating of drums by the Nichiren Sect was a charm to scare away the enemies of Buddha. However, it seemed rather a poor answer. I felt that my definition should be changed to describe the drum-beating as a charm to attract believers.

In the adjoining temple was a bonze over 70 years of age. Every day I could hear him cursing away at the English. There was a hideous quality about his curses. It seemed that his hatred for the British was steeped in him to his very bones. He had seen his temple destroyed by British bombs, while a number of his comrades had been buried alive under debris from the falling temple.

After the entry of the British forces into Thailand, an air strip for fighter planes was hurriedly built on the plaza in front of the royal palace. British planes swooped low over the palace as they come in to land or take off. This also helped intensify the resentment of the Thai people toward the British.

This hatred eventually turned into a feeling of friendship for the Japanese. In later days, this old priest dropped in frequently to tell us the latest news. Even after the Japanese residents were placed in camp and cut off completely from the outside, this old priest acted till the very end as our liaison man.

One night, I heard the sound of hammers and chisels. I woke up and walked around the outer walls of the temple. I found several thieves breaking into the royal vault in the compounds of the neighboring temple.

I realized then that theoretically speaking, there was the possibility of establishing Thai-Japan friendship through Buddhism. However, in this pitiful country of Buddhism, the true significance of this religion had been completely lost. Buddhism had deteriorated into a set of rigid disciplinary rules that were only adhered to for a few short months.

Buddhism had failed to save public morals and Thailand was slowly becoming a nation of indolent people.

Gradually I began losing confidence in my initial resolution to stay underground in Thai for ten years and work for the betterment of Thai-Japan relations through Buddhism. It was this that made me think of plans to slip away to China. Out of this was born my campaign to contact Chinese residents.

Lao-Tai was a native of Canton. His shopping and selling were all restricted to merchants from his own native province. In this respect also, I saw the strong unity of the Chinese people that formed the basis for their amazing expansion overseas. When I thought of the future, I felt the necessity of erecting a bridge, connecting me with Chinese residents.

However the Chinese that we had employed before the defeat at headquarters had completely lost their value. They were being hunted relentlessly by Chungking agents. It was necessary to form contacts with Chinese residents acknowledged by the Chungking Government.

It was our faithful Lao-Tai that made the first contact for me. The greater part of the bedding that we had brought from the Army became unnecessary after the departure of my disciples for Mahatat Temple. I thought that now was the time to make use of the bedding. I dispatched Lao-Tai with instructions to bring to me a merchant from his home town who could be trusted.

I gave him one condition, the merchant must be able to read and write Chinese. At the beginning several merchants arrived with the hope of buying things cheaply. They would come often to the ossuary. I felt that my plans were working out. It was not my aim to sell anything. All I wanted to do was to find one or two men that I could talk to in confidence.

My task was not to set a bargain price but to judge the character and the background of the merchants coming to the ossuary. Finally, I hit upon one youth, whose name was Wang Hsiang-cheng. At the beginning, he would not commit himself. As his visits increased, we came to talk about politics and economics in addition to business.

"Fine. I think he'll be of more use than I had imagined."

I gave these Chinese ten quilts free of charge. The next day, he came into my room and wrote down some interesting information for me. He also told me that he had friends among Chungking agents in Bangkok. He also told me that their headquarters was on Surion Avenue and also listed the main members of the group.

I finally succeeded in obtaining daily a Chinese language newspaper published in Chungking. Through this paper, I was able to obtain a picture of Japan after the surrender, blurred as it was. The papers also told of the actions of the Occupation forces, their confiscation of military, government and private property.

In this way, quite by accident, I was able to: find a loophole through which I could jump into the Chungking fold should any crisis arise. I realized, however, that, if Wang were noticed coming and going from the ossuary too frequently, he would be noticed not only by Thai police but also by the secret members of the underground Free Thai Movement. Thus, I took every precaution to make it so that our relations appeared strictly business. Wang, himself, came as Lao-Tai's friend and slipped in without being noticed as much as possible. In this way, I collected my information on things happening outside the temple yard.

I learned that just recently Lan Tung-Hai, Chief of the Chinese Residents Affairs Section had been called back to Chungking by wire. It was rumored that he had been recalled because he had extorted much gold and money from Chinese residents in the name of purging collaborationists. It was also said that he would be shot to death as soon as he reached Chungking.

I was also able to confirm the exact location of the headquarters of the Chungking Government's underground on Surion Avenue. According to my information, the make-up of the headquarters was as follows:

Lieutenant-General Hsieh, Head, a native of Hainan Island,

Major-General Hwang, Secretary-General, also from Hainan,

Colonel Cheng, Chief Secretary, also from Hainan,

Lieutenant-Colonel Kuo, Chief of the Chinese Residents Affairs Section, a native of Canton.

Lieutenant-General Hsieh, I learned, was at that time in Hanoi. Cheng and Kuo were in Bangkok, while Major-General Hwang was now travelling through both French Indo-China and Thailand. The special characteristic of this personnel setup was that the majority of leaders were from Hainan Island. At the same time, I was disappointed to learn that the leaders were absent from Bangkok.

When I went out on the streets with my yellow robe, I noticed, in particular, the Chungking demonstration corps walking through the streets with megaphones blaring and carrying large placards calling for the arrest of all collaborationists and traitors to the Chungking Government. The aim of this drive was to collect donations from Chinese residents, precariously placed for having shown sympathy for the Japanese.

The only funds that I invested in making my first contact with the Chungking underground was ten quilts.

In the Nichiren Sect, a bonze is called a "se shonin" (Saint or His Holiness). In the Shinshu Sect, this appellation is used only for such preeminent priests, as Rennyo or Shinran, or priests destined to receive in succession the abbotship of a large temple. When I heard of His Holiness Maruyama, His Holiness Koyo, etc., I wondered what great priests they must be. Imagine my surprise when I found that the majority of these men, so addressed, were men still in their thirties. Of course, I realize that age alone does not add or detract from a man's true worth.

Yet, I could not get over the queer feeling of these youth addressed with the same honorific usually reserved for saints.

There was Nagai Shonin, who had been a member of the Hikari Kikan, the organ in charge of underground activities in India and Burma, had opposed Ba-Maung and had been thrown out of the Japanese Army to become a priest. He was an amazing priest. Once in his youth he had participated in the independence movement of India, had been arrested by the British authorities and questioned under third degree methods, had refused to answer till the very end.

He arrived in Bangkok immediately after the end of the war and I had received a telegram from the Chief of the Hikari Kikan warning me to ignore this priest. However, I felt that anyone with as much guts as this man was worth my interest.

He had arrived with several members of the royal family of Burma, successfully breaking through the tight net thrown around him. I did not know whether he would succeed in his mission.

However, with the permission of the Commander, I gave him 100,000 Bahts from our secret service funds, the last remaining funds set aside for special purposes by the Japanese Army. He carried through his task superbly, leaving with his charges for French Indo-China the night before the arrival of the British occupation forces. It was on the day preceding his departure that he came to the ossuary to say good-bye to me.

Though only 34 or 35, he had an enormous physique that reminded one of the rough soldier-priests of sacred Mount Hie. He sat before the image of Saint Nichiren and in a voice loud enough to be heard by a mighty host, he chanted his own praise of Buddha. His perspiration flowed freely and discolored his yellow robes. The 60 year-old Chino, sitting beside him, seemed like a small child by comparison.

Then there was Maruyama Shonin, who was also Chino's senior of the same sect. He was man of the world and very shrewd. Immediately after the war, he flew to French Indo-China.

Kobiki Shonin was only around 30 years of age. He had been the guardian of the ossuary before old Chino, but had become an army priest and had gone with the Japanese forces to North Burma. At the end of September, he returned to Bangkok.

He was an intellectual, a priest well versed in dogma and he often found himself arguing with Chino over the contents of Saint Niching's writings. From the standpoint of age, Kobiki was a child compared to Chino's 60 years. However, in the Nichiren Sect, the one who begins wearing the priestly garb first is the senior and age does not count.

Old Chino gave up his bed to Kobiki Shonin and slept on a straw mat spread on the dirt floor. I could not but be impressed with this practice of the Nichiren Sect to ignore completely the human attributes of age differences.

Old Chino continued to beat his drum. When the entire Japanese population in Thailand was sunk in the depth of despondency over Japan's defeat, this Shonin alone remained unchanged. During the war when American planes blasted the neighboring temple into debris, Chino refused to leave. To the very end, he stood guard over the ossuary beating his drum. Fortunately, the ossuary escaped being bombed. Chino believed that it was the beating of his drum that had repulsed the B29's of the enemy. He also believed that Japan's reconstruction would depend upon how he beat his drum and so continued without letup night and day.

The bonze, whom these men acknowledged as their spiritual leader was the old priest His Holiness Fujii of the Nihon Sanmyohoji Temple. On his way home to Japan from India, carrying with him the bones of Buddha, he called on the Headquarters of the Japanese Expeditionary Forces in China at Nanking. This was in 1940, and it was the first time that I met him.

[In 1933, Nichidatsu Fujii, the founder of the Nipponzan-Myohoji Buddhist Order, acquired small fragments of the Buddha's remains in India with the promise to enshrine them in stupas (Peace Pagodas) across Japan and the world as symbols of peace and nonviolence. Fujii's mission was to promote global peace through the teachings of the Lotus Sutra.]

[The first of these Peace Pagodas was constructed atop Mount Hanaoka in Kumamoto, Japan, beginning in 1947 and inaugurated in 1954.]

In a dirty robe, His Holiness walked the streets of Nanking morning and evening, beating his drum. He was disliked by both Japanese and foreigners for disturbing the peace with his unceasing noise. On top of this, he had the habit of beating his drum without consideration of time and place. Once I was aroused to indignation and spoke to him in a sharp voice. The old priest showed not the least sign of anger, said to me with a benign look:

"You are the victim of reverse fate. There will come a time in the future when you will come to believe in the Sutra of the Lotus."

During my days at the ossuary, I felt a strange feeling of nostalgia for the old priest, at the same time no small degree of shame. His Holiness Fuji, I hear, is still beating his drum in retirement in the foothills of Mount Aso, praying continually for the reconstruction of Japan. Someday, I plan to visit him and apologize for my rudeness of the past.

In the first week of September, just before the arrival of the British occupation forces, Commander Nakamura-came to pay his final respects to the souls of the war dead enshrined in the ossuary. I had not seen him since that memorable day when I said good-bye to him and he had answered "Good-bye" in return, deeply-moved and with tears in his eyes, his hands clasped before him as if in prayer and penitence.

He looked with pity at me, wrapped in a yellow robe, copying sutras on a crude desk, beside a rough wooden bed set on a concrete floor.

With compassion filling his eyes, he said:

"I do not want to make you waver in your determination. However, won't you come back again as a staff officer? I cannot but regret letting you rot away in this manner. Please reconsider your decision."

He spoke as if he were talking to a beloved child. The British Occupation forces had not yet arrived. I had not been designated in particular as a war criminal. It also seemed impossible to continue hiding in such a small city as Bangkok. It was because he thought it best both for the Army and for me, as an individual, that he wanted to persuade me to greet the British forces honestly and openly and do as they ordered.

"I greatly appreciate your advice. However, I cannot abandon the young men who chose this life with me and return all by myself. It would also mean that I deceived the Thai people. When the worst comes to the worst, I shall die bravely all by myself. I would like you to consider me as one who has already died."

This was my answer.

Later, there was a meeting of high officers after which Lieutenant General Hamada again asked me through Interpreter Hatano to reconsider my decision and return to the Army.

However, I had already transcended life and death and had decided to go underground against the express wishes of His Majesty. Now that I had taken a step forward in line with my decision, I could not change my intentions in face of danger and surrender so easily to an enemy that I had not as yet seen.

I knew, in particular, that if I abandoned this life, I would become laughing stock among Thai leaders. I would also be disregarding the efforts made on my behalf by Ambassador Yamamoto.

Of course, I realized that I was not as yet being pursued by the British and that there was a good chance of my returning unharmed to Japan. However, if the fact that I had tried to go underground came to light, I would have to stand trial as a deserter. I knew that I had reached the final stage when I would have to consider my own interests or else.

But, I could not change my decision at this stage, despite the kindnesses of these advisors. I immediately wrote a letter to Deputy Chief of Staff Hamada:

"I accept with deep gratitude your kind advice. It hurts me greatly to have to refuse your kindnesses. However, with things as they are now, I have no other intention than to stay underground until the very end. With regard to the British Army, I have already left my last will and testament, dated August 16, indicating that I have fled elsewhere. Though far removed from you, I continue to pray for your happiness...."

It was the day before the arrival of the British Occupation forces. The order was received that this would be the final day during which the Japanese Army would be' allowed to use their aircraft. Braving danger, Staff Officer Wakizaka came to visit me in civilian clothes.

He advised me with earnest intent:

"We are planning to send off our final plane today. Please make preparations and escape immediately to French Indo-China. To stay in Thailand is to court absolute danger."

"Thanks for your friendship. However, I cannot at this stage abandon my seven disciples and flee all by myself from the, teeth of the tiger. Please resign yourself to my fate. Everything is destiny."

I had refused to accept the last friendly warning. I knew that all these offers were the manifestations of true feelings of friendship for me, arising from the desire to save me from the inexorable approach of danger and destruction. However, all these thing I had thought out on August 16 when I had made my final decision.

Already the, Japanese Army organization had been dissolved. No one was allowed to carry any weapons. At such a time, it was up to everyone to decide his own fate. My greatest worry was whether the Commander would not be held responsible for my desertion. It was because of this fear that I had slipped out of the headquarters without anyone knowing and it was because of this that I had left behind a note stating that my action had been taken on my own free will. I believed that in the last analysis, my crime would not be pinned upon either my higher officers or my colleagues.

Man's fate is already decided the day he is born. Life and death are not within one's own power to decide. If it were necessary for the sake of the race, the gods would offer me a way to life in the midst of the greatest perils and if they no longer felt I was netted they would call for my life at any moment.

I again resolved not to think of life or death but to proceed along the path I believed right for the sake of the rebirth of the Japanese race. I believed that outside of this, there was nothing that should influence me. Spilt water did not return to the tray.

On September 15, that which was due to come did come.

The advance party of the British occupation forces landed at Don Maung Airfield, with the pride and prestige of victors. They were greeted in trembling fear by a group of representatives of Japanese Army and the Thai Government and military leaders. The British Occupation forces immediately took over the actual authority of the country.

Rumors flew that Japanese soldiers and residents alike would be placed in concentration camps that only those residing before the war would be left free, that priests would be safe. Depending on the individual, rumors both optimistic and pessimistic criss-crossed and tangled with each other.

The atmosphere became tense with each passing day, and seemed to press more heavily upon myself. Japanese residents began to be herded together. The concentration camp was a village of barracks, which the Thai Government had built during the war to accommodate the dispersed population of Bangkok, and was located in a farming area in the northwestern part of the city, facing the Menam River. The Japanese in Thailand were bereft of their homes and property the crystallization of years and years of bitter suffering and hard work. Their furniture was looted. Then, carrying a single suitcase or with rucksacks slung over their drooping shoulders they were dragged through the streets under burning sun toward the concentration camp. I could hardly bear to watch them pass by.

A friend of one of my disciples that had moved to Mahatat Temple, a young Nichiren believer, came to the ossuary to say his last farewell. He was a serious youth. In the course of our conversations, he said:

"Aoki-san ([Norinobu] Aoki was my assumed name), do you know Staff Officer Tsuji? I hear that famous officer has disguised himself as a priest and has gone underground. However, he has been wounded severely in both his foot and arm and I don't believe he'll ever succeed in escaping."

I felt my jaws drop in astonishment. But, quickly regaining my composure, I replied with a bland face:

"Yes, I've heard of his name, but haven't had a chance to meet him as yet. He sure is an eccentric person. Is he still supposed to be in Bangkok ?"

"No, a man like that wouldn't be sticking around here in idleness. In all probability he has already gone either to China or French Indo-China. Pardon me for asking, but what sect are you?"

"Oh, I'm of the Shinshu Sect, a green priest!" And we parted in great laughter. I pondered over the fact that even the Japanese residents in Bangkok knew already that I had gone underground. I knew that I could not let down my guard. And I girded my loin cloth tighter around myself.

There was a Mr. Tajiri, who had been a resident for 30 years in Thailand. He was the head of the Japanese residents' organization in the city. A native of Toyama Prefecture, he claimed to be a Shinshu believer and often during our conversations I ran into the danger of exposing my ignorance of the teachings of this sect.

He would come to see me often and angrily declare:

The darn fool militarists. They started a war on their own initiative. And look at what happens. The products of 30 years of hard work all for nothing." And he condemned the military with bitter denunciations.

I told myself it was a natural indignation. The product of 30 years of labor had been completely confiscated. Even his home, his furniture and all his belongings were looted by unthinking Chinese and the Thai people.

Then, there was the youthful owner of the Thailand Hotel, a thin pale man, who came to the ossuary, bringing with him the cremated remains of his beloved wife. For three years, this young man had suffered from tuberculosis and he had married the woman that had sacrificed everything to look after him during those years. Immediately after his marriage, the war ended. His wife tired from the long years of nursing her husband and completely disappointed by the defeat of Japan and the dark days ahead, had died of self-poisoning on the eve of the arrival of British occupation forces [September 9, 1945].

The young man left the brand-new wooden casket carrying his wife's remain in front of the image of Buddha. In final failure, he knelt before his image and clasped his hands in tears. Somehow, I could not but feel that the picture that he presented was that of the whole of Japan. I felt great pity for him as I watched him leave with lifeless steps through the evening rain.

The headquarters of the British Occupation forces was established quite near to the ossuary. The plaza in front of the Royal Palace had been turned into an emergency landing field and fighter planes were leaving or arriving in a constant stream both night and day. Both white and colored soldiers and officers filled the streets and many peeked curiously into the bombed-out remains of the neighboring temple or peered through the hedge at our ossuary.

It seemed as if the drum beaten by old Chino attracted them. I felt that the drums which had characterized "as a charm to repel the enemies of Buddha" would become "an invitation to the enemies of Buddha." In order not to attract the attention of the British, I asked Chino time and time again to stop beating the drum, if it were possible. However, he seemed, just out of meanness, to beat the drum still louder. He refused to budge and would not have budged even if I had used a crowbar. The confidence that he had repelled B-29's with his drum could not be so easily overturned.

Time and time again, I noticed several well-armed British soldiers standing outside the hedge. Time and time again, I thought they had come to arrest me.

"If they have come to get me, then I shall be killed in front of the ossuary," I resolved. In this way, several tense days passed by. One day, I approached the British soldiers; my hands clasped in front in a pose of prayer and beckoned them into the compound. However, none would dare come into the home of this ominous-looking priest, which was indeed a thankful thing.

After the arrival of the British, the surveillance of the Thai police became more strict. Night and day plain clothes detectives would be standing watch around the temple.

One evening after I had finished chanting the sutra and was walking in the courtyard, I noticed somebody hiding behind the trees. I approached the man in order to find out who he was. He was a Thailander. In the moonlight, I saw the muzzle of a pistol peeking out from under his coat, pointed directly at me. In all probability, he was one of the secret police. It seemed as if the Thai authorities were beginning to suspect even priests. After all, we were the only Japanese left free.

I felt that an unseen net was being drawn around this small ossuary in the midst of temple ruins-drawn closer and closer around us from all directions. Waking and sleeping, I thought of ways to break through this uncanny net tightening around me.

The Chinese vernacular press reported the death of Lieutenant-General Hamada.

Following the arrival of the British occupation forces, General Hamada had taken upon himself every bit of the liaison work. The night before the Japanese forces were to be disarmed, he held a final dinner with his aides and his orderlies. Then, when all had gone to sleep, he quietly cut open his stomach in the traditional manner.

A part of the letter he left behind was announced. What particularly struck me were his words of gratitude to the leaders of Thailand.

This General, at a glance, seemed an ordinary quiet and kind person. However, he never lost his composure no matter what situation arose. At the same time, he never sought to avoid responsibility for his mistakes, no matter how great they may have been. Although outwardly soft-spoken and quiet, inwardly there burned a fiery heart. When we had prepared fortifications in Bangkok to enter into siege when worst come to worst, it was this General that first agreed to turn the Headquarters into a fort, and defend it to the last, as part of the entire fortification network.

In contrast to the usual Oriental type of hero, addicted in everyday conversation to grandiloquent speech, who now found trouble, at the end of the war in backing up their own boasting, Lieutenant-General Hamada did not change, maintaining a quiet dignity from the beginning to the end.

It was natural for all the Thai leaders to trust explicitly in General Hamada. There were few men of General Hamada's caliber in the Japanese Army in its heyday. In addition, there were few who were able to judge correctly the true worth of this man.

A man is truly understood only when he is placed in a coffin. A new votary tablet was made for the General. I placed his tablet alongside that of Lieutenant-General Shidei and Subhas Chandra Bose, leader of the Indian Independence Movement who had died of an airplane accident in Taiwan and morning and evening I offered sutras in memory of these men and presented before their tablets grateful and respectful prayers.

If the British Army had come to investigate the ossuary and had seen the three votive tablets lined side by side in front of the image of Buddha, they would have immediately suspected the true identity of this priest. I knew the danger that I was risking, but felt that I would never regret being arrested if for the sake of these three men whom I respected and admired.

I did not have the opportunity to meet Chandra Bose. However, I recall Wang Ching-wei's impressions of the Greater East Asia Conference told to me while I was working in the headquarters of the Japanese Expeditionary Army in China at Nanking. Of all the leaders whom he met there, Wang Ching-wei said, Chandra Bose was the most impressive.

Dr. Wang was indeed loud in his praises of Bose.

Lieutenant-General Shidei had been my teacher at the War University. He was rare man of integrity and character-a true warrior-general.

I pondered over the fate of the three, who had died within weeks of each other and felt an overwhelming sense of sadness.

Despite the fact that General Hamada had, immediately after the war, pointed out the path which I should take and had afforded me every convenience in going underground, I had not been able to reward his high anticipations placed in me. It was indeed regrettable.

When I later I slipped out of the ossuary, I burned the tablets of these three men.

Although all Japanese residents in Bangkok were placed in a concentration camp, Japanese priests were left untouched to the very last. However, a thorough investigation of the real identity of the priests was begun. My seven disciples at Mahatat Temple were first to be questioned.

According to our presentations to the Thai Government authorities the seven had been studying as priests in the Shibow Temple in Burma and had come to Bangkok just before the end of the war.

In reality, however, it was only their leader (myself) who had been in Burma, while none of my disciples had ever been there. But, I had anticipated just such a situation and had taught the seven all I knew about Burma and had put in considerable work in making the seven men tell the same tale. Yet, I knew that once each man was questioned in detail, there was bound to be some discrepancies and there was bound to be final showdown.

The day the investigation of the seven men was held, Dick, the young Thai messenger, brought me a secret message. In this letter, was written in detail the chief questions asked by the investigators and the answers of the seven men who had been questioned in a body.

They had felt it absolutely necessary to have no discrepancies in the answers of both disciples and master. I memorized the answers through the night and wrote out a detailed statement of my past.

The next day, a young Thai lieutenant came to investigate me. The interpreter was fortunately, our good friend, the priest Sasaki, who had also interpreted at preceding day's investigation of the other seven men.

The lieutenant said that he had come not of his own volition but at the express orders of the British Occupation authorities and immediately began his interrogation of the imposter priest, Norinobu Aoki. He asked me my birthplace, religious sect, my schooling, my teachers and relatives and their addresses and my occupation. The interrogation lasted for roughly three hours.

I had prepared in advance a considerable number of lies, but I had not expected to be questioned in so thorough a fashion. If I were asked to repeat my statement, I would have been immediately exposed as an imposter. Fortunately, however, I was only asked to sign the stenographic copy of my statements, made up on the spur of the moment as I went along.

I had Priest Sasaki to thank for the leisurely way in which he interpreted. This gave me time to think out the next phrases and helped greatly.

Old Chino had been in Burma before and fearing that any reference to his life in Burma would bring about the exposure of his entire past life, he had planned not to mention this phase of his life. However, in face of the piercing questions asked by the investigator, he lost all command of himself and seemed to realize that he couldn't lie successfully.

As a result, he confessed his entire life as it had been. Of course, it would not implicate anyone else, only his own right to stay in Bangkok was at stake. He spoke of his youth as a pearl buyer in northern Australia, his laundry in Rangoon and all the unsavory jobs that he had done.

In the very end, he even confessed that he was born a twin child. This last confession made everyone laugh to burst his seams.

This innocent confession by old Chino helped greatly in softening the feelings of the Thai police officer.

Chino's child-like confession, following my severe interrogation, completely changed the atmosphere of the investigation and created a good impression on the officer.. As a result Kobiki Shonin was questioned for only about 30 or 40 minutes. Having been connected with the Hikari Kikan, he was not altogether an innocent priest, but managed to pass the grilling successfully.

After the investigation was completed, I brought down the cakes and the fruit, placed as an offering in front of the image of Buddha and served them to the Thai officer. To my surprise, the officer confessed.

"This investigation was ordered by the British authorities and is not of our own wishes. Please do not hate us for it. Although Japan lost the war, the Thai people have a high regard for the Japanese. The greatest people in the world are undoubtedly the Germans and next come the Japanese."

I asked him why he thought the Germans were the greatest people on earth, and he answered:

"Only German-made dyes, used to color the robes of priests, do not run or fade.. German razor blades are the sharpest and last the longest. There were about 20 or 30 Germans in Bangkok, but they were the finest people among the foreigners. They never did anything wrong and never were arrogant."

The reason for calling the defeated Germans the best people on earth lay in the fact their dyes did not fade and their razor blades were sharp. To the people of Buddhist Thailand, these two things were felt most pressingly in their daily lives. At the same time, I could not ponder over the high moral character of the 20 or 30 odd Germans chosen to be stationed in Thailand.

I felt that I had obtained a valuable clue to the future course an unarmed Japan must take. The practice of greedy Japanese industrialists and merchants who dumped cheap goods, made roughly in large quantities, in an effort to make easy money and the poison spread in the past half century by self-seeking groups, that had exploited the prestige of the Army, must surely have done much to destroy the trust of the Thai people in Japan.

It was probably flattery that made him place the Japanese second to the Germans. Judging from the words of this Thai police, the resentment against the British forces had taken the shape of a longing for a Fast master.

After the officer completed his day's investigation, he passed by the ossuary. There he knelt before the image of Buddha and clapped his hands thrice. I was surprised at the closeness of our feelings through the grace of a common religion.

It was thanks to the religious restraints of the Thai people that my military sword, hidden in the hollowed back of the image of Buddha, never discovered to the last. Thanks to the karma relations of Buddha, I was able to pass my second test by the skin of my teeth.

The advance of the British forces into Thailand began on September 15 and was roughly completed by the end of the month. I felt that I must take advantage of the uproar and began walking around the streets for the outward purpose of shopping. At the beginning Lao-Tal was my guide. But as I became used to the streets, I went out by myself.

In comparison to what I had seen in the past, when I had sat back relaxed in a limousine, driven here and there with red flag, (denoting a field officer) flying valiantly in the wind in front of the car, there were many new and startling sights. I saw groups of Thai beggars fighting with mongrel dogs over scraps of food thrown out on the streets. I saw the virile haggling of the Chinese merchants, arguing with militant bearing over a fraction of a cent profit with British soldiers and officers. It seemed that I had come to an entirely different world.

British troops patrolled the streets on small carriers, mounted with machine guns, and helmeted soldiers armed with pistols strode down the sidewalks arrogantly. It was only thanks to my yellow robe, that I could walk unmolested through the streets. I could walk through a crowd of fighting men and not be drawn into the quarrel. For all this, I thanked the mercy of Buddha and the religious background of the Thai people. But, I knew full well that Buddha would never have protected such a war criminal as myself. It was an act that I could never have done except in this country of Buddhism, Thailand.

It is a historic inevitability for the victor to become arrogant. In the streets of Bangkok, I saw the windows of restaurants smashed to bits and clusters of British solders quarrelling with Chinese and with Thailanders.

Immediately after end of the war, I had dropped into a Chinese shop to have my glasses fixed. The youthful clerk immediately saw through me and knew that I was a Japanese. His eyes shone with hatred and enmity and I had been only able to come away unharmed thanks to the presence of Lao-Tai with me. This atmosphere gradually disappeared with the arrival of the new visitors and I began to notice a growing sense of a friendship toward the Japanese.

Both Thai and Chinese, who came to visit the temple, had grown daily more friendly and the atmosphere of resentment and hatred toward the Japanese had completely changed.

It is not easy to wipe away the resentment of a differing race of people, once aroused. I could not but feel deeply in my present position the bad feelings toward Japan planted among the simple people of Manchuria and China, as a result of long years of stationing of Japanese troops in these countries. In time, the sight of beautiful ladies disappeared from the streets of Bangkok. Only the priests and the dogs continued to fill the streets as in the past.

Immediately after the end of the war, the blue and white flags of Nationalist China were sold in Chinese stores throughout Bangkok. The Japanese Army could not halt the sales of these flags and at the same could not openly condone these sales. However, together with the arrival of the British, Chinese homes, schools, hospitals and factories all blossomed forth with the flag of the Nationalist Government with its white sun against a blue background. In all probability, every Chinese home or business of the entire 1,500,000 Chinese population of Thailand, must have begun flying these flags.

It was no wonder that the Thai Government should feel this demonstration of loyalty to a foreign government unwelcome. Suddenly, the Thai Government issued a proclamation and ordered a ban on Nationalist flags unless flown together with that of Thailand.

The temper of the Chinese people, who saw their motherland now placed among the four great nations of the world, would not allow them to obey the orders of such a small country as Thailand, in particular, a defeated Thailand. Arguments and clashes rose in all parts of the city.

The Chungking overseas agents jumped to take advantage of the situation. Flaunting the slogan, "Death to Traitors," they busied themselves forcibly collecting funds from Chinese residents.

The Thai police also took advantage of the weakness of the Chinese residents, who had cooperated to a degree with the Japanese during the war, and used force in banning the Nationalist flags unless flown together with that of Thailand. Because of this, bloody clashes arose everywhere. These clashes spread not only throughout Bangkok but throughout the rest of the country.

September 28 proved the climactic day. The clash between the Chinese and Thai police developed into full-scale street fighting. The Chinese issued a manifesto throughout the whole of Thailand and shooting began between Thai police, armed with pistols, and the Chinese. Bangkok's streets and avenues were turned into alleys of death. The Buddhist city became dyed with blood and dead bodies were to be seen lying around all over the city. It was an ironic phenomenon that this city, which had succeeded, even in the dangerous last days of the war, in avoiding bloodshed, should now become the scene of massacre on the very day of victory celebrations.

Only priests were unmolested by each side. Wrapped in the protective yellow of a priest's robe, I slipped out of the ossuary and mixed with a group of Thai priests to catch a glimpse of the massacre, The Thai police mobilized armored cars and mowed down resisting Chinese youths. I saw half drunken British soldiers and officers pass laughingly by a Chinese youth lying moaning in the streets from bullet wounds, as if all the bloodshed was not of their concern. Their attitude was one of watching with interest a fight between dogs.

Street fighting continued for three days and three nights. Since the ossuary was situated in the center of this fighting, I was awakened time again at night by the sound of fierce machine gun and rifle. Spent shots pierced with ominous noise the roof of our barrack quarters. I, who had felt no fear on the battlefield while in a uniform, now felt fear in my yellow robe. I wondered what I would do if the Chinese sought refuge in the ossuary. I wondered which side I should favor, the Chinese or the Thai. Or should I, like the British troops, look on with arms folded. After much thought, I reached conclusion I will hide refugees in the underground vault under the ossuary.

There is room enough to hold 20 people. Once within, I'll lock the door. If the Thai police should come, I'll clasp my hands in prayer and chant sutras. They would never point a gun at a priest. Both Chinese and Thai are Orientals. I must try and mediate between the two.

Once I had decided upon this, I felt no fear or anxiety. I slept alone in my barrack bedroom, oblivious to the spent bullets piercing the roof. Fortunately or unfortunately, I did not have to put my plans into effect. The September 28 Incident began to loom as a big diplomatic issue between the Nationalist and Thai Governments.

However, victorious as the Nationalist Government was, it did not have the reserve to send troops to Thailand to protect the Chinese. Thailand, on the other hand, seemed to realize that it had gone a bit too far. The thing that surprised me at that time was the fact that the British Occupation army, which held the real reins of power in Thailand, took the stand of a disinterested spectator. The Chinese expressed no small degree of dissatisfaction at this attitude. It was even rumored that the British Army had fanned the Thai Government in its suppression of the Chinese. The Chinese residents in Thailand, holding in their hands, the major economic wealth and power of the country, could be an unwelcome existence to the British.

I greeted the first birthday (October 11) of my new life in such an atmosphere. Somewhere in Japan, my wife and my children would perhaps be thinking of me, still setting a table for me at meal time. They would be wondering whether I am alive or dead. Perhaps my children would be praying for my health. And my 75-year-old mother would also be waiting anxiously for me.

I handed 10 Bahts to Lao-Tai and asked him to buy as much delicacies, as possible with the money. He brought back three big lobsters almost a foot long and a bunch of bananas. This was all he could get for the money to grace our supper table. Old Chino insisted that he would make some "gomoku zushi" and he brought out from a hiding place in the bottom of his trunk a portion of treasured "sushi" rice. He put heart and soul into the making of the "gomoku zushi" and fairly outdid a professional cook. Since my seven young disciples would be unable to celebrate this occasion with me, I decided to send a portion of the food to the children of a Thai Government official who had moved only recently into neighboring house.

In clumsy Thai, I said: "This Japanese rice. Please eat," and I handed the two a bowl filled with the "sushi."

These two children, brother and sister of about five and seven years of age, were extremely pretty and looked as if they contained European blood. They had often come to play in the temple yard but they had never dreamed of receiving a gift from a priest.

In Thailand, it is the custom of Thai priests to go early in the morning with their begging bowls to the gates of believers and to receive their day's food. They never give anything away. It was therefore natural that this Thai home should be surprised at receiving unexpectedly some food from a Japanese priest. Soon, the two children carried to us, with an effort, a heavy dish heaped with papayas. The two children were about the same age as mine and somehow these strange children of another race seemed as sweet and lovable as my own whom I had left behind in my native home. I felt deep gratitude for the kindnesses of the next door people, knowing that they were risking a great deal in these perilous times to show favor toward a Japanese.

The year before, I had spent my birthday in the front lines in Northern Burma. I had picked bracken sprouts for my birthday dinner in our headquarters in the bamboo groves of Monyeu. I wondered where would spend my next birthday.

I asked my Thai language teacher to have dinner with me. But he politely refused.

I gave him a bundle of dried fish that I had brought from the Japanese Army. I saw him off to the gate and clasped my hand in prayer as he left.

Thinking that I would have to live in this temple for a long while, I had stripped the grass from a plot of ground in the back yard and had planted a vegetable garden. The seeds were already sprouting, pointing two delicate leaves each toward the sun. The baby birds in a crow's nest built in the branches of a willow tree growing in the compound, gradually matured and four small baby crows would flutter to the ground in this peaceful temple in search of food. The parent bird would stand close guard while teaching its young the ways of a bird's life. This love, no different from that of a human being for its offspring, touched me deeply. Despite the bloodshed and the uneasiness filling the Buddhist city of Bangkok, beyond the temple hedges, this compound was all peace, a paradise of quiet where small living things and plants nurtured their will to live.

In contrast to the brutality and sanguinity of the first half of my life, my first birthday in my new life was spent in peace. I had changed my uniform for the robe of a priest, my sword for rosary and I watched in peace the growth of birds and plants amid an ocean of danger.

I read of the misery and tragedy that had overtaken my homeland from the incomplete and vague accounts in the Chinese vernacular papers. And I wrote a verse for my seven disciples in Mahatat Temple and placing it in a rice bag sent it to them care of the young Thai boy, Dick.

This was the verse I wrote:

Ominous are the clouds that darkly whelm the eastern skies. A hundred devils daily trod our homeland hills, Gallons are the searing tears that wet this yellow robe But for a while must I sit within this crucible, steeling my soul.

Immediately after the tragedy of the September 28 Incident, the Chinese residents in Bangkok launched a city-wide strike. All leading stores closed and barred their doors. Rice shops, fish markets and vegetable vendors refused to deal with Thailanders, Only well known Chinese customers were served from the backdoor. There were many days that Lao-Tai returned home empty-handed, disappointment written over his entire frame.

Fortunately, we did not want for food thanks to the rice, the dried fish and the dehydrated vegetables that I had brought from the Army. But, we could no longer buy any fresh foods. Lao-Tai's common-law wife (of mixed Chinese and Thai extraction) had never entered the temple compound in deference to the Buddhist ban on women, and had always met her husband outside of the temple gate. But, now, she disguised herself into a vendor and entered boldly into the compound to sell her wares. Through this route, we were able to obtain fresh fruit and fish during the strike.

For five full days, the Thailanders suffered. Farmers and fishermen saw their bananas and vegetables and fish rot away. The great Thai leaders could no longer indulge in luxuries, while the ordinary citizen, deprived of rickshaws and samro, were paralyzed.

There could have been no better demonstration of the deep underlying strength of the resident Chinese. The five-day strike showed convincingly that the economic basis of Thailand rested with the Chinese people. The British occupation authorities could not rest unconcerned in face of Thai's paralyzed economic life. They ordered the Thai police and army to ease their suppressive measures.

The ability of the Chinese, without the protection of national prestige or military power, to expand to the outermost corners of the world and there to send down rugged economic roots, calls for unconditional respect. I realized that if these Chinese residents could be led along correct political lines and their scattered strength ably organized, China would be able to gain control of the whole Southeast Asia area, without using a single soldier.

As tension increased in Bangkok, I knew the danger I could be placed under arrest at any moment. Not a single Japanese both civilian and soldier, was to be seen on the streets of Bangkok. I knew that I would have to be on constant alert, gathering what information I could from Thai priests and the Chinese.

At the end of September, non-commissioned officers from the headquarters arrived to take away the list of remains of Japanese soldiers and officers deposited in the ossuary. I realized that unless I took advantage of this opportunity I would lose my last chance to contact the headquarters. I wrote a last letter of farewell to the Commander, placed the last 100,000 Bahts of the Headquarters secret service funds in a wooden box for the ashes of the dead, covered it with bones and sent the whole with the non-commissioned officers back to headquarters.

It was apparent that the special privileges of priests could not be maintained in face of pistol muzzle. It seemed my very morrow was in danger–let alone the ten years of underground activity that I had first planned upon.

I knew that the officers and men placed in the concentration camp would need every Baht in their restricted, and inconvenient life. And felt great relief when I had sent back the greater part of the money entrusted to me for hiding in the ossuary. I sent what little remained to my seven disciples and only left behind for my own emergency use 10,000 Nationalist yuans and 1,000 Thai Bahts.

The only other thing that I had left was my sword. I retained this to the end in anticipation of being attacked and kept it hidden in the hollow back of the image of Buddha. On my own person, I retained the prayers beads given to me 20 years ago by my mother.

At the beginning of October, the Free Thai Cabinet changed the name of Thailand to Siam and began suppressing the Japanese. I could feel the stealthy approach of danger. I wondered whether I should move to a Thai temple and disguise myself as a Thai priest and move from one temple to another. I called Kubo, one of my disciples staying at Mahatat Temple and asked him to negotiate with the Chief Abbot. The Chief Abbot readily agreed to take me in. However the day before I was scheduled to move from the ossuary, Kubo come frantically to me, his face sallow with fear and said:

"A terrible thing has happened. This morning, when the Chief Abbot got into contact secretly with the Chief of the Thai Bureau of Religion, he was told that all Japanese priests would also be placed soon in detention camps. The plan is to carry out our wholesale arrest without prior notice early on October 29."

Kubo could hardly speak for excitement.

I too felt my heart thump. The Chinese vernacular press had for some time reported rumors that Japanese army officers had gone underground in disguise. Chinese sources had also reported to me that the British authorities had changed their attitude and had decided not to treat priests as exceptions. All Japanese priests were to be detained and questioned until their true identities were established.

I know that if I were placed in a concentration camp, I would soon be recognized by some of the Japanese inmates, who had known me before the defeat. On the other, there was too strict a surveillance for any possibility of my escaping out of the country. I felt that the end had come.

However, it was by heaven's grace that I had been able to learn one week in advance through the kindness of our Thai friends, that the British planned a surprise raid on us early on October 29. I knew that I would have to utilize this valuable time to its fullest extent and see a way out of my trap.

That evening a group of uniformed Japanese officers and non-com's suddenly appeared at the ossuary.

"The senior aide-de-camp has ordered us to take home all the remains of Japanese officers and men deposited here in the ossuary."

There had been no forewarning at all. Neither did any of them seem to know that I was Norinobu Aoki. No, they were pretending that they did not know me. As long as they were officers attached to the headquarters, they should have known my face.

It was evident that they were extremely afraid of being involved in my affairs. The remains of the soldiers and officers were kept in the underground vaults of the ossuary. It went to where two of the men were getting the remains ready to take home and asked in a small voice:

"Say. What in the world has happened? Why do you have to take away all of the remains so suddenly."

Something quite unexpected has taken place. Last night, the British headquarters sent over an urgent order, demanding Staff Officer Tsuji's immediate appearance. The Chief of Staff told the British that Staff Officer Tsuji left a note and disappeared on August 16 and that Tsuji, judging from his character, had probably committed suicide. "On orders from above, we are taking back these remains as soon as possible to the headquarters for the purpose of making a clean cut between the Army and the temple."

Everything was now clear to me. The worst had come. There had not been a single letter either from my commanders or from my colleagues. If I had not taken the initiative to ask, I would have remained completely at a, loss to explain this new development. It was certain that everybody was extremely afraid of having anything to do with me.

“Alright. Such being the case, I won’t rely on anybody. And, I must also not cause anybody any trouble. Today, I cross the dividing line. If the British only knew that I disguised myself as a priest, am living incognito right next to their headquarters, they might come tonight at the latest to get me. If this happens I’ll fight to the last with my sword until the rivets holding the blade to the haft break.”

And I steeled myself to meet my doom.

Ever since my desertion, I had never felt as oppressed and stifled as this day. In reaction, I found my indomitable will to fight flare up within me like a fierce flame. In the bloody fighting in North Burma, the units defending the fortifications at La-Meng had fought even after they had lost their legs and arms and not a single man surrendered in this fight to the last. I resolved to stay underground in spite of all dangers. I thought the Imperial Rescript terminating the war is undoubtedly a Rescript, but so is the Imperial Rescript declaring war. If one Rescript is obeyed the other is automatically disobeyed. In such an age as now, the only thing to do, I felt, was to throw away all thoughts of self and to choose, on one’s own initiative, the path one believed correct in working for the eternal future of Japan. From the standpoint of the British Army, it was natural for them to hunt to the corners of the earth for me. Ironically enough, Churchill had said “Undaunted in defeat, magnanimous in victory, liberal in peace and resolute in war.” If Churchill were in my place, he would undoubtedly have done what I did.

I told the departing senior officer:

“Thanks for telling me everything. I have already made my final resolution and shall die bravely when the worst comes to the worst. I will not cause anyone of you any trouble. Please convey my feelings to all.”

When the time came to see off the 150 odd remains of the war dead, which I had faced in my morning and evening prayers each day for over two months, I was filled with sadness and a sense of impending tragedy.

As long as these remains had rested in the ossuary, I felt as if the spirits of the dead, for whom I prayed, watched over me. But, now I felt as if I had lost my last support. I was now completely alone in a hostile world. The tragic determination that filled my heart seemed to freeze the blood in my body. I realized fully how easy life had been in the great organization of the Army with supports both above and below and both left and right of me. Even when I had stood at the head of my troops in a hail of bullets with the gold braid of a staff officer slung over my shoulder, life was easier than this utter sense of aloneness that now oppressed me.

All the strength I had was poured into the performance of the religious rites that evening. Old Chino was scared and visibly agitated. He felt that he would be inevitably involved. Kobiki Shonin, on the other hand, retained his composure.

I felt as if I were a fish caught in a far-flung net that was slowly being drawn around me. The net seemed too high for me to jump over and too strong for me to attempt a bodily break-through. I am ashamed to admit it, I couldn’t sleep a wink that night. I felt a certain sense of uneasiness at the sound of Lao Tai’s walking stick as he went his nightly round of inspection. Even so, the sound gave me a certain sense of comfort to know that someone was watching. I mistook often the light of passing automobiles reflected in the windows the light from the flashlight of men come to arrest me.

On the morning of October 23, I awoke early. I poured water over my whole body in purification and began to pray in earnest before the image of Buddha. When I had completed offering the "introibo" and my praises of Buddha, in a state of complete detachment from self. I imagined I saw as if in a dream Saint Nichiren [a Japanese Buddhist priest the Kamakura period whose teachings form the basis of Nichiren Buddhism] appear before my eyes. His big eyes seemed about to pop out of their sockets and in a voice like a thunderclap said:

"Decide!"

I could not tell whether it really happened or whether it was a hallucination. However, on the inspiration of this suggestion, I resolved to smash through the surrounding trap.

The only way to gain life in the face of death is to crash head on toward death itself. It is only the propulsive power toward death, so great as to cow the god of death himself that is the means of opening up a passageway to life. When I resolved to trust my fortune to heaven and stake everything in one determined charge at the very breast of the god of death I felt the suffering and anxieties that had beset me from the day before fade away as does a summer shower. I felt as if I had gained transcendence over all worldly cares. After deep thought and the weighing of all factors, I reached the conclusion that I should throw myself into the headquarters of the Chungking Blue Shirt Society [a secret Chinese organization formed March 1, 1932, in response to the Japanese invasion of Manchuria by members and graduates of the Whampoa Military Academy] on Surion Avenue.

The Chungking underground had continued its secret activities through the war in the streets of Bangkok. However, with the end of the war, they came out into the open and put out a sign:

"The District Office of the Overseas Section of the Kuomintang Party of the Chinese Republic."

They had moved into a former Japanese company residence on Surion Avenue and were tasting to the full the joys of a victor nation.

The Chinese underground expended every effort in negotiations with the Thai Government to bring the September 28 Incident to a satisfactory solution. At the same time, they did not halt in their work of carrying on the exposure of collaborationist and traitors among the Chinese residents of Thailand. The former work was accompanied by many headaches and dangers, while the latter offered much profit.

I hesitated in throwing myself directly at the headquarters and thought of sending a letter, stating my case, through some Chinese. But, in the atmosphere of the times, I feared that the Chinese I would use on such an errand stood in danger of being arrested as a collaborationist, if he had ties with the Japanese. Thus, I decided that I would I have to proceed by myself.

I knew that if I could offer sufficient money, the Chinese would risk their life to take me to China. However, what funds I had in hand were insufficient even for travel expenses.

It was a great help when Kobiki Shonin, of his own volition, said that he would accompany me to the Chinese headquarters. He was familiar to a certain extent with both Thai and Chinese languages. We went together and shopped at a certain Chinese store and asked the proprietor whether he would carry my letter to the headquarters. However, the store keeper was unwilling to accept the task. Finally, he said:

"I'll take you to the headquarters on Surion Avenue and you hand them your letter personally."

That was the most he would do for me.

The proprietor of the Chinese store led the way on his bicycle, while Kobiki Shonin and I followed, sitting together in a "samro." In this way, we passed the scrutiny of British sentinels.

Next door to the headquarters of the Blue-Shirt Society was a British officers' club. Fortunately, it was early Sunday morning and only one or two British officers were around.

I left the "samro" waiting with Kobiki Shonin aboard. I went into the gate, acting like a Thai priest begging for food.

To the thin, white-faced youth at the reception desk, I asked:

"Is Mr. Cheng in?"

"Yes," he answered.

I handed the youth my letter which I had hidden in my inner sash. The youth did not seem surprised and went inside smiling after bowing to me. I watched him go, all the while clasping my hands in a posture of prayer, and then I left. Later, I realized that I should have waited for an answer. However, at that time, I was relieved at having been able to hand the letter over and felt that I fully succeeded in my initial venture.

In the letter, I had written in poor Chinese:

> *"I am a Japanese officer. I have gone underground disguised as a priest. I am staying at the Japanese ossuary near the British headquarters. I have secret information on the September 28 Incident and would like to tell it to you."*
>
> *"I would like Mr. Cheng to come directly by October 26 to the ossuary. If that is not possible, I would like Mr. Cheng to dispatch his most trusted secretary. I am a person for whom the British Army are looking. I am to be placed in the concentration camp on October 29, and would like to speak with you before that time, even for a few minutes."*

Coming away from the headquarters, I felt that I had succeeded in escaping from the tiger's teeth. When I returned home to the temple, through streets filled with British soldiers, it was already noon. I felt as if a solitary ray of hope had pierced through to my heart. For several days after that, I waited with impatience, watching the front and back entrances to the temple compound, wondering, in alternating spasms of joyful anticipation and fearful expectancy, whether Cheng would come first or the British.

October 24 passed. October 25 ended uneventfully. October 26, the day that I had waited for with such hopes from early morning, passed without a word from Cheng. I thought that he might come late at night and waited up until midnight. Neither friend nor foe came.

After all, I told myself, China and Great Britain are allies. There was every possibility of that letter being sold to the British. The more I thought of the possibilities, the more anxious I became. That night, again, failed to sleep a wink. I tossed all night on my hard bed and greeted with a heavy head the arrival of October 27.

But, October 27 also passed with nothing happening. Minute by minute the final day was approaching. My disciples at Mahatat Temple began to show anxiety on their faces. If everything failed, I knew that I would have to enter the concentration camp, from where I could attempt a breakout. Thinking of this possibility, I got together a few items to prepare for a civilian life. However, I could not get over feeling regret after having established my first contact with the headquarters of the Chungking underground movement.

Old man Chino began to get nasty. He started saying that he could not bear living together with an ominous time bomb like me. I also knew that I could not cause Kobiki Shonin any further trouble. On the night of October 27, I dreamt for the first time in a long time of my dead father. My pale-faced father scolded me.

"What are you hesitating about?"

And I woke with a start. The morning sun of October 28 shone red through my window. It seemed to bless and encourage me.

I left the temple alone, filling my heart with prayer. I picked up a samro and hurried toward Surion Avenue passed through a number of check-points by making out that I was a Thai priest. Once again I entered the Chunking headquarters.

The youth, who had received me the other day, greeted me with a smile and led me into waiting room. After a short while, a difficult looking youth swept into the room. In written exchanges, I learned that he was Section Chief Kuo. A little later Chief Secretary Cheng turned up for work. He was no more than 30 years of age. A thin and rather pale man, he was nonetheless handsome looking. Who would have imagined such a youth as being the representative of Lieutenant-General Hsieh responsible in actuality for the 1,500,000 Chinese in Thailand. He resembled Colonel Hattori in his youth. I was seized with the illusion that this youth had come instead of Colonel Hattori to save me from danger.

With the big paws of a country bonze, I gripped the delicate hands of this youth in a grasp of steel. I could not help notice the slight blush that colored his face. For roughly an hour we conversed by written exchanges. He, being a native of Canton, could not understand my broken attempts to speak the official language of Peking.

My real name, my career, my part in the East Asia Federation Movement, the memorial service for Generalissimo Chiang Kai-shek's mother and my relations with Tai Li, I wrote down everything that I could.

I want to go to Chungking, meet General Tai Li and Generalissimo Chiang Kai-shek and open the way for Japan-China collaboration. If this is impossible, please arrest me immediately and hand me over to the British authorities. I am not a man to fear for my life and run away."

This was the conclusion with which wound up the lengthy conversation. The two youths nodded as they read my answers and I noticed their faces gradually color with excitement. Their eyes were shining. When had finished, Cheng then wrote:

"Please wait. We want to hold a conference," and then left the room.

"I wondered whether it would be good news or bad. I waited for 30 minutes and when they came back, both were smiling. I shouted to myself:

"You've made it."

Cheng wrote with delicate fingers just two words:

"Very good."

He told me then to arrange everything with Section Chief Kuo.

Kuo wrote that he would come to call for me at nine o'clock the same night with a car.

I had finally succeeded by my decision to throw myself bodily at the problem.

Yet, to think that these youths, still not quite 30, should handle, the affairs of 1,500,000 Chinese residents, as representatives of Lieutenant General Hsieh, negotiating on more than an equal basis with both Thai and British authorities on the problems arising from the September 28 Incident I was filled with envy. Looking back, I, at their age, was still a lieutenant in the War University and the best that I could do in comparison with these men was to quarrel with my teacher on tactical warfare.

After all, these men were civil servants of China. To protect a man wanted by the British authorities could well cause international complications if handled clumsily. If they had been Japanese bureaucrats or members of the diplomatic corps, just what measures would they have been able to take? In all probability, the best they could answer would have been: "We'll ask Tokyo. Just wait a while!"

Filled with emotion and excitement, with a serene exterior, I returned again on a samro to the temple.

Old Chino breathed a sigh of relief. Kobiki Shonin congratulated me on my success from his heart. I also told my disciples in Mahatat Temple the good news. With my arrest scheduled for the next day, I had succeeded in planning an escape only the night before.

From several days ago, British sentinels had begun to be stationed at night a short distance from the front gate of our temple. Inside the temple gate, plain-clothes Thai police began to appear constantly. It was evident that we Japanese priests had been placed under close surveillance. It was not an easy job to get through this barrier.

On the afternoon of October 28, I hurriedly cleared up all my personal matters. I felt that neither the British nor the Thai authorities (except ex-Prime Minister, Apaiyon) know as yet that the priest, Norinobu Aoki, was in actuality Staff Officer Tsuji. I concluded that the best thing to do as to make out that Aoki had committed suicide. Then, Chino, Kobiki and the seven youths would not be held responsible or be grilled. I shut myself into my own room and wrote my last letters.

> *"To the Chief of the Bureau of Religion of the Thai Government, and to the Chief of Police:*
>
> *"I am deeply moved with gratitude at the special consideration accorded me by you and the people of Thai, in the midst of a difficult international situation.*
>
> *"Faced with the fate of a defeated national have lost all confidence in my own future aged, as I am, I regret extremely that I am causing you trouble, the longer. I live without being able to repay any of your kindnesses. I have, therefore, decided to throw myself into the waters of the sacred Menam River and find everlasting life in the bosom of Buddha. In the life to come, hope to become a link in the bonds of Japan-Thai friendship."*

"I pray that in consideration of my life, now to be ended, you will continue to help my comrade priests and my seven disciples."

"Realizing that I would be causing you great trouble if were to throw myself in the river while still a priest, I am returning my priestly robes and my certificate of priesthood (identification card). I shall die as a worldly man."

"I pray for the prosperity of your country."
"October 28 (night)"
"Japanese Priest"
"Norinobu Aoki"

"To Kobiki, Chino and Sasaki Shonin:"

"I have no words to express my gratitude to 'your for you special kindnesses during the long past. Please allow me to a become a Buddha one step ahead of you have lost all hope in my old age and hope to find everlasting life by throwing myself into the Menam River."

"I pray that you will all remain in good health and return safely to Japan. I ask that you guide and care for the young men that I am leaving behind.

Norinobu"

"To My Seven Disciples: "

"I have no words of apology for leaving this world one step ahead of you. I have, however, chosen death, unable to bear the thought that I will be causing you more trouble when I in my aged condition, am made to enter a concentration camp, with all its inconveniences. I have been an unworthy leader of our group. Please do not consider me as dead: Each time you see the Menam River, think that you are seeing me and study hard and grasp the nucleus of Buddhism. I shall be praying in the other world for you, so that you will form in the long future to come the spiritual tie through Buddhism between Japan and Thai. "

"Norinobu"

False wills as these letters were, I could not help the tears from forming in my eyes as I wrote on. This was perhaps because I thought they might become really my last will and testament. I cleared up everything personal. All that was left was to say good-bye to Lao-Tai, his common-law wife and his children.

For 70 odd days this old Chinese had worked for me with unchanging fidelity. His wife, too, had come often to bless our table with gifts of fruit and vegetables. I simply could not pass them by as would people on the street had watched with pity how Lao-Tai, afraid of old Chino had not allowed his wife to enter the temple compound and had kept her waiting at the gate to take home the left-overs from each day's meal. Thus, when I allowed her to come into the compound, the family almost worshipped me in gratitude. With the passing of time, I had come to love Lao-Tai's two children, brought into the compound by their mother. When they heard that the final parting would come on the morrow, they raised their voices in tears. (They did not know that I was to commit suicide.) They had believed all along that the Japanese ossuary would be their sanctuary in old age, and now from the morrow they would be thrown out in the street without any support. I gave Lao-Tai half of the money I had and let him take home all of my poor worldly belongings and a large part of the food that was left.

To Kobiki, Chino and Sasaki Shonin, I said:

"I feel that if I were to stay together with you, I will be causing you great trouble in the future. Thus, I am making out that I am going to commit suicide and am leaving on a long travel. When you wake up tomorrow morning, please report to the Thai authorities as if you had just discovered my death notes."

The three listened in silence and wiped their tears.

I called my seven disciples in two groups and told them also of my intentions. We had only been together for two months, but we were people that had decided to die together, if die we must, and thus our parting was indeed painful. As dusk set in, I saw them off to the gate.

Some could no longer hold themselves and cried with loud voices, while others gritted their teeth and uttered low groans. When we thought that this would be perhaps our last parting, we felt as if my bones were being gouged.

In a wooden box, inscribed "Articles belonging to the late Colonel Masanobu Tsuji of the Japanese Army," I placed my sword and my blood-stained binoculars. Then wrapped the box in a yellow robe and handed it to my disciples to be placed in the custody of the Chief Abbot of Mahatat Temple.

I saw my yellow-robed disciples off with clasped hands until the dusk swallowed them from sight.

Around eight o'clock that night, the three of us finished our last dinner together. I knew that Lao-Tai would shut the temple gate at ten o'clock as he started out on his nightly rounds. I had to slip out before that time, unnoticed by anyone. I also had to fool the secret police loitering about.

It was completely dark outside. Every home was lighted up, as the family got together to enjoy supper. I turned off the light in my room to make it appear as if I had already I gone to bed. On my desk I placed a figure of Buddha, given to me by the priest in the adjoining temple, and offered a flower before the statuette. I then lit some incense. I placed my three death letters, my watch, my wallet, my identification card, my club bag and my priest's robes on the desk.

On the floor, I put together my leather sandals. Everything was arranged in an orderly manner. I could not leave any trails behind.

I slung a carpet bag over my shoulder, the bag in front and the blankets behind. Then dressed in the white shirt and trousers of a typical Chinese, I slipped into the dark from my room. I stopped for a moment in front of the main temple and offered a silent prayer. Then like a thief, I walked silently toward the main gate.

It was exactly nine o'clock, October 28, 1945.

The main gate, to describe it more accurately, was the front approach to the adjoining Ryab Temple. Left half-destroyed by bombing, it was now overgrown with grass and weeds and because of the infrequent worshippers, it had become the gathering place of thieves at night. In one corner of approach was an old poplar tree. It was a suitable spot to hide. Cowering behind the tree trunk, I waited for the promised car.

According to Kuo, an automobile was to come for me around nine o'clock to pick me up somewhere near the main entrance. With eyes as wide as saucers and ears alert, I waited. But, the car did not come.

It was slowly getting toward the hour when thieves would begin to gather here. I decided to hide in a nearby air-raid shelter. As I stepped from behind the tree. I felt myself tread on something out of the ordinary. The sharp smell of human faces hit my nose. I had really put my foot into it! I wiped my shoes on the grass, but the smell remained. I rubbed my shoes in the dirt, but the odor still persisted. I did not know what to do, when suddenly I felt the presence of someone behind me.

Thief? Police? Was I going to be caught?

I quickly stepped into the air-raid shelter, but no sound followed. I peered through the darkness to see who it was. It was neither an adult or a child. It was a lad from the adjoining temple. A young Thai lad who had been employed as a waiter by the Japanese gendarmerie during the war, he understood a little Japanese and also spoke a few words. I had spoken to him on two or three previous occasions and had become friendly with him. He recognized me despite my new disguise and in broken Japanese said:

"Bonze, bonze, you go home, you run way?"

There was no show of hostility as he approached. I thought that everything was at an end. I had been discovered despite all my precautions just because I had stepped on a clump of human dung.

"Bonze, bonze, dangerous, dangerous, English dangerous," and he pointed toward the street. There a helmeted British sentinel stood with, the lamp light glinting on his drawn bayonet. I was afraid that the sentinel would be able to hear the boy's voice. If this youngster felt any enmity toward me, this was the end. Without thinking I grabbed the boy's hands–the master of my fate.

Should I let him go? Should I hold him?

I decided that if he raised his voice I would kill him.

"Life?" "Death?"

One voice called to me, "Kill him." Another said: "Let him be."

Good and evil fought in my heart over the life of this small lad.

"Let him live." I had been a make shift priest for only two months. But, in that time, I had chanted sutras praising the infinite mercy and sadness of Buddha. Even if I snuffed out the life of the youngster, I would never be able to escape if Buddha did not will it. I could not kill this innocent child, just to find life for myself. The fact that for a moment, I had thought of murder, made me shameful and penitent.

"I like Japanese. I find samro," he said and showed kindness toward me. I gradually relaxed my hold on the child.

"Boy, save me. I'm going to Saigon. Don't tell anyone," and I placed my hand over his mouth. Against the possibility of his telling on me, I had on the spur of the moment thought of hiding my trail by telling the boy I would go to Saigon.

Then, I realized that I must also not let him think I was a soldier. As if afraid of the British sentinel, I shrank away from the side nearest the street and in a small voice said:

"I afraid English. Afraid. You help me. You stay here. Afraid, afraid."

I felt that I could not let this hostage go until late at night when people had gone to sleep. I placed a hundred Bahts in his small hands, then bitten by a cloud of mosquitoes we sat down under the tree. The automobile did not come to pick me up. Only carriers, mounted with machine guns came and went.

Every 30 minutes, these patrols passed by and every hour the helmeted and armed sentinels were relieved. The muzzle of their guns were always pointed toward the Japanese temple. The watch that I had bought for my escape showed one o'clock. The street outside emptied itself of passers-by. Only once in a while, I could hear drunken British soldiers shouting with loud voices as they chased Thai girls. The youngster began dropping off to sleep, despite the plague of mosquitoes.

Bonze, bonze, I sleepy go temple. I no tell anybody. No tell." He spoke pitifully asking me to release him. I decided to let him go, He looked so weak and tired.

"Okay. Thank you. Go home. Don't tell anyone," and I clasped my hand in supplication as I let him go. I left my fate to the gods and all alone I kept glaring at the British sentinels and at the British sentinels and the Thai police. It was heartening to know that they had not caught sight of me.

Two o'clock passed. Three o'clock passed. The streets became completely lifeless. Only the sentinels were alive and awake. Four o'clock came. My arms and face had become a swollen mass of pain from the countless mosquito bites. I thought that all would end when daybreak came. I wondered if my luck had all run out....

Wait, I told myself. The defense detachment at La-Meng did not surrender. I decided to say my final prayers. Turning toward the northeast, I bowed my head in deep prayer. With the rosary, given to me by my mother, in my hand, I said in a small voice the "introibo," the praise of Buddha and the Prajna Paramita Sutra. When I finished and looked up toward the gate, a miracle had taken place. Both the sentinel were seated on the sidewalk, wrapped in deep sleep. The gods and Buddha had not failed me. Hot tears rolled down my cold face.

The eastern skies began to lighten almost imperceptibly. Early rising Chinese passed by in threes and fives, carrying bundles of vegetables, headed toward the direction of the market. I adjusted the carpet bag and the blankets slung over my shoulders and started walking unhurriedly. I passed the gate. The two sentinels were sleeping away, their heads supported by their rifles.

I coughed as if to clear my throat and stamped more firmly on the ground to make steps heard. And I leisurely passed the sentinels. The sentinels, wakened by my footsteps, glanced upward a second, then as if satisfied at seeing a Chinese, went back again to sleep. I felt a big burden roll off my shoulders.

I mixed with the hurrying Chinese vendors and reached an intersection where a Thai policeman stood on duty. I put my luggage down beside the police officer and waited for a "samro."' Before the Thai policeman could suspect anything, a rickshaw passed by. It was clearly a rickshaw reserved for Chinese.

"Surion Avenue. Ten Bahts, Hurry, Hurry," I said in Thai.

The rickshaw man who had been on the lookout for a passenger put down the shafts of the rickshaw. My Thai language studies had been justified.

The rickshaw man must have been pleased to find such a good passenger as I. The regular fare to Surion Avenue was five Bahts. Here I had promised him ten. Completely happy, the rickshaw driver sped on, overtaking several "samro." On the way, he was told to stop by several British and Thai police. But, in an energetic voice, he shouted back, "This passenger Chinaman," and sped on.

There was little traffic along the streets and the rickshaw man outdid himself for speed. At 5:45 a. m., I entered for the third time the gate into the headquarters of the Chinese underground movement.

This hunted bird had finally found refuge in the bosom of an Oriental. The sun rose to the east, as if to brighten my future path. In this way, I passed through my first death barrier.

The Chinese headquarters showed no unpleasantness at the arrival of such an early guest and I was shown into the waiting room. Close to ten young men were sleeping on the office desks and the sofa. They made room for me on this sofa. I had intended to stay awake, but before I knew it I was dead asleep on the divan.

About one hour later, I was shaken awake. I was served rice gruel, fish and pickles for breakfast. But it tasted far better than all the delicacies in the world. I had stayed up all night too highly keyed to even feel fatigue or hunger. However, as soon as I had entered a haven of safety, I had begun to feel the effects of the night and ravenous hunger. The five or six unknown youths, who sat at the breakfast table, reminded me of my own boys, whom I had left behind in Mahatat Temple.

Kuo, whom I had met yesterday, came to work around nine o'clock. He told me in writing that he had come for me the night before at nine o'clock but had not caught sight of me. He expressed surprise and joy at seeing me and gripped my hand in a firm handshake.

Immediately, we left by car for a hideout in the suburbs of the city. Several times we passed by speeding cars carrying British soldiers and officers. Each time, Kuo moved in such a way as to hide my body from their view.

After travelling for about 15 miles, we came to a pretty little house. It looked like the home of a very wealthy Chinese. There I was introduced, to young Wu Chien-sheng, the son of the owner of this home. He was a large man, with a light brown face, resembling that of a Thailander. Just the same his features had a clean-cut tartness about them. He was a handsome and strong looking individual. Before leaving for the office, Kuo told me that this young man would look after me for a while and that I must not be seen outside. I was able to get along someway with young Wu, after I was left behind, with my pidgin Thai and by written exchanges.

His parents had moved during the war to Sahanaket to escape bombing and had offered this home to their son and to the Chungking underground movement. I saw a pair of half-length boots, used by air corps officers. I also noticed an army tunic in the room. I concluded that this home was used as a sort of billet by young officers and leaders of the Blue-Shirt Society.

Wu played Japanese records on the phonograph for me and borrowed some Japanese books from his neighbors to help me pass the time. (I learned that the neighbors were high officials in the Thai Police.) Wu brought out cooled papaya and his kindness was like balm to my wounded heart.

In the evening, seven or eight youths came home with energetic steps. I greeted them as if they were friends. They had completely liquidated our enmity of the past. In a warm and friendly atmosphere, I talked with them through written exchanges of notes.

"When did you fellows come to Thailand and for what purpose?"

"We were sent out from Chungking two or three years ago. Some few by plane to India. Others landed in Bangkok from submarines. Some were parachuted onto secret air fields and others crossed the Mekong River from French Indo-China. Half of us were caught by the Japanese Army and killed."

"What is your work?"

"The gathering of information."

"Whom do you respect the most?"

"Chiang Kai-shek and General Tai-Li."

"What are your ranks and ages?"

"Chief Secretary Cheng, Colonel, Artillery, 28; Section Chief Kuo, Lieutenant-Colonel, Cavalry, 28; Hu I, Lieutenant-Colonel, Artillery, 28; Han Major, Infantry, 25; and Tang Ying, Captain, Air Corps, 23."

With the exception of Chief Secretary Cheng, all were still single. They had all put off marriage until after the war. They had now won. Their brides had been picked or were being chosen. There was no reason why they should not be full of high spirits. They were all General Tai-Li's men, nuclear members of the Blue-Shirt Society.

I had to admit that they had done well. Despite sacrifices that took a toll of half their membership, they had succeeded in hiding out in Thailand and for several years had continued their underground activities under the strict eyes of the Japanese Occupation forces. They were members of the Blue-Shirt Society, a name feared by all. Children were said to stop crying if told that the Blue-Shirt Society would get them. These central figures of the organization, who did not hesitate to kill in order to accomplish their mission, were innocent and full of fun in their private life. They also seemed to be tied together by firm bonds of friendship.

The plans for my escape from Thailand were made carefully and in detail by these youths. It seemed that they had sent a detailed cable to Chungking, asking for instructions. They had asked for a plane, but because of the strict screening enforced by the British, it was decided that I leave by train, I was to pass through Ubon, reach Vienchan [Vientiane] and enter Chungking via Hanoi. It was a long journey that I faced.

The night before my departure, they scrutinized with minute care everything that I had on me. Nothing could be allowed that would mark me as a Japanese.

In the end, two things were left as impossible items, a sake cup given to me by the Emperor and a pair of cuff links from Prince Mikasa. They also must have sensed that these were Imperial gifts. They had served as memorable talismans during my long life on the battlefields. However, to carry them with me from now on would possibly lead to their eventual desecration. I decided to give them up. I gave the sake cup to Chief Secretary Cheng and the cuff links to Section Chief Kuo and I asked them to treasure the articles as eternal memories of our meeting and friendship. They were wild with joy and wrote:

"Invaluable treasures."

It seemed that we had something in common between us.

I also gave away the other dubious articles as keepsakes. A suitcase and a Chinese merchant's suit were procured for me. I put on a white pith helmet and colored glasses. The typical outfit of a Southern overseas Chinese was prepared for me by experts.

It was decided that we leave on the 8:00 a. m. train on November 1. I felt a pang of regret at having to leave this life, though it had only lasted three days. My luggage was limited to one small suitcase. It seemed that the job of guarding this guest would be no easy matter. Two youths were chosen from among the men living together in this hide-out. The two were Wu and Liang. The faces of both men became tense and pallid, when they were named as my escorts.

In order to dodge the police inspection at the station, Liang left the night before our departure. His task was to place the luggage on the train, sit up all night and hold seats for us. Wu became overnight a lieutenant of the Thai Military Police. He crept into the home of a Thai MP, stole a uniform and identification card, He then changed the photo on the card and presto, he had become a real Thai MP. Wu spoke the language like a native and looked like a native Thailander. The idea behind the quick-change artistry was for this bogus MP to escort a sham Chinese through the barriers of death.

Early in the morning of November 1, we left our hide-out. Kuo, Han, Tang and Wu got on the car with me. When we reached the central railway station of Bangkok, Thai, police, under the supervision of British troops, were scrutinizing the identification cards and searching the luggage of passengers. This was to prevent the escape of Japanese. Once caught, it was the end. I left everything up to my escorts. I was told to stay in the car, wearing my colored glasses and my pith helmet, until the moment before the departure of the train. In front of the station, a long line of disarmed Japanese troops were undergoing inspection. They seemed so pitiful, I could not look at them.

Five minutes before eight, the bell warning of the departure of the train began to ring. Tang came running with the tickets. I took my ticket and headed for the turnstile.

The strict examination of passengers, which had thrown the station entrance in chaos, had been ended. I realized why I had been told to wait in the car till the minute before departure. I ran through the turnstile, as if I had arrived late and was hurrying to catch the train. Neither the Thai police nor the British seemed to think the old Chinese, waving only a newspaper in one hand, a dangerous fugitive. Perhaps there was not time to halt me.

I pushed aside the row of passengers clinging to the windows of the train and with an agile jump pushed myself through the window of the already moving train. Of course, I felt a slight fear that someone would spot the amazing change in the old and tottering Chinese merchant transformed suddenly into a lithe panther. However, I couldn't be bothered by such a worry.

The train puffed slowly out of Bangkok station.

Kuo, Han and Tang looked relieved as they waved me, off. Liang, who had got on the train the night before, had succeeded in getting seats close to the toilet. The seats faced the wall and its occupants could not be seen by the rest of the carriage. In addition, it was a third-class coach and workers, merchants and priests were packed in like sardines.

The old Chinese, with his colored glasses, his white pith helmet, his white jacket and black trousers calmly took his seat nearest the window. Then he unfolded the newspaper in his hands and began scanning its line.

"Japanese Priests Placed in Concentration Camp!" The headlines screamed. I read the news item over and over again.

It said: "Even Japanese priests, though they may be priests, are enemy nationals." It was a great mistake to have treated them with magnanimity till now. During the war, many

Japanese spies disguised themselves as priests and engaged in espionage and counter-intelligence. Their crime is thus heavier in one respect than the ordinary Japanese. It was natural, therefore, that on morning of October 29, all the Japanese priests in Bangkok, a total of 10, should have been forcibly placed in concentration camps. The priests arrested are 'Sasaki, Kobiki, Chino, Kobayashi....

The name of Norinobu Aoki was missing. Neither was anything said about his suicide. I wondered whether they believed it was really a suicide and did not announce Aoki's death or whether they had become suspicious and were secretly investigating my whereabouts.

Whatever the case may have been, I had successfully passed through the second death barrier.

More than half of the passengers in our carriage were Chinese. There were also Thai merchants, soldiers and priests, and conversation filled the carriage. The Chinese, in particular, talked and ate from morning to night. If anyone should broach a conversation with me, I would have been in a fix. In accordance with instructions from Liang and Wu, I acted as I a deaf Chinese merchant, accompanied by his servants. No matter who spoke to me I was not to turn my head. It was not a simple act to put on the whole day long.

The burning sun turned the packed carriage into an oven. My throat burned with thirst. The two youths, in turn, constantly plied me with ice water, sweet potatoes deep-fried on skewers and fried noodles. They were supposed to be my servants and I received their gifts without even a nod of acknowledgement.

Each station along the road was lined by vendors with an amazing variety of commodities. A barbecued chicken could be had for one and a half yen. I tore apart the two legs and bit in. Then without hesitation I wiped my oily fingers on the sleeve of my jacket. Fingers were used for everything.

Thai police were also lined along the stations, wearing and carrying with pride confiscated Japanese swords and rifles. Japanese army boots were to be seen everywhere.

Underground members of the Free Thai Movement were stationed at every place, maintaining a network of vigilant surveillance. In the compartment also were several evil looking men, who kept watching every move and every action of their fellow passengers. It was torture to maintain this attitude of hauteur and selfishness and this stony silence. Time and time again, I found I had to catch myself on the verge of expressing gratitude for the care which the two youths showed me, looking after my every unexpressed whim.

Thanks to my profuse perspiration I did not have to go to the toilet.

This was a blessing.

That evening we reached the station of Kohraht [Korat City, Nakhon Ratchasima Province]. We left the train, picked: up a pedicab and sought lodgings in a Chinese hotel. Without a moment's time to wash the day's sweat and grime, I was whisked away to eat supper in one of the town's restaurants. I noticed that all the larger stores were owned by Chinese. We returned around 10 p. m. and despite the bed bugs and the mosquitoes, I passed the first night of my escape uneventfully.

The next morning, November 2, we woke at 5 a.m. and headed toward the station. On the station platform a Thai girl was selling gruel, Together with hot soup, I drank down several bowls and allayed the hunger pangs in my stomach. We entered a train waiting in the siding. Inside we noticed two Thai couples who had bought several tickets and had boarded the train the night before. They had monopolized several seats and despite the overflowing crowd refused to give up the seats. Passengers, just arrived before departure time, began haggling over the price of the tickets held by the Thai couples. Eventually, just as the train was about to leave, a bargain was reached. The tickets changed hands at several times their original value. The couples left, throwing smiles all around the carriage.

The train, scheduled to leave at 8. a.m., did not get off until ten o'clock.

I thought: "Perhaps they've discovered my escape. Perhaps the train is being held up for a surprise inspection," The psychology of the hunted is to worry about the smallest thing. That is the reason why thieves are caught. Their inner anxiety is reflected in their outward attitude.

The day's heat was especially oppressive, Perhaps because the helmet was a little small for my head, is seemed to bite into my skull and hurt I took it off and wiped the sweat. Liang immediately warned me with his eyes never to repeat this action. Perhaps he was afraid the big scar on the right side of my head would attract notice. During the trip, sinister looking Thailanders were constantly watching me, never for a moment taking their eyes off me.

It was in the middle of the night that we reached the border town of Ubon [Ratchathani].

I greeted November 3, the Emperor Meiji's Birthday, without having had a chance even to doze through the night. I prayed that I would be able to pass successfully through the third death barrier. Both Wu and Liang showed a tenseness not evident till that day.

Wu wrote:

"Today, we pass through the greatest barrier. Please be careful."

In front of the hotel, a sleepy British sentinel stood guard. From the second story window of my hotel room, the sentinel looked very stupid. I told myself that he wouldn't be smart enough to cause me any trouble. I hoisted my suitcase on the right shoulder and passed by the left-hand side of the sentinel. The suitcase hid my face from his view.

We travelled by bus from the town of Ubon to the crossing point on the Mekong River, forming the international boundary. We rode for two hours in a rickety charcoal burning bus that was packed beyond capacity. My face, my hands, and my clothes were covered by soot and smoke. My white pith helmet was completely ruined.

When we reached the boundary outpost, a Thai MP came to examine us. Fortunately, all the passengers were so covered with soot, the MP could not differentiate our features. Thanks to this, the examination proved perfunctory.

The biggest problem was the crossing point on the Mekong River. Here British troops and Thai police were stationed in great numbers, maintaining such a strict watch that even stray kitten could not get through. The two youths looked at the scene and I could see their faces set hard with determination they realized that I could not get through as a deaf Chinese and I was to act as if I were dumb. I, resolved that no matter who kicked me or stepped on me, I would not utter a sound.

We stopped at a small restaurant and while I was washing my coal black face and hands, Wu disappeared somewhere. He had decided to risk his life to get me through this most dangerous of barriers. I learned later that he walked unflinchingly into the Thai Gendarmerie outpost, in his disguise as a Thailand Gendarmerie lieutenant. There he had cleverly convinced the men stationed there that he was a real gendarme. He told, them that he was escorting an old Chinese whom the gendarmerie in Bangkok had used as a spy and wanted help in getting him across to French Indo-China without letting the British know. In appearance, language and make-up, Wu did not arouse a bit of suspicion and succeeded through his composed manner in his negotiations. He received permission to take me across the Mekong at a point where there were no British soldiers. He returned to us, smiling triumphantly, as if he had cut off the very devil's head. Picking up my suitcase, I followed Wu to the gendarmerie outpost. They did not bother to examine either my identity or my luggage very closely, and I was placed on a dugout.

I looked at the broad breast of the Mekong, flowing placidly from west to east. The river seemed 1,000 meters or so wide at this point. The current, too, seemed to be quite rapid. The dugout was like a small leaf in comparison to the expanse of the river. It was only about 30 feet long made from large tree, typical of this country's tropical jungles, that had simply been hollowed out. Two large bamboo posts were lashed to its sides obviously to prevent the boat from tipping over. Although I had passed the check point, it seemed another problem to cross this wide river in such a flimsy vessel.

Several times the boat threatened to tip over and each time it was righted. Finally we left the banks and head for French Indo-China on the other side. Minute by minute the distance between us and Thailand grew. Often I was seized by fears that someone would call out at any moment telling us to halt. It seemed that needles were being shot into my back.

I held down an impulse to tell the boatman to hurry and placed my fate in his hands and his long pole. The flow mid-stream was swift and turbulent. One mistake and it would have been the end. Even if the boat were to overturn, I'll cling to its sides, I thought, and tightened my grip on the sides of the dugout. A short while after we had passed midstream, an unknown Chinese fellow passenger handed me a boiled egg. Relieved at the approach of the other bank, I almost forgot myself. On the verge of saying thank you as I took the egg, I checked myself in time and mumbled a few unintelligible sounds in my mouth, in, imitation of a speechless person.

At noon on November 3, the Emperor Meiji's Birthday, the dugout finally reached the French Indo-China side. When we stepped one foot out of the dugout, from the banks of hell to the banks of paradise, both Liang and Wu, spontaneously, shouted.

"Safe! Safe!"

Then they took out their pistols from their pockets and put on the safety catches. I realized that they were fully prepared when caught to shoot their way out, irregardless of who stopped them, Britains or Thailanders.

At the end of October, a proclamation had been issued throughout the whole of Thailand saying:

"Any person, whether a Thailander or a Chinese, who harbors a Japanese and helps him escape, will be summarily shot on the spot, without trial."

Our greatest and most dangerous barrier of death was in this way passed successfully on the auspicious birthday of Emperor Meiji.

I owed everything to these two youths for my escape from Thailand. Yet, both were still in their early twenties. It was an act which no older person could ever have done. Liang told me that he had crossed and recrossed the Mekong River at the same point as an underground worker, to be discovered and shot at by Japanese gendarmes only to escape miraculously each time. I could not but marvel at the workings of fate that an enemy Japanese national should have been placed in the hands of one who had been hunted by the Japanese themselves and should have been saved by a former enemy at the risk of his life.

Once we crossed, I found that this area was Liang's own territory, so to speak. Every Chinese youth we met seemed to know him and signaled greetings with their eyes. At a small restaurant on the river bank, I gorged myself with food. Together with the sense of overwhelming joy at having escaped from my enemies, I found myself suddenly tired as if all the anxieties and worries of the past 80 days since the end of the war had burst over me.

We went to Vienchan by bus. The bus was full, but Liang, on the strength of his 'face' in the district, obtained a special seat for me beside the driver. My two youthful escorts, clutching heavy suitcases straddled the roof of the bus outside and bore the burning rays of the sun as the bus sped toward Vienchan. Time and time again, the bus was halted by flat-tires, because of the excessive heat and the over-capacity passengers. We finally gave up the bus and reached Vienchan by ox-cart and pedicab.

The white-sun-against-blue heaven flag of the Nationalist Government was flying from every home. I saw Yunnan troops wearing straw sandals and covered with skin diseases, and felt as if they were members of an allied army. This is the natural emotion of mutual affection only found among the Oriental races. To commemorate this auspicious day, I composed a bit of doggerel verse:

They thought not of self, to pierce the hostile net
And save this unknown national of another race
Life lasts but fifty years or more
This life I'll sacrifice to pay this debt
And work to bring two races closer knit

The town of Vienchan was once the site of a Japanese military air field. Before the outbreak of the Pacific War, it was the summer resort of the French people. Overlooking the Mekong river, the town consisted of a number of white-walled European-style villas set upon a green plateau. This beautiful town was a paradise of coolness after our hot and stifling journey.

The hotel, where we took rooms, was a former Japanese army club, and still retained the raw wounds of systematic looting after the war. Only a handful of workers swept and cleaned the two or three rooms available. It was therefore a hotel in name only. It was a relief to have chosen such a place, with less chances of being seen by a large number of people. Behind the hotel, I found a tap of running water. I stripped naked and rubbed the dirt and sweat that had accumulated on my body the last three days. After this washing, I felt fresh and alive once again. Wu and Liang took me for a supper to a Chinese restaurant in town. There we had a leisurely lunch. I returned with friends of Liang's ahead of the rest to the hotel.

Wu and Liang said that they would go and report to local branch of the Overseas Department. Worn out to the bones, I lay down on the straw-filled mattress: and fell immediately into a deep and senseless sleep. I slept like dead man, unconscious of the legions of bed bugs and the squadrons of mosquitoes that attacked me.

In the middle night of the night, I jumped from my bed at the sound of a fierce knocking on the front door, so fierce I thought the door was being broken down. I also heard the rough tread of soldier's boots. Four or five people were talking in a loud voice. I glanced around the bedroom Liang and Wu had not as yet returned to the hotel.

"Have I been fooled? Have the two youths been arrested? The influence of the Chinese Communists I know has already extended its long tentacles into French Indo-China...."

Liang's friend, who was sleeping with me, also seemed surprised, He stood blocking the door and was engaged in loud conversation with the intruders. Soon, he was pushed aside by the intruders who strode into the room. There were four of them. I decided my last moment had come but was ready to fight my way out with my fists. I stepped behind the bed, in order to use it as a shield, and took up a stance of self-defense. I had neither pistol nor knife. All I had was my body. Even if I could not be a match for the whole four, I was ready to fight, snatch their pistols and shoot the rest. Suddenly, a flashlight was flashed into my face and I was momentarily blinded. I heard a peculiar brand of Japanese being spoken:

"No need be afraid. You, you Japanese?"

"Yes, I am a Japanese, and who are you?"

"We are Chinese. You come with us. Good thing waiting for you. No need be afraid."

"Enemies? Friends?

Perhaps Wu and Liang are already in prison. Perhaps, these men want to imprison me in the same cell and are trying to fool me into coming with them without using force. Half believing and half doubting, I decided to go with them, leaving my fate to the gods. The four mysterious strangers, all-armed with pistols, immediately surrounded me. I was led to the home of a Chinese resident.

It was already past midnight. In the middle of the room, there was a round table. On the table was a plate of delicious looking sponge cake. I knew that I would be safe. This was no den of thieves.

A huge ugly looking man, over six feet tall, seated himself at the table, dangling his pistol by a cord from his shoulders. The youth who had spoken to me in Japanese seemed like a quiet person no more than 20 years old. The men introduced themselves as members of the Vienchan Office of the Overseas Department of the Chinese Nationalist Government. The huge man was the Chief, Colonel Chin. The others were Lieutenant-Colonel Cheng, Major Lin and the interpreter Captain Wang.

These men were also nuclear members of the Blue Shirt Society and all subordinates of General Tai-Li. Nevertheless, I thought that they lacked in common sense to come in the middle of the night, the whole four of them, all armed with pistols just to get an unarmed Japanese.

They knew in general what had happened to me up to that time, from what Liang and Wu had told them. I felt ashamed to have been seized by doubts.

I talked with them for about two hours through Interpreter Wang, feasting on the sponge cake and drinking coffee. They told me that the next day I would be accommodated in a different place. The whole group again took me home in the middle of night.

Next morning, Wu and Liang returned to the hotel. Having completed their perilous mission, the two had sought solace that night in the company of fairer sex. In the afternoon, I was taken to my new quarters by Captain Wang. It turned out to be the second floor of the Chinese home that I had visited the night before. It belonged to a Chang Kung-ying, a native of Wen-chang Village on Hainan Island. I remembered the village, for I had stayed there for two weeks once while taking part in tropical area maneuvers.

The family had escaped by junk from Hainan Island eight years ago when the Japanese Army occupied the island. Since then, the Chang family had gone through one hardship after another. But Chang had finally managed to save some money and had bought this home cheaply from a Frenchman at the outbreak of the Pacific War. In additional to his aged wife, Chang's oldest son and his wife, and the widowed eldest daughter, with two children, lived with him and helped in the family business. I marvelled at the irony of fate that I should be welcomed with such warmth to the home of a Chinese who had been forced out of his former home by the Japanese Army. I was given the sunniest room on the second floor and lived together with Wang for four or five days. The whole family welcomed me and treated me like an honored guest.

Captain Wang turned out to be a native of Canton, his home being situated on Weichung Road. His elder brother had studied in Japan and he also, in the hopes of crossing over to Japan, had taught himself Japanese. He was an amazing character for so young a man, 25 or 26. He was already a captain but had not received formal military education. The two children of the widowed daughter were five and seven years of age. I called the older brother "Boku-kun" and the younger, " Boku-ten." I felt as if they were my own children and took them out with me on my morning strolls.

It was my greatest pleasure to stroll with Wang after supper along the beautiful streets lined with summer villas. There was a teashop that served wonderful cakes, perhaps to meet the taste of the French, and it was there that we sat and rested ourselves. It was run by an Annamese.

On the night of November 7, Wang and I were seated in the teashop as was our wont. Suddenly, four or five rough Yunnan, soldiers came stomping into the shop. They ate and drank with gusto, but when it came to payment, they began quarrelling with the Annamese proprietor. If these soldiers had been Japanese, I would have knocked their heads together, but I restrained myself and shouted in poor Chinese "Pay." They did not realize that I was Japanese, and they looked with marked antagonism at me, but they reluctantly did pay.

Wang must have sensed that trouble might arise and he signaled me with his eyes to leave. Two or three doors down the street were a barbershop. I entered and was having my hair trimmed, when four or five Chinese soldiers, shouting, at the top of their voices, broke down the front door and rushed into the room.

They shouted:

"He's a Japanese. He's a Japanese," and surrounded me from four sides. They were the soldiers that had tried to get away without paying in the restaurant. In addition, they were now completely drunk.

The four planted their pistols against my half-trimmed bald head. The muzzles of the pistols were no more than five inches away. And there were four of them. The fingers on the triggers were held taut and the soldier's eyes were bloodshot.

The Annamese barber, still holding his clippers in one hand, was shaking from head to foot. Color had run out from Wang's face and his lips were quivering. He could not speak a word. All the other customers had fled from the shop. All the while, the four Yunnan soldiers kept jabbering away in their own unintelligible dialect.

I knew that I must calm myself. If I became flustered, it would be the end of me. Instinctively it flashed into my head that I must ease the tense atmosphere created by these soldiers.

Smiling away, I said:

"Wait, wait. Please sit down for a while."

Without moving a muscle as I sat in the barber's chair, I broke into a smile and as I watched my reflection in the mirror, I noticed some of the tension leave the faces of the soldiers.

"Good. Now I'm safe," I told myself.

I urged the barber with my eyes and chin to continue trimming my hair. With shaking hand, the barber began to cut my hair again.

In a while, he had finished his job. But, I knew he would not be able to hold a razor in his trembling hands. I had the barber wipe my head with a hot towel. I then paid him and bowed to the four soldiers:

"The barber has finished. Thanks for waiting. Let's go."

Interpreter Wang, whom I had counted on as my only help in this crisis, was nowhere to be seen. All alone, I had to entrust my life to the four soldiers to do as they pleased.

They pressed their pistols into my sides and signalled me to go with them. It was already past 10 o'clock. One soldier preceded me. Two took their stand on both sides and the fourth followed. With four pistols pointing at my close-cropped head, I was rushed through the dark woods to an unknown destination.

We were almost running. Often I found myself falling behind. Each time, the soldier behind me would poke his pistol into the small of my back. After about 30 minutes of this, we reached a lone house in the middle of the woods. It seemed this was their headquarters.

We've brought Japanese. We've brought Japanese," they shouted and I was pushed into a room where 20 or 30 soldiers got to their feet. There was a six-foot long desk and six-foot long benches, which obviously had belonged to the Japanese Army. I sat down on one of the benches in silence.

"Give me some tea," I asked.

The noisy group of: soldiers suddenly became quiet. I drank at a gulp the luke-warm tea offered me in a dirty cup.

A lieutenant, who seemed to be the deputy-commander, arrived with a short youngster who spoke Japanese. I knew that the youth, from his pronunciation, was a Korean.

"What's the idea of you, Japanese, still sticking around here in disguise?"

"Sure, I'm Japanese. I have been specially requested by your supreme commander. Generalissimo Chiang Kai-shek, to help him and have come here from Hanoi."

"That's a lie. Where's your proof?"

"Sure, I've got proof. Plenty proof. Call Colonel Chin, Chief of the Overseas Department here. Go and also call your regimental commander. They're all my friends."

"What's your occupation?"

"What do you think it is?"

"I bet you're a soldier."

"Don't joke. I'm a priest, a bonze." (I rubbed by bald head and struck it as if I were beating a drum.)

"Ha, Ha.. So you're a Nichiren priest.... Yes, To think of it, you resemble a face I saw in Shanghai... Ever been there?"

"Yes. I lived in Shanghai 20 years. I've been a very good friend of General Tai Li from that time."

The soldiers and officers who were listening to this give and take gradually held themselves erect. Soon, an officer, obviously, the regimental commander came in, He handed me his name card. On it was written: Chen Kung-han.

Interpreter Wang, who had disappeared in the barbershop, seemed to have reported the incident to Colonel Chin of the Overseas Department and Chin, in turn, seemed to have gotten in contact with Regimental Commander Chen, to try, and save me through behind-scenes string pulling.

The regimental commander apologized to me for the actions of his soldiers. He then called the four rude soldiers and said to me:

"I am going to have these four men shot immediately. Please don't report the incident to General Tai-Li."

The four soldiers began to quake in their shoes.

"Not at all. You should be proud of your men. They are to be commended for being able to see through the disguise of Japanese. Please praise them instead."

In this way, I calmed the regimental commander, while the four soldiers, on hearing my words, broke down and began crying.

They requested:

"We shall escort you on your future travels, guarding you with our lives. In this way, we'd like to make amends for our sin."

In a twinkling, this scene of fear had been transformed into a scene of emotions.

In a short while, Colonel Chin hurried in.

The four soldiers who had gone berserk and had threatened to kill me, turned out to be in the ensuing two weeks of dangerous travel through French Indo-China the most faithful and the bravest of escorts.

In this way, I successfully passed through a fourth death barrier, a barrier that I had not anticipated.

The next day, when I met Lieutenant-Colonel Chen, the regimental commander, for a conversation at the Overseas Department offices, he brought along a Japanese sergeant. His name was Kazuo Kaneko. After the war, he and several of his colleagues took to mountains, resolved to commit suicide together. However, he was found thin and starving by an Annamese and handed over the Chinese Nationalist Army. He had changed his name to Chin Chung-i and was treated as a guest of the commander.

I often wonder what has happened to him since...,

Chapter 3 Annam in Ferment

The Annamese independence movement had spread like wild fire even to this district. In the town plaza, once the parade ground of the French Army and now a racehorse track, young Annamese men and women gathered from early morning, waving their revolutionary flags and making a lot of useless commotion. Without any suitable leaders, all they did was march in step in columns of twos to and fro under the burning sun, to the command: "Mott Hai, Mott Hai' (One two, one two). I recalled that when youth training was first introduced throughout Japan, this was the very sort of thing that was seen everywhere. The occupation of French Indo-China by the Japanese forces had touched off a spark of racial consciousness among Annamese youths. However, only the seeds of independence were sown. No objectives were given in the guidance afforded the Annamese later, nor was there any basic principle.

Four precious years had been wasted.

The history of 100 years of French rule in Annam was the story of military and political enslavement of the Annamese. By manipulating puppet sultans and local kings, the French had divided the Annamese people and had involved them in constant bickering and rivalry. Economically also, the French utilized the Chinese as middle men in their exploitation of the native population. Looking back at this history of misrule, the Japanese occupation forces should have been able to find a guiding principle for the Annam, independence movement. The Annamese were shackled by two steel chains. One was the chain of French military and political rule, the other the economic aggression of resident Chinese. The object of the present independence movement was directed at sundering the former shackles, but in actuality the task of breaking the second chain was much more difficult.

If after the end of the war, the leaders of the Chungking Government had the far sighted vision of eventual leadership of East Asia, they would have helped the Annam independence movement with the financial resources of the resident Chinese. It would not have been a heavy sacrifice if Chinese residents had thrown in a tenth of their financial resources into the attainment of Annamese independence from the French. If left unaided, the Annamese, after chasing out the French, would probably point their rifles at the Chinese and would drive them root and all from Annam. Such a possibility should not have been overlooked.

Unable to understand the essence and the direction of the independence movement, the Kuomintang Government frantically applied itself to confiscating the property of Japanese nationals and the Japanese Army in their allotted area north of the 15th Parallel and allowed looting and raping to take place everywhere. The short-sightedness of the Chinese Nationalist leaders, as revealed in this debauchery, was a prophecy of today's decline of the Kuomintang Government.

It was the Communist Party that took advantage of this situation. The Viet Minh (The League of the Vietnam Democratic Revolution), led by Ho Chih Min, turned its spearhead toward the overthrow of French Imperialism, and succeeded in winning over the majority of Annamese youths. On the other hand, the group centering around Emperor Bao Dai, sought autonomous government, based on close relations with France and was split by jealousies and rivalry. There was no need for a soothsayer to predict which side would succeed.

It was all too apparent.

November 6 was a day of big demonstrations throughout the whole of French Indo-China. Crowds of young and old, men and women thronged the streets. That night, Interpreter Wang and I visited a down town gambling house. Several hundred men and women thronged around several tens of straw mats, watching the falling of dice with bloodshot eyes. The losers were practically always the Annamese, the winners the Chinese.

The Annamese worked like draft animals all day long and were paid pitiful wages by the Chinese. Their farm products, the fruits of back breaking toil, would be bought up by the Chinese for a song. Then, in turn, the Chinese would sell them expensive groceries and cloth. The daily life of the Annamese was a hand-to-mouth existence. They sold their wives and their daughters into brothels and the handful of money received in return would be taken from them wholesale by Chinese gamblers.

The measure adopted by France to enslave the Annamese in the past was to deprive them of education and to encourage gambling. The fate of the Annamese was just like that of the Mongols who were sucked dry of all virility–Yea, to their very marrows–by Lamaism and opium.

Thus, the roots of the disease that sapped the Annamese of the strength to rise in revolt went deep into the very vitals of this race of people.

A large number of Annamese closely resembled the Japanese. In particular, this was true among Annamese women. The sight of Annamese women, wearing large saucer-like sun hats of cypress bark, tied with white cord under their chins and with their long black hair flowing behind them, often reminded me of color prints depicting the women of ancient Japan.

There were many fathers and brothers who encouraged their daughters and sisters to marry Japanese officers and soldiers, saying: "If you give birth to a child from Japanese, I'll bring it up." This was a phenomenon never seen elsewhere. Even at that time, I heard that hundreds of Japanese soldiers and officers had taken to farms in the palm groves, marrying Annamese women and settling down in French Indo-China. I felt a great affinity between the Japanese and Annamese, a closer affinity than with Chinese, Thailanders or Burmese.

Lieutenant-General Hsieh, a subordinate of General Tai Li, was responsible for the leadership of all southern overseas Chinese residents and a leader of the Blue Shirt Society. His base was in Hanoi. However, he wired Colonel Chin in Vienchan [Vientiane] that he would meet me in Sahnaket and ordered Chin to escort me there.

We launched immediately on preparations to sail down the Mekong River. For fear that our intentions would become known if we travelled by ordinary passenger boat, it was decided to borrow the private boat of the King of Laos.

The time had come for me to say good-bye to the Chang family that had treated me with sincerity and kindness during my week's stay in Vienchan. Our party was composed of Colonel Chin, as leader, Lieutenant-Colonel Cheng, Interpreter Wang, and my old friends Wu and Liang.

The four Yunnan soldiers who had nearly killed me in the barbershop accompanied us as escorts.

Seen off by Regimental Commander Chen and members of the Chang family, I bade an eternal farewell to Vienchan. I had only been there a week, but I felt an attachment for the place, as if I had lived there for ten years. The two sweet children, Boku-kun and Boku-ten, kept waving their chubby red hands as we left by car.

It was around noon on November 10 when we reached the river pier. I had expected the King's private boat to be a luxury yacht. To my surprise, it was a small, miserable-looking wooden vessel, equipped with an engine of about five horse power. It was of course, better than, a rowboat or a sailboat. However, when ten passengers got on board with their luggage, it was so small we couldn't move. Just the slightest shift in weight, I felt, would tip the vessel over.

There was a tremendous amount of unaccountable luggage. The boat sank into the water to its maximum draught. I managed to shade myself from the burning sun, but found myself tormented by thirst. The cocoa nuts, which Liang had thoughtfully purchased in Vienchan, were a god send.

I could not believe that this was the same Mekong River that I had crossed one week ago in the teeth of death. Seated safely aboard a King's boat, I watched fish jumping in the mirror-like surface of the river. The white sails dotting the river and the .beautiful scenery on both banks fairly outdid a landscape painting. Because of the twisting course taken by the Mekong, I often lost my sense of direction. Thai territory would be seen sometimes on the left, at others on the right and the sun changed its position momentarily, west, south; east, and north. I told myself that this dizzy change evident in my surroundings was because I was viewing my environs with myself and the boat as the center of it all I- realized that this phenomenon was not limited to this situation alone. Unless I judged with cool deliberation the phenomena appearing around me, always retaining the overall picture of myself as one among many Japanese and Japan as one country among many in the world, I would never be able to grasp the essence of things.

I was told that travelers on this river were often shot at by bandits from both banks of the river. This was particularly so of the French Indo-China side, now in the throes of revolution, with armed bandits appearing from time to time to threaten river navigation. In order to avoid danger, our boat was pushed into midstream. The flow of the river helped push our puny five horsepower vessel and we sped down stream leaving a wake of white foam. We seemed to have no special itinerary whatever. We were to touch shore at nightfall and set sail again at daybreak.

When the sinking sun transformed the Mekong into a bas-relief of gold and silver waves, we tied up our vessel at a nameless hamlet. There we found an old temple in a banana grove, which we chose as our hostel for the night. Colonel Chin cooked and served the rice himself, while the escort soldiers stood looking on blankly, just holding their light machine guns. Such an action would never have been seen among Japanese soldiers.

I asked Interpreter Wang:

"Why don't the soldiers do the work instead of the Colonel?"

He answered immediately:

"They are escort soldiers and not work soldiers."

It was evident that no hostile elements were to be found in this hamlet and it was evident that the four soldiers did not have to stand constant guard. Yet, they took no step to help the Colonel cook the food. Here I thought of at least the good qualities of the Japanese soldier.

There was no difference made at mealtime between officers and men. The whole ten of us sat in a circle on the wooden floor of the temple hall and partook of supper.

The Annamese villagers gathered around us out of curiosity, bringing with them fish and bananas to sell. All the while a chill wind blew through the temple hall, composed simply of a roof, four supporting pillars and a plank floor. Belying the heat of the daytime, the evening was almost cold. I wrapped myself in a single blanket with Interpreter Wang and managed to sleep thanks to the combined warmth of our closely placed bodies.

The next morning, we again took to the boat. We met a large steamer sailing upstream and tied our vessel alongside, while we bought cakes and tobacco: I wondered why they had not bought these articles in Vienchan before our departure. But that was purely a Japanese view of things, I discovered. If purchased in the middle of the river, the goods were that much the cheaper, since the extra freight was not included in the price. For noon we had a' fish almost two-feet long, caught with a bit of meat as bait by the helmsman. That night we slept in a schoolhouse in another nameless hamlet.

Our travel down the river continued the next day. I had tired of the scenery on both banks of the river and the boat seemed to crawl at a snail's pace. I dozed all afternoon, my legs bent double and my arms as a pillow. At nightfall we reached our destination, the town of Takeku [Takeo]. Thus, ended our journey by boat.

This town was also a summer resort of the French. But, as everywhere, the whole town seemed to be blanketed by Chinese Nationalist flags. We were allotted billets in the local Kuomintang Headquarters. People going in and out were practically all members of the Blue Shirt Society. Here again I found the same seething joy over victory, after eight long years of hardships underground. From the window of my room on the second floor, I watched the patient figures of Annamese women selling their varied ware, the thronging crowds of Chinese soldiers, the bicycles scattering Chinese residents here and there and the proud showing of high grade Chinese officers dashing by on beautiful cars. It seemed that I was getting a preview of the whole panorama of French Indo-China above the 15th Parallel.

I was sick and tired of the short three-day boat travel. Wang and I lay down on our beds on the second floor of the Kuomintang Headquarters, resting our weary bones; while all the while the other members of our group were busily engaged doing business in town.

The huge volume of merchandise that had been loaded on the boat at Vienchan was being sold here and with the money so gained more goods to be sold at the next stopping point was being purchased. This preoccupation with side business was unavoidable. Given insufficient travel expenses, not even enough to pay for lodgings; the members of our group had to earn through their own efforts, funds to pay for their transportation, Thus, it was no wonder that they were busy selling and buying after their arrival at our destination.

In the last analysis, this form of money-making was the most honest and conscientious way possible. They first carefully studied the price of commodities in different localities. Then they would take advantage of free cartage, either by boat or automobile, to take goods cheaper by even a cent to the next place, where they could command a higher price and a margin for their efforts. Those who hated to expend this effort took to looting and blackmail to obtain money. I understood for the first time why our boat had been loaded to such a point that I feared it would sink. In the final analysis, the Chinese Nationalist Army was nothing more than a legion of merchants wearing uniforms. At the same time, there were no small number of thieves wearing uniforms.

On the morning of November 14, our group of ten men boarded a bus and left behind the town of Takeku and hurried toward Sahanaket [Savannakhet, Laos]. The road that we travelled was well paved and graded. It was one of the few trunk highways built by the French in order to maintain their hold on French Indo-China and was well built for that reason. After travelling for about three miles, we encountered a road block of felled trees. I wondered what had happened, when I saw French soldiers and officers emerge from the roadside jungle, carrying automatic rifles. Their sunburnt faces were covered by a week's growth of beard. These citizens of one of the most civilized countries of Europe were no better than bandits when transplanted to the jungles of Annam. The refusal to grant the Annamese the independence which the entire race desired, was causing bloodshed and strife in all parts of the country.

The French seemed to show considerable respect for the white and blue flag of the Chinese Nationalist Government and permitted our passage without examining the interior of our, bus. We heard that several days previously, a car belonging to a Chinese resident had been attacked and its occupants massacred. The report filled all passengers with fear as the bus continued on its ominous journey.

The Chinese Nationalists although having troops stationed in the midst of this revolutionary crucible neither helped the Annamese race in their fight for freedom nor cooperated with the French in the suppression of the rebels. All, they did was to engage in a halfhearted attempt to protect the lives and property of resident Chinese or to squeeze as much money as possible from their own compatriots.

The sun-baked road wound on and on toward Sahanaket, As we neared the town, we were halted by 20 or 30 Annamese bandits. Fierce looking Annamese immediately surrounded the bus, pointing their guns at the occupants. They were a motley crowd, the majority wearing tatters.

I saw the killer instinct flash in the eye of our escorts. It appeared as if these were the very bandits that has halted the Chinese car and killed its occupants, as told us by way of warning at the French road block. The bandits and the Chinese kept shouting back and forth at each I couldn't make head or tail of what was being said. Colonel Chin had two of the machine guns pointed out of the bus window and ordered the men to shoot at a word of command. In such a spot, I disguised as an important Chinese resident, and completely unarmed, was in the most dangerous position.

The air in the bus was charged with electricity, ready to explode at any moment. I looked over the crowd of 20 to 30 bandits, searching for the weakest looking man. If fighting started, I planned to dash out of the bus, tackle the weakest enemy, take his gun away and defend myself with the rest of our group.

A single spark would have started bloodshed, when suddenly Colonel Chin bellowed out a storm of abuse. I had never believed a Chinese capable of raising such a voice. To back his thunderous words, was his enormous frame, well-over six feet tall, and his ugly features. Completely overwhelmed by Colonel Chin and the two machine guns, the bandits reluctantly fell back and let the bus pass.

In this tense situation, I was surprised to notice that it was the four Yunnan soldiers with whom I had become acquainted in the barbershop, who had tried to shield me with their own bodies. It was as if they were trying to repay their debt to me, incurred in the barbershop incident, by protecting me at the risk of their own lives. We soon reached our destination, Sahanaket.

The group took lodgings in a cheap hostel operated by a native of their own birthplace, Hainan Island. In the inner room, where we were to stay, awaited a host of bedbugs and mosquitoes even in daylight.

After supper at a restaurant, also operated by a native of Hainan Island, I visited at nine o'clock the home of General Hsieh. It was a beautiful mansion of a wealthy Chinese resident. Wang and I were led into an inner room. There stood General Hsieh, of short and stocky build; he stood, with close cropped hair, wearing cotton Chungsan robe and cotton slippers. I could hardly believe that this man was the person responsible for the entire fate of all Southern Overseas Chinese residents recalled to mind how the Christian General, Feng Yu-Hsiang, ordinarily wore the same cotton uniform and cotton slippers as did his soldiers, but had the fattest American bank book of any one in China. I shook hands firmly with General Hsieh. He skipped the usual ceremonies of a first meeting and, immediately started speaking in a deep bass voice, with foam forming in the corner of his mouth:

> *"Japan's defeat was due to the mistaken direction of her politics. The starting point of this mistake was the launching of war against China and the terminal points her challenging the United States. God is most just. He gave the atom bomb to the most humanitarian country in the world. If he had given either Germany, the Soviet Union or Japan the atom bomb before the United States, the human race might have been wiped out. The Japan of the future must quit her aggressive policies, become a humanitarian country and become firm friends with China."*

Those were the first words of Lieutenant-General Hsieh, head of the entire underground operations of Chungking (the Blue Shirt Society) in Southeast Asia. These were in all probability the first words he had ever spoken to Japanese.

Unlike the average Chinese, he had no use for polite and diplomatic expression. His blunt attitude of thrusting his knife right into the heart of Japanese on first meeting impressed me favorably. I felt that this old man was an honest man at heart, despite his appearances. I immediately answered through Wang:

> *"I know that the majority of the Japanese people realize fully as a result of Japan's defeat, what you have just said. I would like to say a further word from the standpoint of an Oriental. What are your opinions of the Yalta Conference: Did Generalissimo Chiang Kai-shek know beforehand what was going to happen there? What do you intend to do with Manchuria, which you won by defeating Japan? I believe that Japan and China should both reflect on their past, liquidate the hatred that has been engendered between them, stand on the same basis as Oriental peoples and collaborate more closely in the future."*

> *"Japan was defeated in war. However, what she lost were simply superficial things. I feel that spiritual Japan will rise to still greater heights as a result of her trials. As it is written in 'The Destiny of China' (by Generalissimo Chiang Kai-shek), the biggest problems confronting China today are her industrial development and the question of what to do with the Chinese Communists. The question of the Chinese Communist is, at the same time, the question of what*

to do with Soviet Russia: In this respect, Japan will be able to offer China human resources to cope with both questions. I believe that the time has already come for China to liquidate her policy of "checking one foreigner by utilizing another." Finally, I express to you, Your Excellency, my deep gratitude for the way in which your men risked their lives in self-sacrificing effort to save me on numerous occasions from the jaws of death and escort me safely here. I pray that Your Excellency will try to enable me to get to Chungking as soon as possible, so that I may meet Generalissimo Chiang Kai shek and present to him detailed plans for Japan China collaboration."

In this way, I wound up roughly two hours of fervent appeal. Lieutenant-General Hsieh listened with close attention to my impassioned plea, devoid of all flattery, and said:

"I shall be sure to report our conversation to Generalissimo Chiang Kai-shek. Once an answer comes from Chungking, I'll have you taken there."

When we departed, his attitude had completely changed. It seemed that our appreciation and respect were mutual.

The next morning, his chief secretary came to visit me. He asked that I write down the gist of my statements, made to General Hsieh the preceding day. I. wrote down my plea in, what I realized, was poor Chinese. I knew that the old clichés of fat common script (between Japan and China) a common race or "Oriental morality" were viewed with disdain by the new Chinese and that the relations between the new Japan and China would only be possible on the basis of honest action, devoid of all flattery. General Hseih's reactions to my statements seemed to substantiate these long-held views of mine.

I was registered in the hotel as a doctor from Shantung Province. Perhaps for this reason, I was treated with exceptional deference, for the hotel was situated in a town with few doctors.

I could not understand the conversation of the people of this inn, neither could they make head or tail of what I tried to say. However, this did not arouse any suspicions. It was usual for a man from the Peking area to be unable to converse with a man from the Canton, area, despite their common nationality as Chinese.

Despite my fatigue, I could not get to sleep that night, attacked by succeeding waves of blood thirsty bed bugs. I lay tossing on my bed when I heard a soft knock on the door. It was long past mid-night. I jumped out of bed alert to danger and wondering over the reason for such a nocturnal visitor.

To my surprise, my visitor was a woman, a beautiful Chinese woman. She was the younger sister of the owner of this hotel. She came into the room, her face red with shame, and began conversing with me by written notes.

"I have no menstruation. Doctor, will you please take a look at me?"

I was completely frustrated. This was, a more formidable foe than an armed robber. In a flash of inspiration, I. wrote:

"I am not a woman specialist. I am a surgeon and a pediatrician. I cannot diagnose women's ailments."

She looked at me with wonder. Perhaps she thought that all doctors were capable of healing every disease and sickness.

"Is there no way you can help me? My stomach pains me at present."

"Have you a lover?"

"Yes."'

"Then, you should get married as soon as possible. That's the best cure for you. Have you had any bowel movements?"

"No, I have had no movement for two or three days."

I rummaged through the bottom of my suitcase and brought out some laxative pills. I gave her two or three and thereby succeeded in repelling this unexpected visitant.

The next morning, the young girl came cheerfully into my room and wrote down her thanks:

"My stomach is alright now."

After that, the conversation at the dinner table turned always to children's ailments and surgery. The next afternoon, the youngest child of the hotel keeper, a boy about 13 or 14 years old, came crying home. He had been playing on a horse on the streets but had been thrown badly. His knee had been cut and was bleeding profusely.

Here was a true case of pediatrics and external medicine, a case made to order for this quack doctor. I couldn't very well turn around at this stage and say I was a specialist of internal medicine. I was stumped. I decided that I would attempt an escape from this predicament but utilizing my knowledge of first-aid learnt some 20 years ago. There was, however, no drugs on hand. Suddenly, I remembered the spirits the hotel keeper was drinking, as was his custom, with his evening meals.

I asked Wang to get me some. With the spirits I washed and sterilized the wound. The child howled with pain as if he had suddenly sat on a hot stove. Quickly, I sprinkled some tooth powder on the wound.

The old man looked over my shoulder with anxious looks. I noticed a large plaster stuck on his back. I ripped it off and used it to cover the wound. I had no bandages. I ripped apart a towel and wound it around the boy's knees. I guaranteed the wound would heal in about five days.

On the night of the fifth day, I fearfully unwound the bandage and pealed off the plaster. The wound had healed cleanly and a thin new skin covered the boy's knee.

Immediately, my reputation as a wonderful doctor spread throughout the small town. From early next morning, I found a parade of the sick and maimed flocking to the hotel, people with suppurating hands, and rotting eyes. Their smell overwhelmed my nostrils.

"This is terrible. It will mean the end of this quack doctor," I thought and hurriedly left the hotel by the back door with Wang. I continued strolling through the town until late at night.

The next day, the eldest son brought to me an electric treatment machine, which he had stolen from some French home. He asked me how he should use it. This was my first experience with such a gadget. However, I wrote down for him;

"Use five or six times a day, for five or six minutes at a time. This is good for nervous disorders and inflammation of the joints."

He asked me how much the equipment would be worth. I was stumped, but wrote down:

"Around 10,000 yuan."

This was an enormous amount of money at that time and the youth blinked with surprise and joy.

Thanks to all my "medical" services, I was able to spend ten days at this hotel without paying a cent. I decided that if I were to continue my disguise as a doctor I would next become a specialist of internal medicine. But the thought of handling dirty patients deterred me. I decided I would become a government official. I consulted with Wang and sqeezing out 50 yuan from my sadly depleted purse, I had a Chungsan robe made.

"With this, I'll seem on the same level as General Hsieh," and I smiled wryly to myself.

One day a handsome young lad came to call on Wang. He was so beautiful that I felt he would pass as a woman. I offered him some left-over bananas lying on the table. He ate them without hesitation and puffed at cigarettes as if he were an inveterate smoker. However, I could not get over the fact that his waist was small and his "derrier" unaccountably large. I bent my head in wonder at this remarkable youth. The next day, he came dressed as a woman. After that this "neuter gender person" came to take out Wang every night. And they seemed to find pleasure in each other's company. Later I learned that the girl was Wang's childhood friend from his native Canton.

Thanks to the girl, I was liberated from surveillance and constant escort. Night or day, I was able to set out for the down town area without worrying about anybody.

In several places, I noticed loudspeakers inciting the population tor revolt. On moonlit nights, young and old, men and women gathered on the town plaza and stood judging this exhibition of youths appearing in turns to bellow their impassioned revolutionary pleas. The preponderance of young girls struck my notice. The essence of this revolutionary drama was nothing more than the search by moonlight of young men and young women for mates.

The priest of the large French church had been chased out and the place of worship first changed into a training center for youths and later into the rendezvous of young couples.

I could not bear to use the hotel toilet, and it became my daily routine to relieve nature out in the open in the town suburbs.

One day, I inadvertently passed by the outpost of a revolutionary group. I was immediately halted and questioned was dragged to the headquarters. I had hoped to be able to see the interior of the headquarters and was thankful for the unexpected earliness of such an opportunity. I knew, however, that if I were not careful I might get into serious trouble.

I was questioned in unintelligible Annamese. In answer I rattled of all the Chinese words I knew and clutched my abdomen, shouting:

"My stomach's running. My stomach's running."

It seemed they understood the meaning of my words and immediately let me go.

All they had was one Model 38 infantry rifle. This was passed from hand to hand among about 10 men, who took turns at sentinel duty. At the outpost I saw a mountain of delicacies. It was tribute exacted in kind from the suffering people.

Revolution seemed an escape for these youths from the back-breaking toil in the rice fields. Using revolution as an excuse, they extorted money from the people and started their all-night drinking parties from broad daylight.

Brought face to face with the absurdity of seeing revolution, nominally aimed at saving the people from suffering, but actually bringing instead increased hardships on the populace and faced with the sight of youths, at the heights of their productivity, abandoning their fields and drinking wine from noon, I was filled with sorrow. If there had been an actual enemy against whom they had to fight, it would have been different. But, there were no French forces in this out-of-way town. I realized sadly that the Annamese, though a loveable people, was not a race capable of standing by itself on its own feet.

The Chinese Nationalist forces stationed in this town were also from Yunnan Province. I realized that some of these soldiers would include troops whom I had fought against in North Burma.

Wang warned me:

"Yunnan soldiers are like wild beasts! Once you are caught, they will kill you outright."

At each strategic point of the town, were outposts. Each time I passed a sentinel, I turned my face from him and blew my, nose with my hands and then wiped my hands, like any typical Chinese, on the sleeve of my jacket.

I noticed disarmed Japanese soldiers, wearing their combat caps, walking the streets, with "official duty" armbands on their arms. Overwhelmed with nostalgia, I stopped one of them in a dark alley and spoke to him in Japanese. I learned that the Chinese Nationalists treated them with comparative magnanimity and allowed them considerable freedom of action.

The district was noted for its papaya. I will not forget the taste of the ice-cold papaya just taken from the refrigerator that I ate in the teashops of this town. It was wonderful. The price was two papayas for five yuan. This was about the limit I could go in treating the children of the hotel where I stayed. In return the younger sister of the hotel keeper took me to the same teashop and treated me to the same papaya. I wondered why she did not pick another teashop. But, it was always this same place.

The puzzle was eventually explained to me. The owner of the tea shop was the eldest son of the hotel keeper. Because of family ties, the sister always took guests to this teashop. However, the surprising thing was that she always paid the regulation price. It was no different from than any other store. This, I realized, was the manifestation of the thorough commercialism of the Chinese. Once profit is involved, once it becomes a matter of business, it was the Chinese custom to make no exception even if the customer were father, son, brother or distant relative Herein also lay the motive power behind the amazing ability of the Chinese to spread overseas and spring up like weeds unaided in all parts of the world.

The evening sun sank in the west and a myriad silver and golden ripples flecked the broad expanse of the Mekong. As I walked the slowly darkening banks of the river my thoughts turned to the children that I had left at home. Their faces in sleep came to my mind wondered where they now lived and how they now fared.

I imagined them turning in their sleep, dreaming of a father that had not returned and felt envy at the sight of Chinese and Annamese, taking their evening stroll, hand in hand with their children.

Perhaps because of the beautiful younger sister of the hotel keeper, the number of youthful visitors increased with the passing of days. The storekeeper's main boast was the "doctor" from Shantung who had miraculously healed his son's wound. The majority of visitors were natives of Canton and Hainan Island, whom I was able to fool to my pride without much difficulty. However, my pride was due for a bad fall. I was not to keep fooling all the people all the time for very long.

One day, the hotel keeper said that he had a guest whom he wanted me to meet badly. I followed the master to his room. There I met a middle-aged man in uniform with sharp piercing eyes. His features clearly indicated he was a northerner. Brimming with intimacy at meeting a man from his own part of the country he said in fluent Pekingese:

"So you're a native of Shantung. I'm also from Shantung. Where is your home town? Why are you so far from home?"

After I had spoken a word or two, the face of my conversatant began to cloud gradually.

"This is what I get for having snoozed through the lectures on Chinese language at the War University...."

However, there was nothing I could do at this point. I felt my armpits drip with perspiration. I knew I was a cooked goose.

With sparks in his eyes and in rough voice, he said:

"You're Japanese, aren't you? Undoubtedly you're a spy."

He put his hand into his tunic pocket. Undoubtedly, he was gripping a pistol. Danger was increasing momentarily. I pulled up my sleeve and with my eyes signalled that I would converse with him in writing.

"I am Japanese. You have smart eyes to see through me, I have been called on an important mission by General Tai Li. Both Major General Huang Min-hun and Lieutenant General Hsieh are my friends. Please help me keep my secret."

I was covered with sweat as I finished my note. I saw the excitement and his wrought-up feelings gradually subside. There was a headquarters (staff) mark on his breast. He was probably a major-general or a colonel, I had escaped from dangerous predicament.

That night I was called out by Major-General Huang Min-hun. He was General Hsieh's chief secretary, and had been placed in sole charge during the General's absence in Saigon. General Huang turned out to be a good friend of my unwelcome guest of yesterday. He was a likeable person, tall and sparse. General Huang was a long time friend of Major-General Takuya Suzuki, who had been killed on Saipan Island while serving as divisional chief of staff. Although not a member of the faction that had studied in Japan, he had nonetheless friendly feelings toward the Japanese. He expressed deep regret over the untimely death of General Suzuki and expressed condolences with deep grief written on his face.

General Huang told me that an automobile had been prepared to take us to Hanoi and that we would be leaving the next day. He expressed regret at having to part with me so soon. A beautiful girl, the General's private secretary, brought us ice-cold papaya and we conversed for a while in a friendly atmosphere. On our departure, he wrote on a piece of paper:

"The Spirit of the Yellow Races." He wished me God-speed as I left. He was indeed a most pleasant person.

Bidding farewell to the hotel that had sheltered me for ten days, we left early on November 23.

They told us that the area where we would now pass through was untroubled by revolutionary hostilities and Colonel Chin and the escorts were relieved of their mission of guarding me. The four Yunnan soldiers that I had first come to know in that memorable barbershop asked to be allowed to escort me as far as Hanoi. But they were refused. It was the same with both Wu and Liang, to whom I owed my life a number of times since our departure from Bangkok, I had nothing to give them in repayment for their efforts on my behalf. I took off my watch. It was an Omega watch, given to me by a friend in Shanghai, which I had not worn and had kept as a cherished treasure. I had carried it with me through bloody battles in Burma and it was my last and most valued property. I gave this to Wu as a life-time memento. To Liang, I gave a valued box of German-made medicine for malaria.

I had only gratitude and friendship, transcending all boundaries, for these two youths that had protected me day and night. We had shared our pleasure and our hardships and they had guarded me with their very lives for 20 long days. I regretted also having to part with the four Yunnan soldiers that had nearly killed me in the barbershop and I felt sad at parting with Colonel Chin, for whom my friendship had ripened with the passing of days.

Young Wu said that he would return again to Bangkok and promised to report to my seven anxious disciples in the concentration camp there that the erstwhile priest had safely reached Sahanoket.

The new group which was to accompany me was composed of Interpreter Wang, Han (the son of a rich man); Captain Huang (a member of the Blue Shirt Society), a Madame Wang and our driver. Only Huang and Han were armed with pistols, two pistols for a party of six people, starting out on a week's journey. I felt little comfort in being in such a party. Just before our departure, fearing unexpected happenings, I asked Interpreter Wang to purchase some food. He answered:

"You don't have to worry from now on," and refused to listen to my request.

On our departure, our automobile broke down with one trouble after another. We wasted a half a day waiting for the rickety old car to be fixed and it was around noon that we finally got rolling. Because of the unreasonable speed at which the driver drove, the car had to stop once an hour because of punctures. On each occasion, we had to waste both valuable time and money. It seemed that we would never be able to reach our predestined stop before nightfall.

Around five in the afternoon, we reached a small hamlet. The entire village had been gutted when caught in the flames of the revolutionary war. In its ruins, a platoon of French soldiers were encamped in tents. Near the encampment was a food store, built out of burnt sheets of corrugated iron. It was owned by a Chinese. The store sold sausages, tobacco and vegetables to the French troops. This was typical of all French Indo-China.. Whenever an Annamese village was burned out, the place would be taken over by Chinese. The phenomenon of the spread of Chinese, like persistent weeds, was also evident here in this burned over battlefield.

On our way, we also encountered a. French armored car. It was the first time in a long while that I was able to relive the atmosphere of a battlefield created by this sight. We passed a desolate hamlet and reached the bank of a large river. The sun had completely set. Across the river lay our destination, but we were forced to halt our car near the approach to a bridge that had been bombed out. The opposite bank was evidently held by native Annamese forces.. The Chinese Nationalists seemed to be holding the bank in collaboration with the natives. The river barges had stopped operating with night fall and had been taken back to the other side of the river. We raised our voices and called. There was no answer. Occasionally, the sound of rifle shots from both banks broke the uncanny silence.

Obviously, we were caught in between two hostile lines. We realized that this unarmed caravan would have to spend the night on the river bank. Captain Huang let out a fiendish cry and called for the boat on the other side. The only answer was the sound of gunshot.

This was the end. We had only two pistols. There was little possibility of the French forces opening fire on Chinese. But it was the bandits and the packs of thieves that live by foraging the battlefields that we feared. The automobile would make a wonderful prize for these hungry men.

Han turned out to be spineless brat, spoiled by high living. The driver folded up with fatigue and hunger. Madame Wang was all energy, but her energy was limited to her tongue. She was more of a debit than an asset. That left the three of us, Captain Huang, Interpreter Wang and myself. With two pistols between us, we three would have to guard the car. I resigned myself to standing sentinel all night.

Wang and I hid the car in the shade of trees and took up positions to guard the vehicle. Hiding in the shade we sought to utilize the light of the moon to discover the approach of the enemy. This was the first time Wang had been forced into a rough job like this and he trembled each time he heard the sound of guns.

"Don't shoot even if the enemy approaches. Only shoot after I fire."

Leaving Wang with this stern injunction, I stepped into the bushes. In contrast to the extreme heat of the day, the cold night bit into our bones. I covered myself from head to foot in a blanket and stood guard without any hope of relief the night through. I decided that if those French soldiers from the encampment a short distance back should approach, I would try and capture one of their automatic rifles. I feared, however, the Annamese soldiers on the other bank. There is no hiding from bullets shot at random.

Midnight passed. In the distance I heard the twinkling of a bell. I couldn't believe that it could be an enemy. No enemy would approach ringing a bell. With my eyes like saucers and my ears alert, I waited. After a while, I realized that it was a bullock that had lost its master. It was seeking food on the battlefield, the bell around its neck still ringing as in more peaceful days.

The group had no lunch. Of course, no preparations had been made for supper. None had listened to my precautions before departure. Of course, none anticipated that such a situation would arise. I could feel hunger eat into my stomach. But there was neither a crust of bread to be had or a drop of tea.

About the time the eastern skies began to whiten, Captain Huang came to relieve us, rubbing sleepy eyes. I handed him the pistol and returned to where the car had been hidden. But the new sentinel, still clutching his pistol, soon began to snore loudly.

Day finally broke. A Chinese truck arrived on the scene. Its driver was an Annamese, a native of this locality, it seemed. He jumped into the river and quickly swam with light overarm strokes to the other side. He soon came back on the ferry boat, pulling himself to our side by the steel cable stretched across the river.

When we finally managed to get to the other side of the river, we were met by a group of 20 or 30 native Annamese soldiers, led by a young Chinese lieutenant, with hair done in a pompadour and wearing gold-rimmed glasses. Captain Huang and the lieutenant conversed in loud voices for a short while and we were allowed to pass on.

I later learned that the young lieutenant that we had met was a Chinese Communist. According to Wang, the lieutenant had been a former newspaperman. I felt that this was a development that could not be overlooked. While the Kuomintang Party was engrossed in a mad race to make money, young leaders of the Chinese Communist forces were throwing themselves heart and soul into the Annamese Revolutionary Army and grasping important positions of leadership.

Our first meal in 24 hours was composed of bananas and tea. I noticed that the Annamese forces was made up of boys no more than 12 or 13 and old men in their 50's, armed only with clubs and bamboo spears. Their only weapons were rusted Japanese rifles and even at that there were no more than two or three for the entire group of 50. This must have been the actual state of the Annamese revolutionary forces. It reminded me of the farmer's revolt near the end of the Tokugawa Regime in Japan.

In our search for food, we came upon a road side Annamese farm house. Despite the poverty-stricken exterior, the home afforded us sufficient food–chicken, fish and rice, to assuage our hunger. The swift stream flowing down a nearby valley reminded me of the mountains and streams of my homeland.

Completely satisfied, we scrambled aboard the car again and headed eastward through a tropical downpour. It was already evening. A short distance along the road, we came upon a truck, stalled at the side of the road with all its lights extinguished. From the driver's seat came the sound of Japanese, which filled me with nostalgia:

"Our lights have gone out on us. Please let us follow you."

It was a Japanese army truck, confiscated by the Chinese Nationalists and the Japanese soldiers were also being kept behind to drive the truck. I killed an impulse to speak and comfort them.

It was ten o'clock when we finally reached the town of Hotung, our stopover point for the night.

We had supper in a Chinese restaurant and stayed at a Chinese hotel. The hotel was supposed to be a first-grade establishment, but inside it was no better than an inn for beggars.

The steamy heat, the columns of attacking bed bugs and the hosts of mosquitoes prevented me from sleeping until morning. I thought that at least I could ward off the mosquitoes if nothing else. But I found the mosquito netting full of holes. I felt the netting, but the holes would only enlarge. It was just as if I were touching the skin of rotten fruit. The netting seemed melted in one's hand rather than tear. The netting was as good as useless wondered if it had been strung just as a charm against mosquitoes. On the dirt floor of the hall-way, I saw a Chinese and a native woman, carrying a child, sleeping peacefully in a swarm of malaria-bearing mosquitoes. They were the picture of the people of Asia, able to sleep with thanksgiving in their hearts for any roof to shelter them from the rain.

We warmly thanked the keeper of the hotel and prepared to depart before dawn, unable to linger longer. We had to refill our gasoline tank at this point. Just then, five or six young children, jabbering away among themselves, came by rolling a drum can of gasoline. They had obviously stolen it from somewhere. The driver opened the can and tasted the contents. Apparently satisfied, he bought five gallons. We were told that these children often slipped past Chinese Nationalist sentinels into former Japanese Army warehouses, stole whatever they could lay hands on and sold their ill-gotten products to passers-by. In the background I noticed an adult with shining eyes watching the transaction. He was likely the master of these child thieves.

In the afternoon, we had one trouble after another with the car's engine. Careful investigation revealed that the gasoline that we had bought that morning had water added to it. It was to late for us make any protest. When the driver had opened the lid and tasted the gasoline, he perhaps did not know that gasoline was lighter than water.

The highway along the sea-shore was well maintained and ran in a straight line. Paved with asphalt, it was undoubtedly built as a driveway for the French to drive along and enjoy the beautiful seaside scenery. At the same time, it served as a major artery for the French in their management of this colonial territory.

Despite the wonderful road, we were constantly plagued by breakdowns. We seemed to spend more time in fixing punctures than in driving ahead, due to the extreme speed at which the driver pushed the dilapidated vehicle. Even without speeding, the road was as hot as a grill. But, because of the desire to make up lost time, the driver was whipping the car up to speeds as high as 55 miles an hour. I suggested through Interpreter Wang that we hold the speed down to under 30 miles an hour. After that we had hardly any troubles.

It was close to dusk when we reached the town of Vihn. Confiscated Japanese Army horses were being led through the streets. Worn to a mere bag of skin and bone, they were loaded heavily. There were also Chinese soldiers leading confiscated Japanese Army German shepherd dogs, with large patches of scabies on their thin backs.

Neither dog or horses were exempt, it seemed, from the bitter sufferings of defeat.

Vihn was a large city on the east coast of French Indo-China. Its streets were crowded. At each important street junction stood Chinese Nationalist sentinels, in mixed uniforms, but all armed with bayoneted rifles. I was astonished at the tense atmosphere pervading the city–as if martial law had been declared. When the Japanese forces were here, the only sentinels were those at the entrances of army barracks and headquarters. I wondered why such bristling precautions were necessary when war had already drawn to a close. The hotel where we halted was situated on the second floor of a large food store and restaurant. This hotel-restaurant was owned by an Annamese. Han was supposed to buy new tires for the car and also gasoline. But he had gone out somewhere and did not return when night fell. Wang and I lay down in our semi dark room to sleep. However, in the next room, four or five Chinese were playing mahjong until late at night. Then they began drinking, brought in street women and kept fighting and screaming and uttering lecherous sounds the whole night through.

Han failed to return again the next morning. Wang alone worried about our departure, but he could do nothing.

A seven-year old boy poked his head into our room in with strange hand gestures, he indicated that he had come to take our orders for something or the other, I finally realized that he wanted to make reservations for us to have prostitutes brought up to the room, pretending that I did not understand, I cocked my head questioningly. This young panderer became indignant. He brought in a girl, no more than four or five years old. He pushed her on the bed, squatted on her as if mounting a horse and gave a miniature demonstration of a sexual act. The purchasers of human flesh were always Chinese residents and Chinese Nationalist soldiers. Those being sold were always Annamese girls.

I could see from my second-story window into the barracks of Chinese Nationalist troop's right across the hotel. There I saw sentinels halting Annamese, searching their pockets and their luggage. An old woman, who had come a distance from the country to sell her vegetables, had been stopped on her way home, carrying with her the pitiful: but precious proceeds from her day's efforts. Here again, she was being stopped. It was not the first check point for her. She had had a number to go through. At each point, the sentinel would say:

"This is forbidden."

In this way, one by one her belongings had been stripped from her. She stood there in a daze, unable to bring to her lips the cry of anguish eating in her heart. She stood there a picture of the eternal sadness of Annamese women.

The hostile resentment of the Annamese, first pointed toward the French, was gradually shifted toward the Chinese Nationalists. So were the rifles of the Annamese independence forces. In all parts of the city, clashes arose between Chinese Nationalist troops and units of the Annamese Independence: Army. Unable to bear the atrocities of the Chinese Nationalist soldiery, the Annamese population of Vihn had decided to launch a city-wide strike from that day. Street-side vegetable vendors, merchants in the city's market places, barbers, and restaurants were all closed, and refused to do business. Only the Chinese stores remained open.

Throughout, the city paced Chinese Nationalist soldiers, armed for action. Everywhere were plastered posters with the following proclamation:

"Anyone continuing their strike after today will be summarily shot. Martial law will be enforced from 5 p.m. today."

The strike in protest against suppression by the Chinese residents in Thai was being duplicated here by the Annamese against the Chinese Nationalist Forces. I realized that there was every possibility of running into shooting before we reached Hanoi. In particular, an automobile flying the white-sun-against-blue-black-ground flag of the Chinese Nationalist Government was apt to meet attacks from indignant Annamese youths at any point along our journey. Because of the undisciplined excesses of the Chinese Nationalist soldiery, the reputation of the Japanese forces grew better. This made me happy and at the same time sad. In actuality, I had anticipated a large number of clean and honest youths in the resistance camp of Chungking, youths, who, sick of the corruption of the officials of the Nanking Government, had fled to the opposite camp. However, the things that I had seen in the one month since my departure from Bangkok were all of such nature as to betray my expectations. I felt it was still too early to make a final and overall judgment, but, my disappointment was daily growing greater.

Opium, whoring, mahjong, thievery, falsehood, irresponsibility and impotence, was this the true picture of the people of Asia?

In the afternoon, I took a stroll downtown. I peeked into the rear of one of the stores that had been closed in strike and asked if I could do some shopping:

"You're Japanese, huh. Alright I'll sell to you. I'll make it cheap."

And the storekeeper sold me articles 20 to 30 percent cheaper than indicated on the price tags.

Of the whole five Chinese in our group, Interpreter Wang was the only conscientious person. He alone sought to leave for Hanoi as soon as possible and set out on his own accord to a friend's home to borrow money to buy gasoline. He filled the tanks and completed all preparations to leave. It was after all preparations had been made that Han returned from his whoring. We set off again at four that afternoon. Constantly worried over possible punctures and fearing surprise attacks from Annamese youths. I had no thought of enjoying the beautiful scenery along the way. Late that night we reached a large town. We had finally escaped from the area of martial law. However, I sensed the same, feeling of suppressed hatred among the Annamese as I did at Vihn.

The night of November 28 was spent in travel. We stopped mid-way for a comparatively clean meal. Then utilizing the light of the full moon, we continued through the night. Although the day had been hot, cool breezes flowed in through the open windows, as we sped along the deserted highway. However, we wasted some time at several river crossings on the way, where we had to wake up the boatmen to take us across.

At the last crossing, close to Hanoi, the engine failed to start. Perhaps, it was due to the steep incline we had to climb from the river bank. The residents of Annamese homes in the neighborhood were aroused and five for six men hired to help us push the car up the hill. But, despite our efforts, the car failed to move. It was heavy. I noticed that the trouble lay in the fact that Annamese and Chinese were not pushing at the same time. The Annamese that we had hired were not trying hard, perhaps because of their resentment from being pulled out of their beds at this time of the night and perhaps because of their enmity toward the Chinese. I told myself that there was no reason why ten men could not move this small sedan.

I got out in the front and began leading the group.

"Mott Hai! Mott Hai! I shouted, imitating (One two. One two)," the command words of the Annamese, as I had heard them while watching their drills. Suddenly, the Annamese took up the cry in unison, and the car was pushed easily up the hill, as if it were a baby carriage was surprised at the totally unexpected efficacy of my efforts, and was struck at the power of a phrase to move responsive chords in the inner workings of men's hearts.

When in Rome, do as the Romans do.

The closer we got to Hanoi, the better the road became. Traffic also increased. Day broke over the countryside and a long file of Annamese, carrying loads of vegetables, snaked northward toward the city. The eyes of the travelers brightened at the prospect of soon reaching their destination. It was as if the fatigue of the long journey had blown away somewhere.

Our tired automobile rolled slowly over the last lap of our journey, over streets still wet from the night's rain. Disregarding our hunger, we headed straight for the Kuomintang Party Headquarters. We came upon next to a church a tall western-style residence, over which proudly flew the Nationalist flag. A large placard in front of the building was inscribed with the lengthy legend:

"The French Indo-China Offices of the Central Overseas Department of the Kuomintang Party of the Chinese National Republic."

We seated ourselves in a large sitting room, decorated with cut flowers. The Chief Secretary, Major-General Wang, soon appeared. He was a handsome youth, who seemed, at a glance, around 34 or 35. Introduced to him by Interpreter Wang, I offered the usual greetings of a first meeting. He seemed extremely formal and cold in his attitude. He was a very short-tempered person a characteristic rare among Chinese people. This type of person, I knew, was to be met often among people from the Canton area. Compared with the people of North China, they gave one the impression of being a completely different race. I was overwhelmed at his differences in comparison to Lieutenant-General Hsieh, Major-General Huang and Colonel Chin.

We reached the hotel reserved for us in advance. It was a three story foreign-style building that had been used by the Japanese Army. However, all the furniture and the fixtures had been had been stolen in the confusion following the end of the war. In the middle of our empty room stood a lone bed. Wang and I lay down together to rest ourselves after our long travels. Han and Captain Huang brought in a bed from somewhere and set up house with us on the third floor.

Captain Huang had the obnoxious habit of spitting from the morning to night on the floor. Soon, the floor became mottled with snot-green phlegm, I called the boy and had him clean the floor. But, no sooner did Huang come back to the room, when he would start all over again. During my stay with him, I found this unhygienic practice of his most unpleasant and the most painful experience of all.

We went to the Kuomintang Party headquarters for our noon and evening meals, where everyone ate together. For breakfast, Wang and I sat on our bed and had "shao ping" and bananas.

The day after our arrival, I visited General Wang, taking Interpreter Wang with me, in the hope of making a detailed report.

"I know everything that there is to know. No explanation is necessary have already wired Chungking of your arrival," was his brusque reply

The usually quiet Wang was moved to anger and said:

"That's not a human being. I won't talk with a guy like that."

And for the ensuing two weeks, Wang hardly spoke to him.

The dining room of the party headquarters had scattered about six or seven round tables. The food was simple Chinese dishes. But, there also I saw the picture of a once starved people. Wang, Li, Chang and others–about eight all told, ate at the same table as I. But, I was surprised at the speed with which they gobbled their food. Even Major-General Wang would pour soup on his rice and in three or four minutes flush down his dinner. His character was exposed even in his eating habits. The only thing that anybody seemed to be in a hurry about in this leisurely headquarters was the meals. I wondered why this hasty general could not at least eat his meals in a more leisurely manner. After all there were only three of them a day. Even if you arrived only five minutes late, all the dishes on the table were usually completely debauched. I frequently saw even an officer as high as a major eating his meals standing up. During supper once I heard the sound of some body being slapped in the face. I peeked through the door and saw Major-General Wang striking a kitchen duty soldier. It would have been unthinkable in the case of a comparable officer in the Japanese Army.

It was my greatest pleasure to stroll through the streets of Hanoi after supper with Interpreter Wang. It seemed the streets of this city were monopolized at night by the luxurious sedans of high-ranking Chinese Nationalist officers and dignitaries. These men had confiscated Japanese Army vehicles for their private use, commandeered the finest residences and gathered together the most beautiful women as concubines. It seemed as if they were intent on satisfying within the year instincts that had been suppressed during eight years of war with Japan.

Stores, restaurants and theaters were flooded everywhere with Chinese soldiers. And everywhere quarrels arose, between the Chinese and Annamese over unpaid for feasts and forcible purchases. Peace and order seemed to be maintained somehow by soldiers with fixed bayonets standing on the main street corners. However, I could not escape the underlying current of sharp discontent and ominous resentment.

Because of the flooding of French Indo-China with Kwanchin notes, the Chinese Nationalist currency, commodity prices soared like kites. At the beginning of the Chinese Nationalist occupation, the Kwanchin was at par with the local Annamese currency. Three months later, the Kwanchin had fallen in value, until three Kwanchin yuan equalled one of the local currencies. Because the Chinese soldiery tried to disregard the current exchange rate and tried to buy things with their yuan at par value, quarrels, ending in bloodshed, were common everywhere throughout the city. The Chinese residents, who had said a relieved farewell to the Japanese forces and had greeted the arrival of the forces of their fatherland with fanatical joy, were also sick and tired of the Chinese soldiers, because of their undisciplined greed and brutality.

General Wang warned me that I should not go out as long as I could avoid it, in view of the disturbed state of public peace. However, even without Interpreter Wang, I often went out on strolls. I noticed one day, Chinese Nationalist forces escorting with fixed bayonets a group of four or five Annamese youths, their hands tied behind their backs. This took place in broad daylight. I had no way of telling what crimes the Annamese had committed, but I could not overlook the burning hatred in the eyes of the Annamese that watched this procession.

Three unarmed Japanese soldiers, wearing their combat caps, passed by, whispering in small voices:

"It's because they're like that..." In all probability, the Japanese were technicians, held behind by the Chinese Nationalists. I felt such a strong attack of nostalgia that I followed the three and ascertaining that no one was around, said:

"Yes. What you boys say is right...You must be having a hard time now.... I'm also a Japanese soldier. Please be persevering...."

They showed surprise on their faces and, at the same time, joy at meeting a compatriot so unexpectedly.

"We're glad to meet you.... Please take good care of yourself...." and we parted, turning around to wave again and again to each other.

Among the countless Japanese officers and men who advanced to the continent after the outbreak of the Manchurian Incident and still further south after the Pacific War, there were no few cases of brutality, cruelty, raping and arrogance. [The Manchurian Incident, better known as the "Mukden Incident," was a railroad bombing blamed on Chinese soldiers that was staged by Lieutenant Suemori Kawamoto of the 29th Japanese Infantry on September 18, 1931, as a pretext for the invasion of Manchuria.] In addition, the evil-doings among merchants and Government officials, who went abroad on the band-wagon of the victorious Japanese forces, were more numerous than that of the officers and men. As a result, Japan lost her hold of the hearts of the peoples of Asia. This was one of the causes of her defeat.

However, the surprising thing is that the armies of the victorious nations, who took the place of the Japanese forces, committed more crimes than did the Japanese. Thanks to this, I saw directly with my own eyes, both in Thai and in Annam, a feeling of longing arise among the natives for the return of the Japanese.

Japanese soldiers also watched with critical eyes the excesses of the Chinese nationalist forces, felt sympathy for the Annamese and reflected in calm retrospection on their own conduct in the past. I believe that this was one of the biggest gains from Japan's defeat.

The second-hand book stores in Hanoi were filled with Japanese books. Since I did not have money to buy the books, I had to go to a loan library. There I was able to get books for five yen a volume per day.

In ten days, I hungrily read roughly twenty volumes. I felt deeply moved by Lieutenant-General Yasokichi Hayashi's biography of Masashige Kusunoki. I felt that it was one of the masterpieces of our time. [Born in 1294, Masashige Kusunoki was Japanese a samurai renowned for his unwavering loyalty to Emperor Go-Daigo. He committed seppuku (ritual suicide) after Go-Daigo lost the Battle of Minatogawa, and with that loss the hope of imposing imperial rule over all of the Japan.]

Roughly 50 youths were working in the Vietnam Office of the Kuomintang Party, under Major-General Wang. Their chief job was to keep: the resident Chinese in line with Chungking. The man responsible for guiding the whole 10,000,000 overseas Chinese in Southeast Asia was Lieutenant-General Hsieh. Thus, this excitable young Major-General Wang was equivalent-to General's Hsieh's Chief of Staff.

Young boys in their twenties were the field leaders, as Chinese Resident Affairs Section Chiefs, Supervising Chinese "elders" in their 60's, who had left their homes in their youths carrying only a rough pole to balance burdens on their shoulders, and through tens of years of suffering, and of sacrificing food and clothing, saved and scraped to achieve their present positions. The strongest weapon, which these youths wielded, was the threat of exposure and arrest as collaborationists and traitors. During the war years, young underground workers had worked their way from Chungking in French Indo-China and elsewhere and had thoroughly studied and recorded the movement and actions of Chinese residents who had cooperated with the Japanese during the Japanese occupation. This information was used as material for extorting money and goods from the resident Chinese, with the age-old threat, "Your money or your life." This was the chief work of the overseas offices of the Kuomintang Party. Men who had come to Hanoi at the end of war, wearing tattered cotton clothes and slippers, sported in less than half a year beautiful suits of foreign cloth and brand-new leather shoes. Despite their low salaries, the executive officers feasted daily from morning till night. The income of the lower-level people, with the exception of those in special positions enabling graft was the same as everywhere. They were given free quarters, food was served at discount to them and despite the drop in the exchange rate of the Kwanchin yuan, they were paid in the equivalent of the local currency. Thus, their "real" salaries did not suffer. They managed somehow to make a living. However, it was the custom to cheat the books from 20 to 30 per cent in the purchase of government supplies and the sale of released government goods.

In the room next to mine, lived a Major Shih and his wife, natives of Changsha. Because of the fact our sur names were the same, I had felt a special attachment for them. (My Chinese name was Shih Chenghsin). He was an honest and blunt youth and complained often to me:

"You can't make a living unless engaged in dishonest deals."

On a major's salary, he could eke out a living for one person, but he could never hope to support his wife. His newly-arrived bride should have been enjoying the happiest days of her life. However, I saw her day after day out in the street in front haggling over prices with passing vendors. And with a sad face she would come back to her room carrying the vegetables that formed their sole auxiliary food. But, no matter how much a couple may be in love, they couldn't just live on vegetables every day. One day, Major Shih came to me and said, "In present-day China, you can't succeed on the basis of your character and ability alone."

One day, they placed their battered suitcase and their bedding, wrapped in a straw mat, on a rickshaw, and walked away to the station. They had started on their way home from a disappointed life. Changsa, I remembered, was the native home of Mao Tse-tung.

The greater part of the people working in the Hanoi headquarters were either natives of Hainan Island or Kwangtung Province. Because Lieutenant-General Hsieh was a native of Hainan and Major-General Wang a native of Canton, people from these two districts were brought into the employ of the office through blood and locale ties in ever increasing numbers. As a result, people from other provinces were being pushed out. This clannish partiality for relatives and persons from the same home province would undoubtedly also be true in the Chinese Communist Party.

About 20 days after my arrival in Hanoi, I was able to say good-bye to Captain Huang and Han, who had accompanied me from Sahanaket. I was finally liberated from the wretchedness of having to stand the sight of snot-green phlegm all over the floor. However, it was hard for me to have to part with Interpreter Wang. I was never able to meet as fine a youth as Wang in the China of that time and in the two ensuing years of my stay in that country.

"Chungking will be cold. Please wear this. My sister knit it for me," he said, as he gave me a beautiful wool jacket, which must have been a treasured property, as a memento of our life together.

"I regret that I am unable to repay you for all the kindness during our many days together. Please take this as a token of my appreciation," and I gave him a small wool sleeveless pullover that I had bought before my departure from Bangkok.

Wang told me that he would try his hardest to retrieve the things that I had left behind on my departure from Bangkok, and send them on to me. I gave him, therefore, a thousand Kwanchin yuan to pay what expenses he might need. When youthful Wang returned to Bangkok, he found that my belongings had been stolen by some unknown person. Indignant at the moral degradation of the Chinese people, he wrote to me in Chungking of what had happened and returned through Lieutenant-General Hsieh, the thousand yuan I had given him when he had left. (I shall write of this in greater detail later). Of all the hundreds of Chinese that I met, this young Wang I-kai was the only truly clean and honest youth. However, it was to be regretted that he lacked strength. He did not have a good background. In such a chaotic world, he was bound to find trouble getting enough to eat. Such was China.

I was left all by myself in the large third-floor room. I knew that other phlegm-spitting men would come in and help occupy the room. I wanted a room all by myself, no matter how small it was. I negotiated with General Wang and was allotted a dirt-floored servant's room next to the toilet. It was damp, mouldy and permeated with odors from the adjoining toilet. Lieutenant-General Hsieh was absent. Interpreter. Wang had gone home. I had no one to take my troubles to.

Major-General Wang lived in two rooms on the second floor of the party headquarters with his young bride and his mother-in-law. From his windows he could see his guest in the servant's room. In all probability, he was under the impression that I was a prisoner of war. His wife was still a young girl of 17 or 18 years of age. The mother-in-law, addressed respectfully. as Tai-tai (madame) was extremely proud of her daughter's husband and ate, together with the office workers, in the headquarters dining room. No one seem to question her eating Government food. But, I found that this was the case everywhere.

Perhaps out of consideration for a Japanese in their midst, Major-General Wang, one day, acidly warned in a long tirade, the members of the headquarters to halt their practice of rushing the dining room, even before the dinner bell, and fighting like cats and dogs over their food. The staff stood at attention, while they chewed at the food crammed in their mouths. Two or three days later they had completely forgotten the general's warning and were again back at their old habit of wolfing their food.

The night life of these Kuomintang members, now at the heights of their prowess was spent first of all in tea rooms and strolling. The second favorite pastime was spending time in taxi dance halls. The next was whoring. Outside of these activities, the young men seemed to have nothing to do. I never saw any one of them engrossed in a book.

In the meantime, I studied during every minute of my spare time five volumes of textbooks on a condensed course of the Chinese language. Thanks to my efforts, I was able to learn enough of the language to do my shopping and to get by in my daily life. However, I could not stand my life in the room next to the toilet.

Lieutenant-General Hsieh had not returned as yet to Hanoi. Winter was approaching. The cold began to bite into my body. Of course, I had no funds to outfit myself in winter clothing. I had to think of some way out of my difficulties.

As winter approached the city of eternal summer, Hanoi, I found my single cotton blanket too thin to keep out the cold. I wanted winter clothing, but had no money. After rejecting one idea after another, I finally decided to go hunt for the cheapest things possible with the Annamese boy at my hotel acting as my guide.

An open-air market had been opened for the poor people on the site of a former race track. It was known as "the Thieves' Market." From morning till night the place was black with people. The sellers were Annamese, mostly women, and the majority of the people, rummaging around for bargains, were Chinese Nationalist soldiers. The goods on sale ranged from clothing and daily necessities to foodstuffs. In all probability, everything necessary to maintain living could be bought there. However, at least half of the merchandise on sale was stolen goods.

Brand new woollen overcoats of the French Army were the center of attraction for buyers. Warehouses packed with French Army clothing, when the French forces were disarmed by the Japanese forces at the beginning of the Pacific War, were broken into by mobs and looted. Thus, French Army clothing was to be found in great quantities everywhere. They were being sold for a song. I had the boy do the bargaining and I reluctantly parted with a thousand yuan for a coat. This I planned to use as bedding, in addition to my thin blanket, at night and to wear it for the purpose it was made during the day.

After that I came to the market often by myself. It was the scene of the life and death transactions of the poor class. Everywhere I saw Chinese Nationalist soldiers cut down the prices of goods sold by Annamese women and forcibly leave with their booty, leaving behind the sobbing sellers.

After giving my last prized watch as a gift to young Wu, I had found myself inconvenienced greatly by lack of a timepiece. I wanted to find a cheap bargain and greedily tramped daily through the market in quest of a cheap watch. Finally, I succeeded in obtaining a silver watch of Parisien make. The watch did not seem very bad but it was old. It kept going ahead ten minutes a day. However, I was able to get a rough idea of the time of the day from the watch.

I needed a mosquito net also, and I found a second-hand one for which I paid two hundred yen. But the netting reeked terribly of the body odor of its former owner. Finally, I neutralized the smell by dousing the netting with cheap perfume.

A major of the Chinese Nationalist Army had his uniform stolen one night. The next morning, with several of his comrades, he went to the Thieves' Market and set up a trap. Unconscious that eyes were watching everywhere for the appearance of the stolen uniform, an Annamese appeared with the goods thinking to sell it. He was immediately caught. The Chinese beat and kicked the man until he was half-dead.

I also noticed disarmed Japanese soldiers strolling through the market. I noticed one Annamese halt a Japanese soldier, and tell him in a soft voice in Japanese:

"Japanese soldiers very good. Chinese soldiers not very good." Then with a show of yearning and friendship, he sold the Japanese soldier goods at discounts of from 20 to 30 per cent of the market price.

Cases of Chinese soldiers raping native women were, contrary to expectation, rare. This was because there were plenty of cheap Annamese prostitutes everywhere in sufficient quantities to meet the demands of the poor Chinese soldiers. In fact, it seemed the supply far exceeded the demand.

In the middle of Hanoi were two revolutionary headquarters. Flying different flags, these headquarters were situated close to each other and seemed to be exchanging hostile stares. Both headquarters were ringed with rows of barbed wire entanglements and day and night armed sentinels stood in front, also glaring with, evident antagonism at each other.

I could not but feel sad when I thought of their futures, to see despite the fact that the objective of the two revolutions were the same, this conflict and strife between the followers of the two revolutionary movements. One movement was called "Vietmin," an abbreviation of the Vietnam League of Democratic Revolution. Under the leadership of Soviet-educated Ho Chih Min, it sought to gain simultaneously racial independence and a social revolution through Marxist principles. The other movement sought racial revolution under the leadership of the Vietnam National Party.

After its advance into French Indo-China, the Japanese Army backed the more right-wing Nationalist Party, providing it with considerable arms. However, after the end of the war, roughly 3,000 armed members of the Nationalist Party, all bearing Japanese rifles, defected to the Vietmin camp. Judging from appearances in the city of Hanoi, alone, it seemed that Ho Chih Min's reputation was overwhelming great among the general populace.

The Kuomintang Party of China supported the Vietnam National Party, while the Chinese Communist Party backed the Vietmin. Bloody clashes between the Vietmin and Vietnam were repeated again and again in the streets of Hanoi. In this way, both sides were wearing each other down, at a time when their respective strengths should have been reserved for their fight against the French. However, the Government did not ever try to give the weak Chungking and puny Vietnam National Party any material aid, not even a single yuan, Of course, not a single rifle was given the Vietnam Nationalists: Neither did the Chinese Nationalists have anything of ideological value that might have afforded inspiration to the Vietnam independence movement. On the other hand, the Chinese Communist Party shared, at least, the ideological fervor of racial liberation, and incited the Annamese to seek independence from the French. It was more than apparent that if matters were left untouched, the pro Communist Vietmin, under the leadership of Ho Chih Min would eventually unify the independence movement of the Annamese. The majority of the Annamese people were illiterate and did not have even the fundamental critical acumen to judge the good or the bad of Marxism. For this reason, it seemed they felt a greater attraction for Ho Chih Min, who sought in a more positive manner and by using brute force to drive out the French. I wondered whether I was prejudiced in reaching this conclusion.

A number of Japanese soldiers, detained in French Indo-China, were in the employ of the Kuomintang. The group included a sub-lieutenant who was engaged in the compilation of army manuals. I also heard that a number of Japanese residents in French Indo-China were working within the pro Communist Vietmin camp. However, I was unable to check the veracity of this report.

It was a fact that the Annamese, who seemed to have been sapped to their bones of all energy under a hundred years of French rule, were being aroused to a "bloody wakening." This desire to overthrow being French domination was the only common idea to the two revolutionary movements.

I realized that the future course of the Annamese independence movement would be swung either left or right, or colored red or white, depending on which of the two countries, the Soviet Union or the United States, most correctly grasped the situation and most positively offered aid. It seemed a common characteristic of the two conflicting independence movements that the majority of those abandoning their own work activities were youths becoming engrossed in revolutionary activities.

One day a certain Chinese came to me with a message from the Ho Chih Min camp:

Please stay (in Vietan Annam) and help our revolution."

In answer, I said:

"If you can unconditionally, magnanimously and dispassionately unite the two independence camps, I would not hesitate to offer my wholehearted services. However, I cannot bear the thought of helping one Annamese kill another."

Because the French Army occupied French Indo-China below the 15th Parallel and the Chinese Nationalists the area above the 15th, the French residents in the northern part of the country lived in concentration camps under the friendly supervision of the Chinese Army. Even at the heights of the revolution, Annamese women would come to the French camps with vegetables and fruits and stealthily sell their goods to the occupants, picking up the money thrown in return from the second-story windows of the French camp. The Annamese women were running all kinds of chores for the French. I realized that racial emotions are weak in the face of the necessity to live. A people that cannot stand alone economically, I told myself, can never achieve military or political independence.

Looking back at the Meiji Restoration [January 3, 1868, the day the Meiji Emperor formally assumed political power], I recalled that the successful overthrow of the

Tokugawa Shogunate was accomplished through the pooling of resources and energy by the Satsuma, Choshu, Tosa and Higo cláns. In addition, the Loyalist factions had brusquely refused all foreign aid in their efforts to end the rule of the Tokugawa's and restore the Imperial House to power. When I viewed the realities of the Annamese independence movement, in the light of these recollections, I could not but help sympathizing with the powerless and peaceful people of Annam. Everywhere the peoples of Asia were divided into two camps in their struggle to win racial independence, and everywhere the common shortcoming was the short-sightedness of relying on outside aid and utilizing outside backing to carry on internecine warfare. I do not begrudge heartfelt respect for our pioneer patriots who completed the Meiji Restoration in the midst of a tense world situation. At the same time, however, the fundamental reason why the tragedy of internecine bloodshed was limited to a minimum lay in nothing more than in the national polity of Japan. We should recognize anew the fact that this was due to the existence of a continuous and consistent center of blood affinity, of ideals, and of faith. For this; we should be thankful.

Even if Ho Chih Min should win out thanks to aid from Soviet Russia or the Chinese Communists, the net result would only mean a nominal change of masters. No improvement in the national life of the Annamese people can be expected. In addition, should Emperor Bao Dai succeed in maintaining his erstwhile influence by obtaining French aid through compromise, he would never gain the full-hearted support of the Annamese people, in particular the Annamese youth.

Even if he were to succeed in freeing Annam from the military and political shackles of France, how could he ever hope to gain independence from the frightful economic exploitation of the resident Chinese? No matter which way the problem is viewed, Emperor Eao Dai's future is a path of thorns.

Those riding around the streets of Hanoi in beautiful sedans were the high military officers of China. Those who went shopping on bicycles were the French. Those who ride the "yang che" (rickshaws) were Chinese Nationalist non-com's and soldiers and those who pulled these rickshaws were without exception Annamese. Those who lived on their savings were the French. Those who toiled and moiled were the Annamese. Those who made money with hands in their pockets were the Chinese residents. And, sitting on top of this pyramid, were the Chinese military leaders who fattened on the kickbacks from all this economic activity.

The sketch of the features of the streets of Hanoi might change with the evacuation of the Chinese Nationalist forces and the arrival of the French Army. However, it seemed that the positional relationships of the Chinese residents and the native Annamese had become semi-permanent and would not easily change.

The true objective of the Annamese revolution should be aimed at changing this relationship. Already shrewd Chinese residents, with prescient eyes, were beginning to place confidence in the growing influence of the Chinese Communist Party. Some had already started a movement to donate ten per cent of their gains, obtained through the exploitation of the Annamese, to Ho Chih-Min's Communist revolution, with the idea of placing the Vietmin in a position of indebtedness. realized that future developments in French Indo-China would be worth watching. At the same time, I wondered what shape the future would take.

While walking all alone one evening on the street behind my hotel, I unexpectedly encountered two Japanese soldiers. Though without their usual bayonet and belt, they seemed almost majestic in their combat caps.

Careful not to attract attention, I approached them.

"Say, you fellows. Don't you belong to the Japanese headquarters?"

They probably did not believe that this figure in a long Chungshan robe was a Japanese. They looked at me with surprise, and slowed their walk.

"Yes. That's right."

"I'd like to speak with you for a minute. Is it alright?"

"Six o'clock is when the barrack gates are closed. We're in a hurry."

Somehow they seemed to be trying to get away from this ominous looking person in Chinese clothes.

"Don't worry. I'm also a soldier. I'm a friend of Staff Officer Iwakuni. Is Iwakuni well?"

"Yes. Do you know him? Immediately before the end of the war, Staff Officer Iwakuni was hit by an American plane. However, he has completely recovered and is a present at the headquarters in Dosong."

"How is His Excellency Dobashi?"

"He's also well and is staying at the headquarters in Dosong."

"Thanks. Whose orderlies are you?"

"His Excellency, the Commander's."

"I'm a friend of Staff Officer Iwakuni from our days together in the preparatory military school. If you meet him please give him my best regards. I do hope you get home safely."

I was surprised at the way my heart went out to these two strange soldiers. Living all alone in a world, composed only of Chinese and Annamese, I found myself yearning unbearably for the company of Japanese and found myself wanting terribly to speak Japanese.

The face of the commander and Staff Officer Iwakuni floated before my eyes like a vision. Thirty years of life together with Iwakuni from our days together in the Nagoya Preparatory Military School flashed like a kaleidoscope through my mind.

Before I lay down in my lonely room to sleep, I clasped my hands together in prayer and prayed for the continued health and the future success of my friend:

"Please, return home safely. Please work for the rebirth of Japan. Oh Iwakuni, I have left behind my country. I have left behind my friends. I now float alone on a vast ocean, as if clinging to a blind tortoise or a floating log and entrusting my life to the giant waves of fate...."

However, that night I couldn't sleep. My brain seemed to become clearer and sharper with each passing minute. It was not till late that I finally dozed off. In all my 30 years of military life, I had never felt such yearning, nay, such respect for. Japanese soldiers before. I looked back on my life and wondered whether I had ever mistreated such fine soldiers. If there were any shortcomings on the part of the soldier, the greater part of responsibility and sin lay with the officers could there be in any country of the world soldiers so pure of heart, so self-sacrificing and so brave as those of Japan?

The end of the year was approaching. The Kuomintang Party Headquarters also began decorating its entrance in preparation for greeting the first new year of victory.

The entrance was adorned with greens in the traditional Chinese manner. A picture of Generalissimo Chiang Kai-shek, almost ten feet tall, was set out with much ado outside the main gate.

"I'd like to greet at least the New Year with Japanese soldiers." Since my meeting with those two soldiers, I was seized with a desire to get in touch with the Japanese Army and find out, at least, what had happened in the interim. I learned that the liaison office of the Japanese Army was situated in the park at the rear."

On December 25, taking advantage of the gathering dusk, I rode to the liaison office in a rickshaw There were no Chinese soldiers stationed on guard in front and so, I went straight into the office.

Suddenly, a major shouted:

"Sir. Staff Officer..."

I clasped my hand over his mouth. I took him to a separate room and there told him my story in detail. This major was a member of the first graduating class of the Nakano Military School. He was Major Suga, who had sat in one of my classes, when I lectured during a joint education course at the Supreme Headquarters at Nanking.

Here I was able to meet an old friend unexpectedly in one corner of Hanoi. I told him I wanted to move my quarters to the liaison office. Major Suga answered:

"Staff Officer Misawa is absent at the present time. It would be best if you came back again tomorrow."

Major Suga went to great pains to make me feel at home the short while I was there. In order to prevent the Chinese from knowing where I had been. I took a detour on my way home. A cold room, smelling of mould and human excreta greeted me on my return.

The next day, December 26, turned out to be raw and wet. The chill penetrated to my bones and the old wound in my right hand began to pain. That night, I deliberately took the chopsticks in my left hand, and making a great show of my clumsiness, ate my supper. Everyone looked at me with odd eyes. Inevitably the awaited queries arose.

"There's still a dozen or so pieces of shrapnel in my right hand. When it gets cold, it begins to hurt terribly," I replied; and I screwed my face in a show of pain. In my poor Chinese I explained my wounds in detail, Even Major-General Wang, cold as a reptilian specimen, expressed sympathy. He asked me whether there were some means of treating my hand. I replied:

"The only way is to bathe it in warm water. But, I can't do that here."

Thus succeeded my preliminary maneuvers.

The next day, I slipped away unnoticed and visited Staff Officer Misawa at the liaison office.

"Oh. Staff Officer Tsuji,' he cried and I had a hard time keeping him from being heard by others. I explained all that had happened to me to date and explained that I would like to await the time of my departure for Chungking in the liaison office among Japanese friends. He expressed joy at having me with them and said:

"Please come over any time you desire."

I couldn't stop the tears that came to my eyes at his reply.

The Japanese Army had been dissolved. But, people were still held together by mutual friendships. I immediately got in touch with Commander Dobashi to obtain his permission and to await his reply. Knowing that he would not refuse my request, I, got in contact with Major-General Wang as soon as I returned to the Kuomintang Headquarters:

"While walking in the park today, I accidentally met Staff Officer Misawa. He happens to be an old friend of mine. He invited me to join him at the liaison office. There's nobody else there that knows me. I am confident that my secret will be kept, if you will allow me to go. I can bathe there any time I desire and it would be most convenient to treat my ailing hand. Please allow me to go there until the date of my departure for Chungking is decided."

He immediately wrote down:

"Very good."

He probably felt that there was no further need of keeping me under strict surveillance. Or perhaps the recent arrival of his young wife had made him more human and had tempered this young major-general's habit of being aroused to quick anger.

That night I was like a young child on the eve of departing on some pleasant outing. I couldn't get to sleep for joy. I had finally succeeded in escaping from my prison next to the obnoxious toilet. The next day, accompanied by my Chinese language teacher, young Li, and the pleasant Major Yeh, I moved to the Japanese Army Liaison Office. Li and Yeh were picked to be my contact men and were to come and see me from time to time. It was the fall of December 29.

I was overwhelmed with the pleasure of being able to speak my native tongue after such a long period of constraint.

I had no special relations with Staff Officer Misawa in the past. But he greeted me from the bottom of his heart and our relations were as warm as that of friends of ten-years standing. I was indeed grateful to this man, my senior.

Commander Dobashi wired back the next day:

"Overwhelmed to hear you alive. Accede to everything you request. Please arrange details with Misawa."

Commander Dobashi had showered me with many favors since my days at the Tokyo Staff Headquarters. I felt ashamed of the many occasions on which I had spoken impolitely to him and I reflected on my past faults. I heard that he had hurt his leg and prayed for his early recovery.

On the door of, my room was placed a card inscribed

"The Reverend Masanobu Uesugi, Army Priest."

That night Major Suga kindly arranged to have a bath prepared for me. It was the first bath in five months since the end of the war. I had contracted scabies without my knowing it during my unhygienic travels and my sleeping with unwashed persons in cramped quarters felt itchy all over my body, especially between my fingers. There were also sores here and there. I soaked myself in the hot medicated bath, smelling the sulphur fumes, until I thought both body and soul would melt away. There was a sergeant-major who shared the bath with me. He had grown a long bicycle handle mustache and was well-built. He looked at the bulging muscles of this priest, better built than the sergeant-major himself, and with a question in his eyes said:

"Bonze, my you're well-built…?"

"Yes. During my school days I took up 'judo,' something quite out of my line. For eight years I served as Army chaplain, accompanying soldiers everywhere. That's why I've come to look like you people."

Thus, did I manage to prevent my true identity from being discovered.

Later another soldier came into the bathroom. I asked him from what part of Japan he came from. He answered:

"Ishikawa Prefecture."

I felt a sudden desire to become intimate with this youth from my home prefecture, but realized that I would have to be on my guard. He conversed with me with kindness in his eyes, but there was also wonder written in his face at this rough and ready priest.

For breakfast I had miso soup and "takuan"–pickled Japanese radishes. The breakfast tasted better than anything I had had in a long while. The Japanese people have an unremitting attachment for their own-style of bath, for miso soup and for pickled radishes. I felt that this was also true of my attitude to Japanese soldiers. In ordinary life, I had not felt any particular gratitude toward them, just as I did not think miso soup and "takuan" tasty. But when I had been separate from both Japanese soldiers and these foods, I felt an overwhelming nostalgia for them.

The quarters allotted to me was a small room, opposite that of Staff Officer Misawa. It was clean and bright. The floor was tiled and the bed equipped with springs. In comparison with my life of the past half year, the difference was as great as that between the life of a beggar and a king.

A second-class private, called Miyauchi, a native of Kagoshima, was allotted to me as an orderly. He was still a first-year soldier, but he looked after me with scrupulous care. Such treatment was much too good for one who had given up all orderlies with the end of the war. This honest-to-the-bones soldier believed implicitly that I was a priest. Perhaps it was because of the sign on my door, "to be accorded the treatment of a major," that he tendered me so much respect.

"Ah Bonze, let me do your washing for you," he would say. And I would refuse:

"No, to do one's own washing is a part of a priest's self-discipline."

And the young soldier would go away with disappointment written on his face.

Gendarmerie Major Oshima and 20 to 30 other officers and men, all under suspicion of having executed Chinese residents, were staying in the same liaison office.

This major, flush with alcohol, caught hold of Major Suga one day and blustered:

"Say you. It's a damn outrage to accord that dung priest the treatment reserved for a major and to let him have one room all by himself. I'm a major on the active list. Chase that priest out and let me have the room."

I often met him at bath time. I tried hard to soften his psychology of a carp on the cutting-board and spoke to him on the attitude he should take when he took his stand in the courtroom, explaining quietly the truth of "Color, this in sum is the sky. The sky, this in sum is color." My talks seemed to have some effect. He went around telling others:

"Say, that guy is no ordinary priest."

I noticed people begin looking at me with different eyes.

" This is bad. Unless l remain a pure hick bonze, a dung bonze, they'll discover who I really am," I warned myself. But, often I would find myself forgetting my present role and wanting to speak from my heart to different people.

I greeted the first New Year's Day since the defeat in the role of an Army priest. At five in the morning, I purified myself and went out on the roof of the two-story building. Not a single person was as yet awake. The eastern sky had begun to whiten. I turned toward the distant skies over my motherland and chanted aloud the "Saddharma Pundarika Sutra" (the Sutra of the Lotus). When I had finished the Sutra 30 minutes later, the New Year's sun had already poked its red ball over the sleeping trees.

"Oh, what a disloyal, what an unfilial son I am. I wonder what the Emperor feels in his heart today? I wonder what thoughts fill the hearts of millions of people bereft of loved ones? I wonder how fares my old mother, with her constant thoughts of her son? Ah, my wife, my children! Where are you now on this New Year's day...."

At 9 a.m, Staff Officer Misawa gathered together all the officers and the men of the liaison office and "delivered homily." This was the final New Year for the gathered men as members of the Japanese Army, noticed Staff Officer Sato. I managed to check myself in time from speaking to him. I also noticed the face of Interpreter Imai, who had been with me in Nanking. Gendarmerie Colonel Kasuga, held as a war crimes suspect, was living under the same roof with us. Each man seemed engrossed in his own thoughts. Although the central force that once bound these men together had gone, there seemed to remain some where the intangible strength of the Japanese Army.

After the address, a humble festive banquet was held. It seemed all the traditional New Year's food was prepared. There was spiced sake, black peas, "mochi" (rice cakes), lightly salted fried sea bream and herring roe. In all probability, the Japanese family back in Japan were not enjoying such a feast as we were having. A total of 40 to 50 officers and soldiers sat down together at the tables. The majority of these people really believed that I was a priest.

Many would try to serve me sake, saying:

"Bonze. Have a little nip. After all today is one day in a year...."

"Thanks, but this bonze can't hold even a drop of sake...."

What a laugh, "can't drink." Here my mouth was watering! The the aroma of warm sake pierced my nostrils, titillated my instincts and I could feel my throat groan for a drink. But, I held myself back, for fear my true identity would be exposed.

After the meal, I was treated to delicious "abekawa" in Staff Officer Misawa's room. In this way, I was able to greet my last New Year as a Japanese in an atmosphere of warmth and comradeship.

I peeked into the Nationalist Party Headquarters celebrating the first New Year of its victory. Two or three petals of cut flowers lay on the floor marking the passing of a storm of boisterous singing, loud talk and Bacchanalian excess.

From my second story window I could see the home of a newly-wed Annamese couple set in a pleasant truck garden. The bride, with French blood in her veins, was beautiful. Their child was also very pretty. The grandfather, a dignified and warm-hearted patriarch, must have guessed that I was a priest, from watching me morning and night praying with clasped hands. We exchanged New Year's greetings, each silently clasping hands in front. He plucked a beautiful flower growing in his garden and sent it to me, handing it over the intervening fence. I was happy to think that the Annamese still held a certain fondness for Japanese, even though they had been defeated in war.

I was able to have a heart-to-heart talk with Sato for the first time in a long while. A promising junior from the same prefecture as I, he was chosen from a number of other instructors at the Military College and appointed aide-de-camp to Field Marshal Terauchi. From this post, he had been transferred as staff officer to the occupation forces in French Indo-China, where he saw the end of the war. He was a characteristic northerner, a man with a warm heart and of deliberate actions.

After lunch, another old friend dropped in, a Sakai. He was an outstanding youngster, long known for his ability in the Tokyo staff headquarters. Now, he was engaged heart and soul in the difficult task of supervising the return to Japan of a vast host of demobilized Japanese soldiers.

Kashiwara arrived from Dosong for routine business liaison and on January 15, Staff Officer Iwakuni came to Handi. He had lost one eye and had been wounded both in his face and feet. His appearance was pitiful to look at. However, his heart had grown more tender. He grasped the hand of this wretched-looking priest and simply gave himself up to tears. That night over a pan of sukiyaki, we talked until late of the defeat of Japan and of the future of our country. Looking back at the past 30 years of our life, the thought of what the future held in store for both of us was bitter gall. At the time when puberty overtook us at the age of 16, we had changed to uniform and worn the plain unstarred epaulettes of students of the Nagoya Military Preparatory School. There in Nagoya, under the shade of the camphor tree, where lay enshrined the soul of Tachibana, we had studied together, gazing often at the golden dolphins atop the Nagoya Castle. We reminisced of those days during. which we had trained for a common objective. The spirit of sacrificing self for the Emperor and for our country, pounded into us in those days, still remained the same despite the gray hairs that had begun to appear on our heads.

We forgot that we had begun to bald, that our eyes had no longer the same eagle-sharp vision to our youth, and we returned to our childhood and talked as old friends. It is good to have a good friend, a true friend. Our country had been defeated. The Army had been dissolved. But our friendship still remained.

The visit of Colonel Kanehara of the Medical Corps and Dr. Tanigawa all the way from Dosong, just to get in touch with me, was also unforgettable. I also had gratitude for Interpreter Imai, a friend of my days spent with the Supreme Headquarters of the Japanese Expeditionary Army in China, who taught me the Chinese language both morning and night.

Sick and tired of the corruption of the Nanking Government under Dr. Wang Ching-wei, I had secretly hoped to find comrades, with whom I could really cooperate, within the anti-Japanese resistance camp of Chungking. However, what struck my eyes during my 100 days of waiting in Hanoi all pointed to a definite "no" to my expectations. Taking advantage of the fear of Japanese soldiers and officers being designated as war crimes suspects on the eve of their return to Japan, the Chinese Nationalist officer superintending all matters concerned with war crimes openly demanded several million yuan in Vietnamese currency. The sum stipulated was equivalent to several hundred million yen today in Japan (around two million US dollars). In particular, Gendarmerie Colonel Kasuga and the 20 or 30 men under him, were the constant targets of blackmail on the grounds that they had killed anti-Japanese Chinese residents and were responsible for missing American pilots who had been shot down. [See *Incident at Ple Tonan, An*

Imperial Japanese War Crime and the Fate of U.S. Navy Airmen in French Indochina, for details on the fate of these pilots.]

Other Chinese Nationalist officers in charge of supervising the confiscation of Japanese firms and property, would reject all inventories that tallied with the actual property or commodities on hand. Only those cases which left out over 30 percent of the goods from the inventory and left this 30 per cent to the disposal of the Confiscation Committee were passed unconditionally. Nay, such proceedings were not passed, but demanded openly and brazenly. I wondered whether this practice was limited to Hanoi alone. I hoped that it would be. Although I had placed all my hopes in the expectations of finding a hard core of honest and patriotic men in the anti-Japanese resistance camp of Chungking, I could not suppress a budding premonition of the approaching end of the Chiang Kai-shek, regime when I assumed that what I saw in Hanoi was a trend common to the whole Chinese Nationalist Party. Major-General Yeh demanded openly through Interpreter Ima, ten bars of gold, bullion and from a certain colonel, through Liaison Officer Aoyama, two million Vietnamese yuan (about US $500,000). He threatened that if these demands were not met, he would execute the Commander of the Japanese force in Indo-China and several hundred Japanese officers and men. It was, of course, impossible for the Japanese officers and men to raise such a large sum of money. There was only the official military funds, which were to be handed over formally into the Chungking Government treasury. The only way to raise the requested money was to take it out of these official funds. To my surprise, the Chinese Nationalist Confiscation Committee did not seem to mind such a dishonest expedient.

These practices were not limited to the question of war criminals alone. Several hundred of the most outstandingly wealthy Chinese residents were arrested as collaborators and secretly urged to pay ransom. Those who agreed were immediately released, (Those who refused were executed regardless of whether they had actually been traitors.) The average market value of the life of such Chinese residents was half a million Vietnamese Yuan (about ¥50,000,000 in Japan today or about $140,000 United States dollars).

Commander Dobashi heard of these demands and made a strong representation to the Commander-in-Chief of the Chinese Nationalist Luhan Area Army, saying:

"I am entirely responsible for the actions of my subordinates. Please be lenient with the war crimes of the men under me." However, the Chinese Nationalist Commander was not one to be moved by such a plea.

Even my friend from my days in Military Preparatory School, Iwakuni, was placed in a totally new world. Realizing the necessity of sending as many officers and as many men back to Japan as soon as possible, he swallowed his tears of indignation, covered his eyes and tacitly condoned the unprincipled actions of the Chinese Nationalists.

I was filled with sadness. The hope that I would find true friends in the camp of the enemy, against whom I had fought for so long, seemed about to fade under the impact of actuality, like wisps of clouds.

Early in January cholera broke out in the Liaison Office. The intended captain, with whom I had gone into the bath together the night before, was carried into the dispensary, vomiting blood. Medical care was given to him immediately, but he died early the next morning. Only yesterday he had talked with longing of his wife and children and had been happy over the prospect of being in Japan a month later. And now, he lay cold and dead in lonely, concentration camp of a defeated army. Staff officers Misawa and Iwakuni and all the members of the liaison office gathered to hold funeral ceremonies for the captain in the main hall of the liaison office. Laid out in state was this man who had been so full of energy two days ago. Now he was a memory, whose cremated remains lay in a five-by-five-inch white box.

The imposter priest was now called upon to fill the role of a real priest. I had not bargained for this task of chanting sutras and saying requiem mass in front of a crowd of people when I had become a priest. Now, from the memory of past funerals that I had attended, I would have to act a funeral priest with sutras and the burning of incense and all.

Despite being accorded the rank of major and despite my avowed profession as an army chaplain, I only had my cotton Chungshan Chinese robe, now bleached white from many washings. The only accoutrements of religion was a single loop of prayer beads. I had neither shoulder piece nor robes I did not even have a book of sutras. I wondered with what eyes the majority of the gathered officers and men must have viewed me. I broke out in cold sweat and my body trembled. My chanting was poor. All I had was the feeling that I would like to take the place of the distant members of this dead man's family and grieve from my heart over his death. This, perhaps, was the only virtue this sham priest had over a regular bonze. For 30 minutes I chanted every sutra that I had learned from the "Hobinhon," the "Nyoraijuryo hon" and the Prajna Paramita Sutra. I lit the incense and bowed myself out in as respectful a manner as possible.

There was a real priest among the soldiers. He was a private and had therefore handed over the honor of saying mass to this Army chaplain, accorded the treatment of a major. After the ceremony was ended, he came to me and asked:

"Pardon me, but I would like to know what sect you belong to."

"I am of the Nichiren Sect. However, because of my long life as army chaplain, I do ceremonies in any sect. In other words, I'm what you might call an 'army sect.'"

With this crazy-quiet answer, I managed to escape from my dilemma.

Later, Major Suga told me that this private had gone to him and criticized my chanting of the sutras:

"There is something funny with the way that priest offers his sutras. He makes too clear a break between each of the phrases."

I myself knew that there was certainly something funny with the way I said my sutaas and so the criticism was not totally without foundations.

"Forgive the poor way in which I said the sutras, but please enter Nirvana and attain Buddhaship," I prayed and apologized to the dead man. In this way, I managed to pass my first, and probably my last test as a priest.

At Sahanaket, I had almost been exposed as a quack doctor. Today, I was almost caught as a sham priest. How difficult it was to live a lie. However, the gods and Buddha seemed to have forgiven my deceptions.

Members of the Kuomintang Party headquarters often came to chat with me. Gradually our intimacy grew and they began to ask for my advice. The greatest number of requests were concerned with the treatment of gonorrhea and syphilis. At least they were ashamed of going to a town doctor and wrote "gonorrhoea" with their fingers on the palms of their hands and asked me to request aid for them from the Japanese doctor attached to the liaison office.

I called on Army Doctor Niizuma in the dispensary and asked him politely. This doctor believed without a single doubt that I was a real priest. It must have been the first time in his life that he had been introduced by a priest to the people with gonorrhea. However, without showing any distastefulness on his face, he gladly took over the task of treating my Chinese friends. Niizuma's sincerity and his medical competence greatly impressed the Chinese.

The main task of a priest is to say mass for a dead person. But, is it not also an important task to cure and guide living men?

There was a young divisional staff officer who used to drop in often at the liaison office. A native of Shikoku, he was a young man of about 30, named Miyoshi. Despite his youth, he was a very excellent "go" player. Watching the way he played this game, he had an arrogance that would not bow to any man, I felt that it was a shame for so fine a youngster to have so evident a failing. On the eve of his departure for Japan, I felt strongly attracted to this rascal.

Staff officer Sato interceded and said:

"Say, Miyoshi. The priest says he would like every much to have a talk with you. Go and listen to him."

Despite Sato's urgings, this youngster blustered away:

"I hate priests. I just don't like their guts."

I just happened to pass by and saw the young man doggedly refusing to move. I said:

"Now, now don't be like that," and I dragged him willy-nilly into my room."

"What class are you?"

The moment he heard my question, this quick staff officer Miyoshi caught on to everything. He snapped to attention and said:

"Yes sir, I'm of the 54th Class."

Beads of perspiration formed on this brow. For the first time I told him who I really was and asked him to restrain himself and keep on fighting cleanly.

"You're Kojiro Sasaki. You are blessed with an abundance of natural-born talent and ability. However, you lack self-reflection. You have no humility. Unless you change your attitude now, you will have your head split by Musashi's wooden sword."

I talked to him man to man. It was without doubt, the heights of presumption for me to speak to him in such a manner. However, the greater the hopes I had in this young man of the future, the greater the pain of having to keep on fooling him and of leaving him with the false impression of me as a priest. I felt that this youth, no matter what path he should choose in the future, was a man capable of shouldering an important task for Japan. I prayed that my impression of him would prove beneficial?

The figure of the young staff officer, apologizing for his impoliteness and wiping the sweat from his brow, seemed like a dear younger brother.

I prayed that this man would progress through self-restraint, reflection, self cultivation, and passiveness in face of humiliation.

"Young man" the task of the reconstruction of the motherland lies on the shoulders of you youngsters still in your 30's."

On February 15; Major Yeh suddenly appeared at the liaison office. He told me that Major-General Wang wanted to see me. Wondering with a fluttering heart what was doing, I hurriedly left for the Kuomintang headquarters. In the presence of Colonel Li, Major-General Wang passed me a secret telegram. At the same time, he told me that he had received a telegram from Chungking telling me to come immediately in company with Lieutenant-General Hsieh. The deciphered code telegram was from a still unmet comrade:

Translation

"To be transmitted to Colonel Masanobu Tsuji, personally and in secrecy. When I was in Shangrai some time ago, I had heard of you from both Miao Pin and Tamura, and always felt a deep sympathy toward you. Unfortunately, Japan's military clique erred in its policies and did not listen to advice. The result of this was to invite today's utter defeat. Your Excellency has always had high and far seeing aspirations. I pray that you'll be careful in the future, deepen your knowledge of China still more, and fully recognize the magnanimous kindness of our supreme leader and General Tai Li."

January 23
Cheng Chang-feng"

I tried to think who this comrade whom I had not as yet met could be? The only important leader. of the Chungking underground movement having connections with Miao Pin and Tamura that I could think of was a Dr. Chang. Could his be Dr. Chang? Yes, it was. We had both been ideological comrades in the East Asia Federation movement. My relations. with Chang went back to my days in Shanghai, The Japanese gendarmes. in Shanghai tried to arrest him but let him slip through their fingers. Instead, they arrested his family and tried to hold them as hostages. Miao Pin approached me and asked me to help. I went to see Lieutenant Colonel Yamazaki, Chief of the "Tokko" Section of the Shanghai Gendarmerie who immediately released the members of Chang's family and cleared them of all complicity. Chang turned out to be a lieutenant general, but undoubtedly, he was in reality, the very Cheng Chang-feng that had wired me a nuclear member of the Blue Shirt Society (the Military Control Bureau).

Judging from the fact that he had taken the trouble to send me a telegram, after learning in Shanghai that I was waiting in Kunming, to enter China, it was correct to assume that Chungking took a fairly grave view of my case and was investigating both my past actions and words throughout the whole country. If that were the case, I would have to view Dr. Chang, in other words, Cheng Chang-feng, as being a fairly important person in the Chungking hierarchy. Major-General-Wang's attitude had also suddenly changed, after the receipt of this telegram, and he was now extremely polite towards me. It seemed that he had finally gotten around to realizing what this old colonel of the Japanese Army was. Colonel Li rejoiced from the bottom of his heart for me. Thanks to the fact that I had fought for long years by example and in practice an ideological warfare to realize the ideals of the East Asia Federation, against the fierce opposition, hostility and obstructionism from a part of the Japanese people, I had been able to gain a friend in the camp of the enemy.

I wanted to fly to Chungking as soon as possible and begin on the first step toward cultivating relations between the Chinese and the new Japan. My departure was now promised within a few days. At Chungking, the second meeting after the war of the Central Committee of the Kuomintang Party was about to be held. General Hsieh had been called back as a member of the Central Committee. The telegram telling me to go to Chungking with him seemed to shed a bright light that I was to go to on my future path.

I told staff officers Iwakuni and Misawa the good news. They rejoiced with me from the bottom of their hearts. My departure had now been decided. Staff Officer Misawa felt that I would be cold in Chungking with only a thin cotton robe and he kindly arranged to have a worsted Chung shan gown made for me and tried to have Imai accompany me as an interpreter but he was unwilling to come with me. I decided to take Yamatani and presented a request to the Kuomintang headquarters to let him come with me. However, they said that interpreters would be made available to me in Chungking. Yamatani was greatly disappointed. I looked around for a tailor to make me a robe of worsted cloth. However, because it was the Annamese lunar New Year, all the stores seemed closed. After a long search, I finally managed to find a tailor. The bill for the robe was paid from Army funds. I had long shivered in my thin cotton robe because of my empty purse. I was thankful when the warm gown was completed.

From the beginning of February, the war crimes investigations by the Chinese Nationalist Army entered its last lap. The Sixth Area Army, occupying French Indo-China, was to evacuate by the end of March and return to Kunming. At the same time the War Crimes Supervisory Offices were to be closed.

Sub-Lieutenant Aoyama, dispatched to the Sixth Area Army as liaison officer turned out to be the only son of lawyer Aoyama, a native of my home town. Sato introducing me to him, said:

"Here's a bonze that knows your father well." I hated having to fool this young officer with a future ahead of him and wished that I could leave this youngster something to serve as a guide in his life to come. But, I bore with lies as a means of achieving an end great enough to counterbalance the expedient.

Major-Mitsui flew into Hanoi bringing final orders from the "Japanese Army in China." I was favorably impressed by this youth, formerly attached to the headquarters in Canton and learned that he was a good friend of my young brother-in-law, Tsumiaki Yamada. Overwhelmed with nostalgia I asked him a number of questions about my brother-in-law.

However, I found myself at a loss, when he asked:

"By the way, do you happen to know Colonel Tsuji, Staff Officer Yamada's brother-in-law? "

"No, I've heard his name, but haven't had a chance to meet him yet. Where is he now?"

"In all probability, he's already returned to Japan," he answered, indicating that my plans were still secret and safe. However, I was tormented by having to dupe this pure-hearted youth and I couldn't look him straight in the face.

Mitsui was accompanied by two non-com's and Iwakuni invited us all to sukiyaki. One of the non-com's by strange coincidence happened to come from the same town as my bosom friend Hisakado, who had been killed in Hokkaido. I felt as if my entrails were being ripped apart to talk about a friend, no longer on this earth. Here too was the pain of one who had to lie even in such circumstances.

The orderlies, viewing this priest participating frequently at sukiyaki parties, whispered behind my back:

"That bonze is a "raw smeller. He eats meat."

Lieutenant General Hsieh, for whom I had waited so impatiently, finally came back to Hanoi from Saigon on February 23. I immediately went to meet him at the Kuomintang Headquarters. He told me that I was to fly with him by plane to Churtgking on February 29. On the afternoon of February 28, I gathered together my luggage and went to the headquarters. To my surprise, I was told that General Hsieh, accompanied by General Wang, had left on that morning's plane. I was surprised to think that an important man like General Hsieh would not hesitate to break his own word. There were no grounds to think that he had mistaken the time and date. My gaping mouth refused to close.

"You will accompany Hsieh Tai Tai (Madame) on the next plane on March 9," they told me.

That night I met Madame Hsieh for the first time.

The wife of such a distinguished person as a lieutenant-general of the Chinese Nationalist Army and a member of the Central Committee of the Kuomintang Party, I expected, would be of fairly advanced age and a refined and dignified lady. To my surprise, I was introduced to a short and stout woman of around thirty, in the prime of her womanhood. It seemed a common practice among Chinese dignitaries to marry women young enough to be their daughters, then keep drinking aphrodisiacs to continue a life of pleasure into old age.

Major-General Wang's wife was a girl young enough to be still in high school. However, she was strong-willed and was a match for any man. Looking at these women, I felt again the goodness of Japanese women, even as I had felt the goodness of Japanese soldiers.

On the night of March 8, I had our final dinner with Iwakuni, Sakai, Sato and Suga. Despite the humble repast, my hosts were sincere in their joy at the success of my mission so far. This brief meeting in a brief life would soon be over. The feeling of that perhaps this might be our last meeting together, pervaded the dinner. But, the warm friendship of my friends dispelled any feeling of gloom and could be felt even in the sake cups that we exchanged. I had met friends whom I had longed to see, and had conversed as I had wanted to converse. There was nothing more that I could desire. From tomorrow, I was again setting out on my solitary travels, throwing myself alone into a lion's den. Nay, I myself had chosen of my own accord to enter the lion's den. That night, a feeling of joy commingled with that of sadness, and I could not get to sleep easily.

I woke at dawn on March 9 and completed all my preparations. I told second-class private Miyauchi and others:

"I'll be gone for a while on a trip to Dosong."

They did not doubt in the least this lie of mine.

"Bonze, please take good care of yourself. Hope we meet again."

To the very end, they looked after me with brimming kindness. It was painful to have to keep on duping such soldiers. I ate with gratitude the final breakfast they had prepared for me at great pains. Would I never again see the figures of Japanese soldiers, such good and fine soldiers.

Iwakuni had slept in. He woke in a hurry, put on his uniform and came to say good-bye. Tears filled his one remaining eye. Words failed us as we grasped each other's hands firmly, and we parted,

Interpreter Imai was also present to give me a heartfelt sendoff.

I got off the Japanese Army car at the Nationalist Party Headquarters.

I said to the driver:

"I'm going with some Chinese friends on a Chinese car," and I had to force the driver to go back. I watched the parting car in a cloud of dust until it turned from sight.

I was now leaving all alone on a long journey.

Good bye friends! Good-bye Japan!

Chapter 4 Proud Chungking

It was still early when I got to the Kuomintang Headquarters. But, I found already gathered there a crowd of General Hsieh's subordinates and their wives, raising a big commotion over the departure of Madame Hsieh. I heard that they had gathered before dawn. These party members, who ordinarily, showed no interest in their work whatever, nonetheless were willing to risk their lives in looking after the needs of Madame Hsieh. They knew they would never be fired for incompetence or negligence of duties, but their fate could he sealed by a word from the Tai Tai. As a result of all these cooks spoiling the pie, I was forced to get up before dawn and arrived in a packed car at the airfield at 6 a.m. Imagine my surprise to find not a single passenger there, let alone the airport personnel. I soon learnt the reason. The plane scheduled to take us to Kunming was still there and would not leave that city until 8 a.m. After arriving in Hanoi, the plane was only due to leave in the afternoon. When we arrived at the Hanoi Airport at 6 a.m. by car, our plane had not even completed preparations to leave Kunming could not understand why these Chinese, who would argue as if their very lives depended on it over a single yuan, were so wasteful of their time.

Around 11 a.m. the first airport personnel arrived. Then began the inspection of our luggage. The suitcase that I had entrusted to Lin safely passed the customs.

Our plane arrived from Kunming at noon. It was a large four-engined aircraft. Mixed with the forty or fifty passengers that alighted was a giant American pilot, who stood head and shoulders above the Chinese. The airport was a Babel of confusion with those come to greet the new comers and those come to send off the departees.

In the midst of this chaos, the passenger list was read out. Next to Hsieh Tai Tai, my name was called:

"Shih Kung-yu."

"Yu (here)," I answered and got aboard.

I took a seat near the rear. There was an inspection of passengers just before the plane took off.

The women in front of me was carrying a seven-year-old child on her knee, for whom she had bought no ticket. She had succeeded in bringing the child aboard in the surrounding confusion.

"Get off!" The officials would shout.

"I won't get off!" She would shout back and soon a big quarrel had started. In the end, they dragged the howling child off the plane. Aboard the plane, the mother sobbed her heart out, while on the ground outside, lay the child wailing its heart away. Because of this commotion our departure was delayed until 1 p.m.. The large plane began to move and pushed into the air lightly. In all probability, Staff Officer Iwakuni was wishing me God-speed, with tears in his eyes, looking out from his second-story window.

We passed the French Indo-China border and entered Yunnan Province. I was deeply moved at seeing this province far below me, for once I had studied this terrain on a map for the purpose of planning an attack on Kunming from French Indo-China. The majority of passengers immediately began getting air sickness and started opening paper bags, into which they emptied their stomachs. With everybody so sick, I had no worries that my identity would be discovered. Uneventfully, we soon reached the skies above Kunming, capital of Yunnan province. Kunming viewed from the sky proved to be a stout fortress, centered around the airport. During the war, it had been built up as a base of the United States Air Force and the signs of extensive construction work were still evident. We landed at 4 p.m. Thanks to the host of student-servants that came to greet us Madame Hsieh's large trunk escaped customs inspection and was immediately loaded on a car.

On my departure from Hanoi, I was given three gold bars, to tide me over any emergency period. These I had thrust carelessly into my pockets. They were straightway confiscated by the customs. At Hanoi, the inspectors had said nothing and I had completely forgotten I had the gold pieces. The inspectors at Kunming were extremely strict, except in the case of special people. Because of my carelessness, I would now have to continue my travels with poverty always hovering over my head. In my wallet I had roughly 100,000 yuan (about 30 US dollars). However, after I arrived in China, I found that I did not want for food as an invited guest.

Hsieh Tai Tai took me that night to a first-class restaurant and the next day to a shop specializing in Yunnan "paotz," a rare and delicious food. The same night, she took me to see a moving picture and a play. It seemed like a marathon bout of entertainment. The play was a raw melodrama, having as it theme the hunt for collaborationists. In the scene showing the arrest of traitors, two real bullets were fired from a pistol. Only in China could one see such an actual killing taking place on the stage.

I walked the streets of Kunming. The goods flooding the night stalls were practically all American-made. I was impressed by the truth of the advertisements, "Beautiful goods at cheap prices." I was also surprised that such a large volume of every-day commodities had been transported from India by air together with vast military supplies. Six months had already passed since the end of the war, but American goods were still abundant.

The Lieutenant-General's quarters was a house in the suburbs of the city. It contained also the branch office of the Kuomintang, the local offices of the Overseas Department and private quarters. There I met three sweet children and the General's aged mother. The family atmosphere was one of tender affection. A dozen or so youths lived together with the family. They were all natives of Hainan Island. It was still the custom in the officialdom of China to mix home and government office, government officials and student-servants and official and private matters. Here again I saw the same strong blood and locale ties binding a conglomerate mixture into one unified whole. After supper I exchanged views with the student-dependents. Sub lieutenants, no more than 20 years old, argued pretentiously of World War Three. They were indeed embryonic members of the Military Control Bureau. The person they most respected was General Tai Li and next to him Cheng Chieh. It was not "isms" or principles that bound these youths together. It was the feeling of one mutual dependence and mutual self-defence among people related in blood or from the same birthplace that formed the cement of unity in politics, in military affairs and in economics.

To think that the Hsin Hai Revolution was brought into being by the Three People's Principles is a purely Japanese view of things. The revolution going on in China, under the placard of Marxism, when scratched below the surface, was nothing that could change fundamentally the age-old characteristics of China.

After seeing the revolting play on the witch hunt for collaborationists, I felt goose pimples break out all over my body. I lost all appetite and somehow managed to stay my stomach with an orange that I bought on my way home. Soon I was seized with spasms of vomiting and my pulse beat exceeded 120 a minute. It was another attack of acute malaria.

For two days I did not eat a thing and although I did not have a thermometer I knew my fever was over 40 degrees Centigrade (104 degrees Fahrenheit). My urine turned red as blood. My symptoms were the same as that when three years previously I was told at the Army Medical School I would die. I wondered whether I would ever get well. The only care I had was the housekeeper bringing me rice gruel. I did not have a doctor and Madame Hsieh did not come to visit me. When a person becomes sick, a Chinese fears approaching for no reason at all. Perhaps they feared they would catch the same sickness. If I were to die, I wanted to die in Chungking, the destination that I had set out for. I called the lieutenant-colonel who was secretary of this party headquarters branch office and had him send a telegram to General Tai Li.

On March 15, Major-General Wang suddenly returned from Chungking. On orders from General Tai Li he had reserved for me a seat on a plane leaving on March 19. I did not know what change had taken place in this cold and detached general but he brought me a large basket of delicious oranges from Chungking. I wondered whether he had become aware for the first time of the "importance" of his guest as a result of talks with his superiors. At any rate, I was interested to notice the change in his attitude.

On March 16, my fever subsided. My urine, which had been blood red, changed to a brownish color. I was now able to drink down with difficulty a bowl of gruel. The succeeding hardships that I had gone through seemed to have made me suddenly old.

My luggage was limited to 30 kilograms (about 75 pounds). I cut down my luggage to the barest essentials and gave the rest to the student-dependents. I felt that as long as I could take this body of mine to Chungking, I would not need anything. I had no attachment whatever for material things. All I prayed for was that I could in some way or the other transmit to Generalissimo Chiang Kai-shek the feelings of His Majesty and to take the first step toward Japan-China collaboration.

On the morning of March 19, I got out of bed, groggy with weakness. I left Madame Hsieh's home, seen off by the Tai Tai and the students. I got on the plane with Staff Officer Hsu. My legs were weak. My eyes spun. I knew I still had a fever of around 38 degrees Centigrade (101 degrees Fahrenheit). My pulse was over a 100. I managed to gain some energy from oranges I bought at a wayside stall.

The Kunming Airfield was a chaos of welcomers and people come to say farewell. It seemed that an important Allied guest was due to arrive that morning. In the airport terminal a long table was laid out. The sparkling tablecloth was covered with eatables, foreign wines and whiskeys and fruits of all kind. The Chinese looked on with envy at this mountain of food. I wondered who was coming.

Two remodeled four-engined B-29's arrived from India after crossing the Himalaya Mountains. Two or three European passengers alighted and stopped at the table. But after drinking only a cup of coffee, they departed somewhere.

Staff Office Hsu and I boarded one of the two newly arrived planes. My wicker suitcase was deferred until the next day's flight. None of the mortals gathered ever guessed that a Japanese, disguised as a government official, had mixed with the passengers and gotten on the plane. The American pilot lightly lifted the giant aircraft into the air. We had left Kunming Airport behind. Once we were in the air, I felt cold. It was perhaps because the plane climbed to an altitude over 3,000 meters. The fortifications surrounding the airport looked like a miniature garden. Whether they had feared an airborne attack by the Japanese or not, I had no way of knowing, but, the airport was surrounded by formidable infantry positions and a large number of anti-aircraft artillery emplacements.

From my experience aboard small-aircraft, I could not have imagined the way in which a large plane rolled. It was the same movement as vessel at sea, rising on the crest of a large wave and falling into the trough. It was as if I were taking a boat trip.

Soon the majority of passengers were sick and they were more engrossed in looking after their queasy stomachs than noticing their fellow-passengers. Thanks to this, I was again alone to myself. I stared at the terrain of Southwest China. This, I realized, was a habit that had acquired over long years as an operations officer. On the least provocation, I would be viewing almost instinctively a peaceful terrain from the point of view of military operations, and it seemed that I would never get over this habit. The charges of militarism and imperialism against me could, therefore, be justified to an extent.

From about 5 p.m. the weather began to worsen. In such weather a Japanese transport plane would have turned back. Fog began to cover the earth below. The pilot was now flying blind over a rugged mountain country. I was afraid that the pilot would not be able to find the landing field once we reached the skies above our destination. We landed once at Liangchou for liaison purposes, but we soon took to the air again. This time we flew close to the ground. The granary of Szechwan Province spread under my eyes. I was astonished to see land cultivated right to the tops of mountains. In places I even saw rice paddies on the mountain tops. I wondered at the remarkable feats of irrigation that had been accomplished to enable such thorough cultivation.

It is a great mistake to think that there are great areas of uncultivated land in China because of her vastness. With the exception of the desert expanses of the Northwest and Mongolia, every available plot of arable land is under the plow.

As dusk began to fall the fog deepened, cutting off the path of the plane. I wondered whether we would be able to land. The passengers, now over their air sickness, began also to worry.

When I made out the confluence of the Kianglin and Yangtze Rivers, I felt a surge of relief.

A little past six, our giant plane began a sudden descent and spiraled lower and lower. I felt as if my body were being twisted, around and around.

"We've come." "We're there." "I can see the airport," were the cries of joy that arose within the plane.

We finally landed on a narrow air strip situated in the middle of a dried river bed.

I recollected the numbers of times, since the outbreak of the July 7 Incident [Marco Polo Bridge Incident, July 7-9, 1937, at Beijing between the 29th Army of the National Revolutionary Army of the Republic of China and the Imperial Japanese Army] that I had worn my heart out, trying to fly to Chungking and by own efforts bring about Japan-China peace. Now I had finally succeeded in reaching Chungking, but war had ironically already ended, and that in an unprecedented defeat. How deeply I felt at that instant that heaven was without pity. Why had I not been allowed two years ago to land on this airfield? The runway was built on a dry part of a river bed and paved with rocks. The airport buildings were practically all barrack-like structures of wood and woven straw mats. Probably only flimsy structures were built because of the possibilities of floods two or three times year. The city of Chungking, viewed from the river-bed airfield, seemed to be perched on a high cliff. Up the face of the cliff, perhaps close to 900 feet high, zig-zagged two or three roads. Several Europeans, greeted by government dignitaries, walked stiffly away from the plane.

All along the other side of the river stood several tall chimneys from which curled thin wisps of smoke. The number of Chungking's munitions factories, transferred westward during the war, could be guessed from these few chimneys. The small volume of smoke seemed to suggest, that the factories were catching their breath after the busy war years.

Because of insufficient liaison, there was nobody from the Military Control Bureau to greet us. Staff Officer Hsu, himself, seemed surprised. However, thanks to Hsu's efforts, I was saved the bother of undergoing an examination. Despite the fact that I had only gotten out of my sick bed, I did not feel a bit tired because of my excitement. My fever also seemed to have come down. However, I did not have the strength to climb this steep hill road. A drizzle began to fall and I had no rain clothes. The sedan chair carriers, seeing our predicament, demanded 4,000 yuan per person (around $1.25 US) for taking us up the hill. We bargained with a number, but none would come down a yuan.

Finally, we shout, "We don't need a sedan chair, we'll walk," and we started, as if brimming with confidence, toward the hill road. We knew that they would come shouting after us soon.

They did. We had won. They agreed to take us up for 2,000 yuan a person one half the original quotations. Like the carriers that shouldered people across the Oi River on the Tokaido main route from Osaka to Yedo during the Tokugawa Period, the vehicle that took us to Chungking, was a covered chair set between two poles, shouldered at each end by coolies.

The steep hill, with a grade of over 45 degrees, was traversed by a series of thousands of stone steps.

Wet with rain, the steps seemed to offer little foothold for the men. A single slip would have meant being plunged, carrier and all, down the cliff to our deaths. Each time the carrier swayed, I felt as if death were clutching at my buttocks. Looking down the cliff, I could see the silver wings of the large plane on the river bed being gradually swallowed by the gathering dusk. If the paths were not built in a zigzag, these cliffs could never be climbed. The cliff evidently of aqueous rock, was wet at its base with absorbed water and extremely steep.

I thought:

"If the city of Chungking were to be captured, a frontal assault on these cliffs would never succeed. The attack would have to be made from the hilly area to the west."

Once again, such instinctive thoughts were unconsciously filling my mind. Near the top of the cliff were built innumerable air raid shelters cut horizontally into the aqueous rock face. "Slimy mud covered the whole surface of the path. One of the carriers slipped. I gave a start, but luckily we were already at a resting point on top of the cliff."

Once on level ground, I breathed a sigh of relief. I looked around and saw rows of dirty shops. All were selling food, pork, fowl, fish and fruit. I was surprised at the abundance of everything.

I remembered the old Chinese saying:

"Those who base themselves in Szechwan are strong."

Hsu hired rickshaws at the side of the wide road on top of the hill. The road bed, composed of crushed rock, was solid, but the surface was covered with thick slime, making walking extremely difficult.

Chungking, I found, was a city of hills, barrack-like shops and homes, newly built after the bombing of Japanese planes, were covered with gaudy posters done in cheap paint.

Everywhere were still to be seen the raw signs of victory celebrations, but the expression on the faces of people walking the streets was one of coldness. They seemed to be tense. I noticed a large number of people wearing long quilted robes because of the piercing cold of the drizzle. But despite this rain, the streets were crowded with people. I wondered what it would be like in fine weather. I had expected to find a people still drunk with victory. I watched with care the sights along the road. However, all I saw were people hurrying by glumly. There were no hilarious crowds and no couples walking arm in arm.

We reached our quarters located in the biggest hotel in Chungking, pushing through tremendous crowds that surged through the main street of the city, on which the hotel was situated.. This hotel was in Chungking the equivalent of the Imperial Hotel in Tokyo. In front of the hotel hung a huge sign, still new, written with gilt paint:

"Victory Mansion."

At the same time, the camouflage painted on the building during the war still remained.

We were led to a room on the third floor. It was Lieutenant-General Hsieh's room. There was practically no furniture. There were two wooden beds, each, three feet wide, a small miserable-looking office desk and two wooden chairs. Yet, the room rent was 10,000 yuan a day. (At that time the private's salary was 7,000 yuan a month). I could imagine how austere life had been in Chungking during her eight years of war with Japan. We were told that Lieutenant-General Hsieh was attending a meeting of the Second Central Committee of the Kuomintang, now in session, and only the General's untalkative secretary was in the room. We were taken to the dining hall and had a late supper. For the first time in a really long while I tasted European food and found it delicious beyond words. Up till the end of the war, six months previously, this was the only hotel where United States Army officers and men could be invited and, at that time, the hotel was the most luxurious and elite establishment of its kind in Chungking. No wonder the cuisinery was far beyond anything acceptable in such an out of way place.

General Hsieh returned home late that night. He showed considerable fatigue from the continuous round of meetings and banquets. However, his broad round face, indicating an uncommon virility, was still shining with a lustrous sheen of natural oil.

General Hsieh did not have a single person who could understand Japanese. The smattering of the official Peking dialect that I had crammed in with the help of Interpreter Imai during my 100 days in Hanoi and the broken Hainan Island dialect that I had picked up on the streets failed to be of any use. Usually, I had to get by through written exchanges of notes. That night, this exchange of notes blossomed into a long conversation and when I finally got to bed it was midnight. I lay my completely worn out body on the narrow plank bed, with Hsu sleeping together with me, his feet pointing toward my head. However, I had not realized that the night cold would be so fierce. I had only the jacket given me by Interpreter Wang at Hanoi and two blankets and Hsu and I tried to hug each other as closely as possible to keep warm. Near dawn, General Hsieh, sleeping on the next bed, got up especially to cover us with one of his quilts. Although it was only a thin cotton covering, the sympathy behind the gesture transformed the quilt into something finer than finest of eiderdowns.

I dreamt of my wife and children for the first time in a long while. My wife appeared with a pallid and malnourished face and gazed at me as if to chide me for my long absence. None of my children were missing in my dream and I concluded that none had died and were all healthy. I realized that they must be waiting my homecoming with deep longing. I found myself telling them in my dream:

"Please bear a little longer, for the sake of your country, for the sake of your Emperor."

Then I awoke. The east was just beginning to lighten, but I had no idea what time it was. The fog seemed to be blocking the interval between night and day.

Breakfast was prepared and served us by General Hsieh in our room. It was, however, simply coffee with a raw egg dropped into it and then mixed. I was thankful for the abundant steaming hot water which the boy brought to us, for the hotel had no baths. The next day, General Hsieh gave the room to the two of us and moved to the private residence, of General Cheng Chieh-min. Both were natives of Hainan Island and both top members of the Military Control Bureau (The Blue Shirt Society).

Hsu and I strolled through the streets, wet from last night's rain. Because of the thick fog, we could not see more than 30 feet ahead. Automobiles were creeping about continually blowing their horns. They travelled at the same speed as an ordinary pedestrian. Despite this precaution, there were crashes everywhere. The narrow hill streets seemed to groan under the press of vehicles and pedestrians.

We realized that we could not continue our stroll and returned to the hotel. I walked around the second and third floors. All rooms were filled, and all the occupants were noted dignitaries. The Second Meeting of the Central Committee and the Chinese Communist Kuomintang Trade Conference were being held at the same time. As a consequence, every important Chinese Communist, Kuomintang and American leader throughout the country was gathered in Chungking to talk over the future of China following her victory. In the next room was General Fu Tso-i, who had come by air all the way from Inner Mongolia, his six-foot frame clad in a long cotton robe. In an annex to the hotel was the young King of Thailand [Ananda Mahidol] (later assassinated in Bangkok) who had flown to Chungking incognito. If one large-size bomb were to land squarely on this hotel, the whole fate of China would be changed, I thought, then checked myself for my militarist thought, which would have called forth a thunder of denunciation if it had been heard by anyone.

[King Ananda Mahidol (Rama VIII), born March 2, 1935, was discovered on June 9, 1946, in his bedroom, shot in the head. In 1955, three royal aids were accused, convicted, and executed by firing squad for the killing, despite deep doubts about their guilt.]

The King of Siam's dining room was filled to over-capacity from morning till night. The ruling class of the victorious nations spent their days in conferences and their nights in banqueting, while the dance hall in the annex was crowded with pleasure-seekers all night long. Hsu and I bought tickets and looked into the dance hall. Once in the hall filled with a mob of men and women, dancing madly to jazz, I was hit by an overwhelming odor that almost made me vomit and I ran helter skelter outside.

The lieutenant colonels and the colonels who came to call on Hsu in our room paced the floor with questionable steps, practicing the fundamentals of ballroom dancing. I did not see a single person, reading a book or thinking quietly to himself. They would skim through the economic columns of the newspaper, especially the price quotations, but there were none who read the columns explaining the world situation or the music and art sections.

In contrast to the glum faces of the people on the street, only a frivolous atmosphere, and signs of decadence pervaded the ranks of the higher strata of government and military circles and the nuclear leaders of the country.

General Hsieh came home on the night of March 20, his face alight with joy. He told us:

"At today's session of the Second Central Committee meeting, I personally met Generalissimo Chiang Kai-shek and reported to him that you had arrived. He was very pleased and said in a loud voice, 'Hen Ho' (very satisfactory) and praised me for my efforts."

"The Generalissimo then said: 'I am one hundred per cent behind the principles of joint Japan-China collaboration, which you entertain. However, diplomatic questions cannot be solved in one or two days. The attitude of the United States toward Japan will be one of extreme strictness at the beginning and one of magnanimity later. There is every possibility of Japan-China collaboration. I will decide everything after consulting with Tai Li, when he returns. Please tell him not to worry and take a good rest.'"

"In addition, the Generalissimo asked me to make a detailed report on you, so please write for me a detailed outline of your past life. I believe you'll be granted an interview with the Generalissimo with General Tai Li also attending."

That night I stayed up all night and wrote out the requested auto biographical outline.

"The future looks bright. The only thing now is to await the return of General Tai Li."

General Tai Li was at that time in the Peiping and Tsingtao area, helping with the work of taking over Japanese Army and civilian property. I was told that he would return to Chungking in two or three days' time. His return would give me greater assurance of the success of my grave mission.

General Hsieh brought out a letter from his suitcase. It was from my dear friend, Wang, whom I had parted with in Hanoi.

"You will forgive me for my inadequacies as a guide during long days together. After parting from you, I immediately went to Bangkok and met the persons whom you designated and transmitted to them your messages. The luggage that you had left behind was stolen and not a single item was left. I also could not find out what had happen to the remains of Japanese soldiers. I am ashamed at having you, a foreigner, find out the present state of China. I am sending back the money borrowed on my departure from Bangkok. Please accept it. I would like to see you again. Please take good care of your body."

This was the rough gist of a letter filled with sincerity and, at same time, with indignation at the moral degradation of Chinese Government officials.

On my departure from Bangkok, I had gathered together the few remaining bones of Japanese soldiers entrusted to the ossuary and, together with sufficient money for mass to be said over the remains, had entrusted these bones to the Chungking underground headquarters. Those who knew of my request were only trusted members of the Blue-

Shirt Society. There could have been no one but the members of this trusted circle that had stolen the money and had thrown away the remains of the Japanese soldiers. These youths who had fought with such courage and at such danger to their lives during the war had succumbed to their animal instincts after achieving victory. They had probably returned to their true selves as Chinese, with no morals and no sense of honor when money was involved. The sense of comradeship that I had felt toward them faded with the realization of this situation and disappointment overwhelmed me.

I forwarded the letter of Lieutenant-Colonel Cheng, with whom I had become acquainted in Vienchan, to his son studying in Chungking. A few days later, a young cotton robed sub lieutenant, just bursting with energy, visited me in my hotel. He was the split image of his father. I told him news of his father. He cried for joy to hear of his parents and said that he had already been six years away from home. This young officer told me that the majority of the teachers in the Chinese Residents High School in Chungking, where he studied, were members of the Chinese Communist Party and that he had quit school after a big quarrel with his teachers.

Later he volunteered for the Youth Army and had been commissioned sub-lieutenant. He said the majority of the professors and teachers in the various schools in the capital city of Chungking had become Communist Party members from several years back.

The great household of the Kuomintang Party had somehow managed to weather the storm of her eight years of war with Japan, thanks to aid from the United States, but already not white but red termites had eaten into the very foundations of its structure and the supporting pillars were beginning to wobble.

The son of my old friend Miao Pin, whom I had come to know through the East Asia Federation, I learned, was staying at the home of the Minister of Finance, Yu Hung-Chun. Out of a desire to meet and encourage the youth in some way or the other, I asked General Hsieh to get in touch with him. But my hopes failed to materialize. Although in good health, Miao Pin, already branded a collaborationist, was in a prison in Hsuchow, awaiting trial.

Staff Officer Hsu Ying, who accompanied me from Kunming, was a native of Canton. An extremely intelligent youth, he had worked in the headquarters of various area armies as liaison staff officer. His sense of justice also was comparatively strong. When I told him of the corruption of the supervisors of war crimes affairs, Wang and Yeh, of which I had obtained knowledge in Hanoi, Hsu clenched his fists in indignation He was probably appointed to act as my guardian and warden, but he placed no restrictions on me whatever. I was, of course, free to go outside the hotel anytime I desired. However, feeling that I must never cause this youth any trouble, I was very careful not to abuse my liberty.

Night and day, I strolled the streets of Chungking with Hsu and diagnosed the feelings and the psychology of the post-victory Chinese people. It was only the upper levels that were living in extravagance, while the masses continued to toil and live simply. The oranges of Chungking were especially tasty. I also dropped into small and dirty one dish restaurants and ate the entrails of pigs together with the coolies. The table tops shone with ingrained dirt and much wiping, but the food was delicious.

Against the bone-piercing cold of Chungking's night, my blankets were totally inadequate. I scraped the bottom of my purse to buy a quilt. Two quilts made from Szechwan Province's famous cotton cost me 20,000 Yuan. One I gave as a token of appreciation to Hsu. One night I took myself, now recovered to nourishing European food, to a Chinese bath house and there I rubbed off the grime imbedded in my pores since my departure from Hanoi: It was as if a thick layer of dirt had encrusted my post-illness body. The bath charge was 1,000 yuan per person.

During my rummaging around night stalls, I came across a dozen or so Soviet military magazines, which I immediately bought. Published during the war years, they were worth the money I paid for them from the point of view of reference on technical matters. I also will never forget finding in an old book store in a neglected alley a copy of the Chu Shuai Piao, a volume of Chih Pei Fu (ancient Chinese poems) and the lithographs of Huang Shan-ku. Despite my rummaging around in my poor Chinese, no one noticed that I was Japanese. This was due to the fact that the Szechwan dialect has a peculiarity of its own and because of the great influx of people from all parts of China, nobody felt any suspicion over my halting Chinese. I noticed the large number of beggars. I was flabbergasted at the way they pestered me, not knowing that I was a poor man, despite my robe, a robe worn only by the higher class people.

One night, I was invited by Hsu to see a motion picture. The tickets were all sold out when we got to the theater, where a vast crowd of people were gathered. However, Hsu came back in a moment with tickets he had bought at twice the regular price from some scalpers. The picture, which was already showing when we got in, was a Soviet movie. It was a film which glorified the Slavic Race. It clearly indicated that Communism had changed to nationalism, racism and a worship of heroism.

The prices of foods were much cheaper than in Hanoi or Kunming, but the prices of American-made goods were more expensive than anywhere that I had been. I was invited by a certain colonel friend of Hsu's and had dinner at his home. The life of a Chinese Colonel during the war I learned was below that of a sergeant-major in the Japanese Army.

In face of the approach of the return of the seat of Government to Nanking, there were signs everywhere, telling of houses to rent. The rental charges had dropped to a tenth of that prevalent during the war. All shops were also filled with second-hand household goods. Because only dignitaries could take their luggage with them, the minor officials were selling their belongings at throw away prices in order to make up travel funds. High class furniture flooded even the night stalls: There were practically no purchasers. In all probability, all these officials who had sold their household belongings would have to buy new furniture and accessories once in Nanking, at prices hundred times that at which they sold their things in Chungking. I felt deeply how inconvenient, and how uneconomical could be the undeveloped state of transportation.

The population of Chungking immediately prior to the Japan-China Incident [Marco Polo Bridge Incident] was estimated at around 500,000 to 600,000 people. Just before the end of the war the population had topped the 1,000,000 mark. The people of Shu (the ancient name for Szechwan Province), proverbial for their attachment to the soil and their diligent cultivation of land, also were blessed by the prosperity brought by the influx of people from other provinces, but this was short-lived and almost collapsed with the coming of victory. The "racial" industries which were moved west from the Shanghai and Nanking area during the war almost ceased operation after victory and only thin smoke rose from their chimneys after the war. Now, the owners were competing in the liquidation sales of these facilities. That was the reason for the glum faces of the common people on the street. Behind the joy over victory, lay a deep fear at the waves of economic recession rolling toward them.

I was not the only one waiting impatiently for the homecoming of General Tai.

I shall never forget that day, March 24, when Hsu and I were both walking along the street when we saw a crowd standing in front of wall gazette. I wondered what had happened and peeked over then people's shoulders. I saw a large banner head:

"Liang Nung Crashes!"

Damn it! I thought, as I was drawn by an irresistible power toward the wall gazette. This was a true case of "a blue a bolt from the blue."

General Tai Li's nome de plume was "Liang Nung."

There were not a few among his subordinates in the underground movement who were sympathetic with the ideals of the East Asia Federation. I had accepted General Tai as an old friend, as yet unmet, through my friendship with his subordinates. General Tai himself had also accepted me as a comrade. The fact that I was able to pass through a number of death barriers was due to the death-defying efforts of his men and the greatest reason for my overcoming innumerable hardships to come to Chungking was the existence of this man. On receipt of a telegram from Chungking that a guest was waiting for him, he had hurriedly taken a plane from Tsingtao. Because of foul weather, the plane had tried to land at Nanking and had crashed into a mountain side. The bright light of my whole future in Chungking seemed as if it had been blown out by one gust of an inadvertent wind.

General Tai Li was a native of Fenghua Prefecture, Chekiang Province, and the same native place as Generalissimo Chiang Kai-shek. He was the Generalissimo's beloved pupil at the Wangpoo Military College. A member of the sixth graduating class at Wangpoo, General Tai-Li was only a major-general in rank but had under him a number of members of the second and third graduating classes of the rank of lieutenant-generals. He was a man transcending military rank. Not one to fawn before authority, he was the only man in a position to give bitter counsel and straight advice to the Generalissimo. He was the only man allowed to enter freely into the Generalissimo's room, when he was all alone with his wife, Sung Mei-ling (Madame Chaing Kai-shek's maiden name).

During the Chinese Revolution, that ended in the overthrow of the Ching Dynasty, he accompanied the armed forces, still a student, and was lauded for his abilities in information gathering activities. He won immediate recognition. During the Sian Incident [Xi'an Incident, December 12-26, 1936], General Tai Li risked his life to save the Generalissimo from death. His activities during the war, when he took complete charge of the Military Control Bureau (the Blue Shirt Society) to the admiration of both friend and foe, are too well known to merit repetition.

During the early years of the war, when Kung Hsiang-hsi was Chairman of the Administrative Yuan, Tai Li discovered Kung engaged in large-scale smuggling by plane from India and, by strong representations to the Generalissimo, forced Kung's removal. It was also Tai Li that almost had Sung Tzu-wen (T.V. Soong) killed. Practically all those who had guilty consciences feared this man, and despite the slander and calumny heaped upon him, he lightly brushed these aside and mercilessly shot corrupt officials and liquidated traitors.

During the war, he often wriggled under the strict curtain of surveillance maintained by the Japanese Army and came from Hongkong to Shanghai in disguise: Once he worked as a rickshaw puller while supervising his underground workers in territory under Japanese occupation. His entire network, composed of 150,000 paid government workers and 200,000 men under his immediate command, stretched over the whole of China. While gathering information damaging to the Japanese Army, he kept strict watch against corruption, prevented treachery and became completely the main pillar of the dictatorship of Chiang Kai-shek, as well as generalissimo's ears, eyes and nose. Of all the numerous generals on the Chungking side, he was the only one entrusted with such enormous powers. Not one to seek wealth for himself, he always wore shirts of blue cotton material and cotton Chinese slippers. It was said that the name Blue Shirt society arose naturally among the people because General Tai Li's subordinates, imitating their master, took to wearing cheap blue cotton shirts and trousers. There were, of course, Chinese generals who ordinarily wore the same cotton clothing as their soldiers and put on a show of simplicity and unpretentiousness. However, they all had hidden away in their private life, fat bank books, gorgeous western-style mansions and numerous young concubines. However, Tai Li was different. When he died his family immediately found difficulty in making a living. Tai Li was that clean and honest. Yet, he did not begrudge showering vast rewards on brilliant subordinates and on those who had accomplished meritorious deeds. The workers in the Military Control Bureau were paid much higher wages than those in other Government branches. Through this expedient, he prevented dishonesty in the Bureau and allowed full concentration on the achievement: of the Bureau's objectives. Then, if he ever found anyone engaged in corrupt practices, the man was immediately sentenced to death before the firing squad. One day when he inspected a unit under his command in Chungking, he noticed that the faces of the soldiers in the unit were pale and lacked colors. The next morning, he made a surprise visit, and found that the amount of food allotted to the soldiers was far less than the regulation quantity. He immediately called out the unit commander, who had been making money by reducing the soldiers' rations and selling the difference on the black market. Then, the general, himself shot the commander in the full view of the entire unit. This was only one of the many famous stories that were reported of this legendary figure.

Perhaps, there were none who killed as many people in the whole long history of China as General Tai. However, surprisingly, the entire population of China respected and admired him. The reason for this war that the general himself was absolutely honest and those whom he killed or had killed were men steeped in corruption or traitors to the country.

It was rumored that the death of Liang Nung was brought about by Chinese Communist subversive elements and the saboteurs of a certain country. Chiang Kai-shek, who received the news of this tragedy, while still in the political conference with leaders of the Chinese Communist Party, immediately left the session and flew the same day to Nanking. Taking the hands of the mortally wounded General Tai, the Generalissimo was reported to have cried bitterly. The dozen or so passengers who were on the same plane with General Tai died instantaneously when the plane crashed into the mountain side. General Tai alone had parachuted from the plane before it crashed. Despite the serious wounds inflicted on his legs and head, General Tai managed to linger on for two or three days and finally drew his last breath under the loving care of the Generalissimo himself. Here too was a feat typical of General Tai Li, an almost superhuman feat. A large grave was built for General Tai Li from the private funds of the Generalissimo in the Ling Ku Temple at the foot of the Chungshan Mausoleum. The needs of the bereaved family were also looked after, especially by the Generalissimo himself.

Following the death of: Tai Li, there just wasn't another Tai Li. It was expected that Lieutenant-General Cheng Chieh-min, as deputy chief of the Military Control Bureau would be promoted to take over the post, made empty by Tai Li's death. However, the Generalissimo himself took over the position on a concurrent basis. The job was that important.

General Cheng was from the same native province as General Hsieh, They were both about the same age and on extremely friendly relations. However, General Cheng, based at Peiping, was at that time also head of the Military Adjustment Department, the on-the-spot organ for mediating relations with the Chinese Communist Party! At the same time, he was Chief of the Second Section of the Defense Department (at that time still known as the Military Operations Department). Thus the direct work of supervising the Military Control Bureau was in the hands of Mao Jenhuang, younger brother of Madame Tai Li. I had asked for an interview with Generalissimo Chiang Kai-shek through General Hsieh, but he handed over this question, still unsettled, to the Military Control Bureau and left for Hanoi in a hurry.

It was March 28. Unexpectedly a fierce looking man in cotton uniform came to the "Victory Mansion" to call on me. He was an ugly and sinister looking brute of a man. At first, I couldn't get what he was driving at, but finally I found out that he was a Chang that had been sent by the Military Control Bureau to call for me, I bade a sad farewell to Hsu and to my ten day life as a state guest. I got on a car with Chang and was driven through the streets of Chungking towards an unknown destination. I noticed a pistol peeking from Chang's pants pocket. He sat glum, without saying a single word, as the car sped along the banks of the Kiangliang River in a northwesterly direction. I wondered if they had finally decided to place me in custody. I could not believe that Chang could be the emissary of good fortune. However, all I could do was to leave everything to fate. After an hour's drive, we came to a small industrial town in the suburbs, which seemed to be the center of a number of munitions factories.

I noticed a bus stop, marked "Tzu-Chi-Kou." I got off the car and climbed a flight of stone steps. Then, after passing through a grimy and crowded alleyway, I was led to a residence with a large gate where I noticed soldiers carrying rifles, but it was not an army barracks. I was led to a room on the second floor and introduced to several youths. The liaison chief was called Li Chien-kung, but he was absent when I got there. Chang wrote on a piece of paper:

"Li Chien-kung will return on horseback."

I nodded, thinking that he had gone for a short outing on his horse. If that were the case, Li must be a high-grade officer. However, learned that "on horseback" in Chinese meant soon and the Li, whom I had thought would be an important dignitary was simply the head of this unit.

This residence housed the anti-Communist Propaganda Department of the Military Control Bureau. The chief of the Department was Major-General Meng, who together with his men, Colonel Liu, Colonel Li and Lieutenant-Colonel Hung and their several families lived their combined official and private lives in this compound. During the war the departments and units connected with work dealing with Japan were the most important but now those dealing with the Chinese Communist Party had the come to the fore. I was greeted as a temporary guest to one of the nuclear organs of this anti-Communist structure. This home had belonged to a rich landlord and had been confiscated during the war. Commonly known as Chang Chia Hua Yuan (the Chang Family Flower Garden), it was a famous manorial compound of this district.

The chief, Major-General Meng, had engaged for a long time in underground activities against the Chinese Communist Party in Sinkiang Province during which he had lost one leg. Colonel Liu had only recently returned from Yenan. For several years he had been in Yenan, disguised as a Communist Party member, and had undertaken a detailed study of their tactics. Colonel Li had engaged for several years in underground activity in the areas occupied by the Japanese North China Army against Colonel Hidewa Arakawa. All of these men were nuclear experts of the Military Control Bureau in their own field.

I was allotted a fine, lovable young soldier, called Nieh, as my orderly, who looked after me with loving care. He was a First Class Private. Immediately, he went out and bought for me a mountain of bedding, toilet articles and daily necessities. I was allotted a room of about 18 by 12 feet in size. It was too good a room to be a prison, yet in the adjoining room were around 20 guards. I resigned myself to fate, midway between that of a prisoner and a guest.

Major-General Meng had a seven-year-old daughter and five-year-old son. Colonel Li's two children also soon joined us after an arduous trip from Yenan via Sinking Province. The four innocent children were my greatest pleasure in this restricted life. These children of another race soon became as lovable and confiding as my own children. Every afternoon at three they came, to me for their afternoon sweets. Through these children the parent's feelings toward me gradually changed for the good. I conversed with the parents, using a mixture of Russian and Chinese and argued often with them over the trend of the times. However, the biggest contribution to the budding of friendly feelings among these family members was medicine, army drugs, which I had brought with me from Hanoi. They cured the children of stomach ailments and the colds of the officers' wives. No matter how much flattery you may use in explaining Japan-China collaboration, and no matter how loud you may shout "a common script, a common kind," nothing could compare with the efficacy of medicine to cure their wives and children's ailments. Even the hateful GPU could win over these people if it used the same tactics.

The only person that I did not care to get on friendly terms with was Colonel Li. He was about 35 or 36 years old. In all my two years in China. I never met a man so unpleasant, so insidious and so impolite and unthinking of others as this Colonel Li.

The night I was taken to my new home, I got up to go to the toilet in the middle of the night, perhaps because of the piercing cold. However, I found the door locked from the outside. I was astonished. It seemed that they feared I would try and escape. There was even a guard placed outside my door. Unable to do anything else, I relieved myself in the spittoon placed in my room. Early next morning, I protested to Deputy-commander Li that such was not the way to treat a guest. He apologized with humility but told me that in this great mansion there was not a single toilet. Chamber pots were placed in the upstairs and downstairs corridors and these were used indiscriminately by adults, children, both men and women. In addition, the spittoons in the various rooms were also used for the same purpose. The next morning the duty soldiers would take these pots and spittoons out to the hills behind and empty them. In my life in the flower garden of the Chang family, this was the hardest ordeal I had to bear.

During the first one month, I spent the greater part of the day playing with the children or reading newspapers. The watchfulness of the people toward this stupid looking Japanese gradually subsided and the number of guards was lessened and the surveillance relaxed. During the middle of the night on my way to the toilet, I often noticed a case of secret documents left unlocked. This alone was indicative of post-war laxity.

Every moment I could spare, I spent in climbing the hills behind this mansion. Antiaircraft artillery emplacements, dug during the war still could be seen; while all the hills around were dotted with machine gun nests and strong points made of piled rock.

These defense works had been thrown up in haste in fear of a Japanese attack. Now they lay abandoned, left to the inroads of nature. I saw beautiful flowers blooming where once soldiers had paced. The flowers seemed to be praying pleafully on behalf of Asia and the world that these defense works would never again have to be used. I plucked a single flower, took it home and placed it in an empty soda water bottle and I gazed at it daily to beguile the tedious hours.

Through the good offices of Lieutenant-General Hsieh, I had reported by mouth the reasons and the objectives of my visit to China to the Generalissimo. In addition, I had been scheduled to make a detailed report directly to the Generalissimo upon General Tai Li's return. However, now that General Tai was no longer living, what person could ever break down the formidable barrier lying before me and enable me to meet the Generalissimo.

I had never realized so deeply before that in China nothing counts more than a "somebody." You had to know "somebody" and you had to do everything through "somebodies." In a modern country, the structure does the work. Even if a chief of a section changes, there is always a regular written relay of business matters and business proceeds automatically, whether you are a "somebody" or not, or whether you know a "somebody" or not. However, this did not seem to apply in China.

Colonel Li transmitted to me the wishes of Major-General Mao, the deputy chief of the Military Control Bureau, that I write down my report for presentation to Generalissimo Chiang Kai-shek. I felt a slight suspicion that this was one means of rejecting my requests for a direct interview with the Generalissimo. However, I felt that it was a mistake to worry before I had done my part and presented my report written with brush in the Japanese language on March 31. I had stayed up nights to make two copies, the original I presented to the Generalissimo and the copy, wrapped around my body and safely brought home to Japan:

The gist of my report was as follows:

"Report to His Excellency Generalissimo Chiang Kai-shek"

March 31, 1946

"I humbly offer my congratulations to you for the successful fruition of eight long years of suffering and the steady achievement of the work left behind, by the, Father of the Chinese Republic (Dr. Sun Yat Sen). At the same time, fully realize the anxieties that beset you night and day, over the swiftly changing situation both at home and abroad. In particular, I extend to you my heartfelt condolences for the recent loss of Dr. Tai Li and sympathize fully with your deep sense of bereavement.

"Having heard from General Hsieh of your deep consideration of such a man of lowly position as I, despite the present grave period of busy occupation on your part, I am filled with deep emotions. Thanks to your kindness, my long-standing wish, spanning several years, has been granted, and I am now able today to report directly to Your Excellency. This I feel as an unprecedented honor and I thank you most sincerely.

"What I have to say is contained in the following four articles:

"First Article:

"I would like to speak first on the august will of His Majesty the Emperor of Japan. I am no more than a lowly soldier. However, as stated to your Excellency the other day through General Hsieh, I have been granted special treatment and received in audience in the past. In particular, I served as aide to His. Majesty's younger brothers on several occasions. Thus, I was able to listen to the expressions of His Majesty's wishes both directly and indirectly on many occasions. As a result of a most tragic turn of events, the august Imperial will, has today become greatly misunderstood by many countries and peoples of the world. This is the sin of His Majesty's subjects, in particular, we soldiers. For this, we are filled with heart-breaking shame.

"In explanation, I am setting forth in the following pages a few actual examples of His Majesty's will:

First Example: In August of the 32nd Year of the Chinese Republic (1943), I was appointed staff officer to the Japanese Expeditionary Forces in China. When I arrived in Nanking to take over my post, Prince Mikasa was already working there, in the status of a staff officer major of the Supreme Headquarters in Nanking.

"Around the beginning of November, Prince Mikasa came all the way to my room bringing with him a beautiful photograph album. This album contained the pictures of Your Excellency's birthplace.

"The Prince said to me: "If His Excellency Chiang Kai-shek's mother [Wang Caiyu, December 19, 1863–June 4, 1921] were living today (1943), she should be 80 years old. I have heard that, it is the custom in China to celebrate the 80th birthday in a grand manner. His Excellency, now in Chungking, has already been five years away from his home town and must feel pained to have been unable to visit his mother's grave during all these years. Isn't there some way in which we could hold, in place of His Excellency, a memorial service for his mother?"

"I was deeply moved. I immediately reported His Highness' words to Supreme Commander General Shunroku Hata and obtained his willing assent for the undertaking. Hoping to make the memorial ceremony a joint Japan-China affair, I spoke to Dr. Wang Ching-wei, who was deeply moved and said:

"I would like very much to attend personally too. However, people would misunderstand my true feelings, and might thereby cause His Highness unnecessary trouble. Therefore I shall send Lieutenant-General Chu Chin-chuan as my proxy and, in that way, pay my respects."

"As a proxy to both His Highness and General Hata, I left with Lieutenant-General Chu Chin-chuan. With Dr. Tao Hsiao-Chieh, who was from Chekiang Province, we arrived on November 25 at your birthplace, Hsi-kou-chen, in Fenghua Prefecture. His Highness wanted to attend the ceremonies personally, but because of the poor state of public peace and security in that area, he reluctantly resigned himself to staying in Nanking.

"The memorial services were held for two weeks, one week before and, after this date (November 25). The Japanese defense units removed all patrol and defense lines and both Japanese and Chinese became as one in participating in the festivities. The report of the festivities was carried far and wide. People from neighboring villages and from distant centers came in droves to attend the ceremonies. Your Excellency's birthplace was black with people.

"As I watched the throngs kneel in front of Your Excellency's mother's grave, sobbing and weeping, as they bowed and paid their respects, I too could not help crying. This was because Your Excellency' reputation and prestige had spread throughout the length and breadth of the country. This day was one filled, with deep emotion for Japanese and Chinese without distinction and for both the peaceful occupied areas and the areas still continuing warts on leaving Your Excellency's birthplace, I told the gathered masses of people:

"Please keep clean through your own efforts this grave of a mother who gave birth to a hero of China in the midst of these beautiful hills and rivers and please do not fail to have continual mass said for the repose of her soul. I then took several pictures and returned to Nanking. When I reported on the memorial services to His Highness and to the Supreme Commander, both were extremely pleased. His Highness immediately took my films, enlarged the pictures himself, took them back to the Emperor and made a personal report of the ceremonies held in honor of Your Excellency's mother.

"His Imperial Majesty was particularly pleased and said: 'This is the best thing that has been done since the start of the China Incident.'"

"In addition, Her Majesties the Empress and the Empress Dowager were also very pleased. I later heard directly from His Highness that the Emperor long after cherished the pictures taken of this occasion.

"The Chinese who expended the greatest efforts to make the memorial services a success were Tao Hsiao-chieh, Wang Chang-chun (at that time advisor to the Shanghai Municipal Government) and Dr. Miao Pin, (at that time deputy-chairman of the Deliberative Yuan).

"Hearing that I had returned to Nanking, Dr. Wang Ching wei, though ill in bed, said that he would like to hear directly from me what had taken place. I immediately visited him and showed him the Record of the Sad Thoughts of Mother Chiang, which the Japanese defense units at Hsi-kou-cheng has carefully treasured and guarded.

[In 1934, Chiang Kai-shek published (Muqin de Jiaohui) — "The Teachings of My Mother." The book recounts the moral, religious, and practical lessons Chiang learned from his mother while growing up.]

"On the first page of this album were pictures of Your Excellency and of your mother, on the back side of which was a message of condolence written by Dr. Wang Ching-wei. Dr. Wang was in a very critical condition. However, he lifted up the upper part of his body in bed sat stiffly and erectly and with tears flowing from his eyes cried:

"My friendship with Chiang Kai-shek as an individual remains today as unchanged as it was in the past. But, what a sad fate."

"I too, could not help crying. Behind the eight long years of war, I believe that there are many such sad stories."

"Furthermore, today, there are many people being sentenced in the courts as traitors and collaborators. Everyone is a Chinese. They are all Orientals. When we look deep into their hearts, they all love China, worried over the future of the Chinese race and remained unchanged in their longing and respect for Your Excellency. There are a number of barriers lying in the path of both China and Asia's future. These barriers can be overcome and difficult problems solved, only when the 400,000,000 peoples of China, nay the 1,000,000,000 peoples of Asia, become as one in both body and soul. I humbly pray that Your Excellency, with the largess of a heart as wide as the sea, find fit to forgive with magnanimity the Chinese of whom I have spoken. If it is said that the peoples in the Peace Areas (under the Wang Ching-wei Government) have sinned, I believed then that all their sins are the responsibility of the Japanese Army."

"Second Example:"

"In the fall of the 21st Year of the Chinese Republic (1932), General Honjo, Commander-in-Chief of the Kwantung Army, returned to Japan: from Manchurià. He was called in audience by the Emperor to give a report on the military situation. The first question asked by His Majesty was: "Are the 30,000,000 people of Manchuria really happy over the establishment of Manchoukuo?" His second question was: "Have measures to cope with the flood damages in North Manchuria been formulated?" (At that time there had been an unprecedented flood in Northern Manchuria). Finally His Majesty asked: "Are the soldiers and officers of the Kwantung Army all well?"

"I heard that General Honjo retired from the audience with awe and trepidation in his heart. The fact that His Majesty had given precedence in his anxieties to the happiness of the 30,000,000 peoples of Manchuria over the health and well-being of the officers and men of the Kwantung Army enables us best to realize the actual Imperial Will."

"The Third Example:"

"In the spring of the 27th Year of the Chinese Republic (1938), His Imperial Highness Prince Chichibu inspected various parts of Manchuria. We members of the same graduating class of the Army University as His Highness held a banquet in welcome of Prince Chichibu in Hsinking (Changchun). When the wine had owed freely, a certain staff officer, in a jovial mood, said: 'We must make Manchuria a Japanese territory.'"

"His Highness, when he heard this, was extremely angered and in a solemn manner answered: 'What are you saying? You must not act against the will of His Majesty. His Majesty's will is like the sun. The light of the sun does not change regardless of the race it shines upon. You must not make any mistakes on the basis of your own arbitrary judgments.'"

"We had never heard His Highness so enraged and the whole gathering heard his words with solemn gravity and deep emotion."

"The words of Prince Chichibu remain deeply engraved in my heart even to this day, as the most direct explanation of His Majesty's will."

"The fourth example:"

" In the spring of the 33rd Year of the Chinese Republic (1944), when Prince Mikasa took charge of the education of young officers below the rank of major, he said:"

"Neither the Manchuria Incident nor the China Incident are holy crusades. The front line officers create so-called incidents of their own accord, in contravention of the Imperial Will and they allow these incidents to expand. Then when they can no longer handle these trouble themselves they come and ask His Imperial Majesty to find a solution. This is the height of disloyalty."

"In addition, while His Highness was in China, he received many letters from His Imperial Majesty, which stressed that Japan must maintain its New China Policy (a policy of ethical measures) and put into practice what it preached. His Majesty stressed in these letters that Japan had no other path of progress than to win the heartfelt confidence of the Chinese people. These manifestations of the Imperial will, I heard directly from His Highness on many occasions."

"The four examples that I have outlined above are the episodes that impressed me most deeply among the many that I directly experienced. There are many other examples falling into this category. But they are too numerous to mention one by one here. To round up all these examples into one conclusion, the Imperial will of the Emperor is indeed like the sun. Sometimes, black clouds cover the sun's face, sometimes the moon eats into the sun and darkens its face with eclipses. However the Imperial will that seeks peace and pities all living things and all peoples never changes However, we soldiers have committed a great sin in disobeying His Imperial will, by allowing the flush of our youth and our energies to take the better of us, by falling prey to the human desire to win fame or by letting ourselves be seized with indignations arising out of momentary emotions. This sin of having run counter to His Imperial will is worth ten thousand deaths. I know that the entire responsibility for defeat should be borne entirely by us."

"All together I have served for eight years on the battlefield. I am indeed a war criminal. During this period, I was wounded seven times and have lodged in various parts of my body more than 30 odd pieces of shrapnel, both large and small! Although I deserved to die long ago, I have continued to live till today. Now, escaping from the close siege laid upon me in Bangkok, thanks to the help of Chinese comrades at the risk of their lives, I have finally been able to achieve my wish of long years and have been allowed to come to Chungking. I know I am being presumptuous, but I believe that I have not been able to come here by my own virtue alone, but by the guidance of the gods and that I have been entrusted with the mission of meeting Your Excellency directly to report His

> *Majesty's imperial will and to take the first step in establishing eternal Japan-China friendship and collaboration. What the entire Japanese people wishes to request through me, is to have, at least, Your Excellency of all people in the world, understand the true will of His Imperial Majesty. If this can be done, then I would be willing at any place and at any time to contribute joyfully such a worthless life as mine as a foundation stone." (The rest of this lengthy report is omitted).*

Finally, I dipped my chop in my own blood and sealed the letter. Deputy-Chief Li immediately took this report to his headquarters in Chungking. He came back with the reply that it would be submitted to the Generalissimo after translation.

If only General Tai Li had been living, I would have been able to meet with him the Generalissimo and I could have talked in more thorough and greater detail. I was filled with deep regret. Immediately after V J day, the great Chinese papers were filled with editorials slandering His Majesty and denying the Emperor system. In extreme cases, some papers carried cartoons showing the Emperor, his two feet tied together, being hung upside down from a "torii." I was deeply impressed with the fact that these developments were all the responsibility of us soldiers.

When the memorial services were held for the dead mother of the Generalissimo at the suggestion of Prince Mikasa, how greatly the Emperor had been pleased.

My greatest objective in coming to Chungking in disguise was to sacrifice my life and have the Generalissimo understand the will of the Emperor. Nay, this was the only objective of my dangerous journey. After I had written out my report, I suddenly remembered that tomorrow was the day when my four children would be advancing to new grades in Japan. I did not know which school they would be going to or in what way they were continuing their studies. I felt as if I could clearly see the pitiful figures of my children, viewed coldly by teachers and follow pupils alike, as children of a war criminal. But, this too, I had to bear as the inevitable punishment of heaven, placed even on my children, through the ties of blood. I clasped my hands together and worshipped in the direction of the skies over my distant motherland.

Two weeks passed after I had submitted my report, but nothing happened. I wondered whether they would present the entire report as I had written it. I also wondered with what feelings the Generalissimo would receive it. During an idle conversation after dinner, Colonel Li said:

"The Generalissimo is extremely busy. Thus, all reports are cut down to within 300 words and all interviews are limited to within five minutes. The report, you submitted the other day is altogether too long and in all probability the Military Control Bureau is rewriting and making it much shorter."

I could not keep still when I heard this. If my report were rewritten by some unfeeling third person and boiled down to within 300 words, the objective of my coming to Chungking would be practically nullified. I must try in some way or the other to have the entire report read and to obtain some sort of an answer.

It was the night of April 27. The faces of my white-haired father and mother appeared before me as if in a dream with stern faces and said: "Why don't you go ahead and risk your very life?"

I woke with a start and jumped out of my bed. My watch told me it was past 3 o'clock. I washed my face and purified my body knew that this was a prophetic dream. I resolved to risk my life and show my final determination once again.

I washed the lid of a tea-cup, prepared a new brush and then with a straight-back razor cut deep into my left thumb. Crimson blood began to well from the wound and soon filled the vessel. Holding back the surging blood, I dipped my brush fully into the cup lid and, then wrote to the Generalissimo:

> *"I am sorry to bother you when you are so occupied and busy, However, I would like to request you, in exchange for my life, to read the entire report which I submitted the other day, understand fully the will of His Imperial Majesty and open up the way for more than a hundred years of Japan-China collaboration. At a time when the future of Asia is still dark and unforeseeable, I pray from a distance for the continued health of Your Excellency and the prosperity of China.*
>
> *"April 18, 1946*
> *"His Excellency Generalissimo Chang Kai-shek*
> *"Masanobu Tsuji"*

It was the first time in my life that I had undergone such an experience and probably the last. The brush at times failed to write as the blood coagulated. Then, softening the brush with fresh blood flowing from the cut in my thumb, I had continued writing. My heart and soul went into each of the strokes that I put down on the paper and when I had finished writing, dawn had begun to break. The morning sun of April 18 shone through thick mists and lighted up the earth. I called Deputy-Chief-Colonel Li and handed him my letter without saying a word.

Completely astounded, his hands shook and the blood ran from his face as he took my letter.

He must have realized everything. With tense features, he hurried to the Military Control Bureau, without even stopping to take lunch. He returned a little after noon and reported that General Mao has said:

"Please tell Tsuji that I will transmit his intentions to the Generalissimo together with the entire report as submitted some time ago. Please tell him that I will take it upon myself to do this. Also tell Tsuji to request without hesitation anything that he should need: I am sending him some cakes, regretting though that they are very few."

Colonel Li held back his tears as he made this report to me. I, who should have apologized to the Emperor and my compatriots with death for the great crime of losing the war, now felt that all the shame that I had borne and the dangers that I had risked in coming to Chungking would not be in vain. I had neither wisdom, or power or money. All that I possessed was my own life. If necessary, I was ready to give it up at any time. I worshipped in the direction of the east and with clasped hands offered the Prajna Paramita Sutra. And as I prayed, I resolved:

"If it be the will of heaven, I pray that the will of His Majesty be instilled deeply into the heart of the enemy general, in exchange for this life of mine. If the Generalissimo fails to understand His Majesty's will through my report, then I shall cut open my, stomach in his presence and tearing out my entrails shall fling them into his face."

Roughly one week later, Major-General Mao had transmitted to me the following report: "The Generalissimo has read your entire report. He was extremely moved by your feelings. He asked that you restrain yourself and wait a little while longer."

It seemed that this report, written with my very life staked upon it, had had some reaction. My bullets had fallen near the mark, but they had not yet hit straight into the Generalissimo's heart. If only, if only, General Tai Li were alive.... How I had wished that he were here. Then I could have accomplished my great mission.

Although I had explained the high moral character of Dr. Wang Ching-wei and had suggested that he save without discrimination not only the life of Dr. Wang Ching-wei's wife and children but the great and small collaborators in the Nanking Government with a magnanimity as great as the sea, my pleas had not, to my great regret, been accepted. Miao Pin, who had worked with me to bring about peace with Chungking, had been a comrade of mine in the East Asia Federation of ten year's standing. He had been in constant contact with General Tai Li. Thus, I had believed that he would not be killed. However, the newspapers of May 22 suddenly reported his death. I was able to picture the last days of this man. At 5:15 p.m. on May 21, he was shot to death as the first of a long list of collaborationists to be executed, after receiving the death sentence from Li Shu-tung in the Shitzukou Third National Penitentiary at Suchow.

In the courtroom, Miao Pin had explained the East Asia Federation and the behind-the-scenes efforts to bring about peace between Chungking and Japan with Prince: Higashikuni. He had brought forward as evidence to back his explanations, letters and telegrams from General Tai Li. The judges, however, had brushed him off lightly with "This is not a lecture hall. This is, a courtroom," and straightway had handed, him the death sentence. He was only allowed to live for one week before the sentence was executed. The Chinese Communists exploited his statements in the courtroom and launched an extensive propaganda campaign, claiming:

"The Kuomintang negotiated the surrender of China to Japan, through General Tai Li and, Miao Pin. We, the Chinese Communist Party, alone fought the war of resistance against Japan from beginning to end." The Chinese Communists knew full well that the people living in the Peace Areas (under the Nanking Government) did not in their hearts hate the Wang Ching-wei regime. However, they demanded strict punishment of all people connected in one way or the other with the Wang Government, with the aim of alienating the hearts of people from the Kuomintang and thereby lessening the confidence of the people in the Chinese Nationalist Government. Despite its outward cry of strict punishment for traitors, the Chinese Communist Party did not hesitate to utilize to their own advantage collaborationist living in the Peace Areas under their control. The Kuomintang, who accepted the demands of the Chinese Communists at face value, and carried out the wholesale arrest and execution of members of the Wang Ching-wei Government, was thereby cleverly fooled by the Chinese Communists. As the Chinese Reds had planned, the Kuomintang by its indiscriminate execution of collaborationists was alienated from the affections of the people.

If Tai Li had been alive or if Miao Pin had been able to guess the pervading atmosphere and had not spoken in court of his efforts for peace along the lines of the East Asia Federation, his life would probably have been spared.

At that time, the great Chinese newspapers deliberately distorted the true intentions and objectives of the East Asia Federation and branded it as the "aggression in disguise" proposed by Lieutenant General Kanji Ishihara.

Again, it was Miao Pin's efforts in bringing about the defection of Li Chang-kiang that proved fatal to him. Once he had risked his life to bring about the surrender of the mixed forces under Li Chang-kiang in the Delta Area. Whatever the case, the Kuomintang Party feared criticism for having engaged in peace negotiations with Japan and had executed Miao Pin without giving him a chance to appeal, to prevent further details from being brought to light.

"When the sly rabbit dies, the good dog is cooked," is an old Chinese proverb. It was true in the case of Miao Pin. However, he went to his death with magnificent courage. Just before his execution, he wrote with great composure, sitting at the desk in the courtroom, these last words:

"The broad expanses of my heart at one with nature rests,
"Sincere my thoughts, they light all human breasts.
"In life lived I in words and deeds the teaching of the sages,
"In death I find myself transformed a god of peace for ages."

Living he had worked for peace and in death he became a god of peace.

Chen Kung-po died on June 3. When he was called out to an extraordinary trial at 8 a.m. in the morning, he realized from the extremely tense atmosphere of the courtroom filled with armed police that his last moment had come. He asked to be allowed to go back and change to his long robes. Then he shook hands with Chu Min, imprisoned in the next cell, and asked that the bowl which he had used while in prison be placed in his coffin. Then with a simple, "Good-bye," he parted from Chu Min.

Chen Pi-chun (Madame Wang Ching-wei) was a strong-willed individual, but she was after all a woman. As she watched Chen Kung-po being led away, she burst out crying and in a large voice said:

"If they are going to kill you, why don't they kill me?"

When Chen Kung-po was asked by the courtroom police whether he would like to make his last will and testament, he quietly answered:

"I was writing my last letter to Chiang Kai-shek. I had finished one third of it. It would not be right to keep you people waiting until finished the letter. In addition, there is no longer any need to do so."

"What about your family?"

"I don't know at all where my family is or what my family is doing, I have nothing left, not even a home on earth and not even a single yuan in the bank. What then could I write in my last will and testament?"

To the court police who tried to tie his hands, he said:

"Please don't let me have to go through that at least. Won't you respect my 'face' even a little bit?"

Then he took a brush lying in the desk and with flourishing hand wrote his last poem for posterity:

"I balked not at steep obstacles nor bruising pain,
"I held life cheap a promise to sustain,
"In this I joyed and now I pledge to friends again,
"With heart unchanged, I'll meet my greatest strain."

Pitiful was this man, who had taken over the Nanking Government on his entire shoulders, following the death of Dr. Wang Ching-wei. The only thing that he left on this earth was this beautiful seven-character, four line verse. He walked with firm and deliberate steps, to the place of execution, then, to the officer in charge of carrying out his sentence, he said with a smile on his face:

"So you are the officer in charge. I'm sorry to put you to all this trouble. Please forgive me. As a last favor, please shoot me in such a way as not to leave my face too mutilated to look at."

At 8:30 a.m., at the order to shoot, Chen Kung-po slightly turned his head to the side (to prevent a wound on his face) and fell from a bullet that pierced the back of his head and emerged from his right cheek. In this manner, he went calmly to his death. Chu Min obtained permission from the guards to buy 50 pounds of ice, and placed it in the temporary coffin where Chen Kung-po lay, to prevent the body of his friend from rotting too quickly. The family of this dignitary, Chen Kung-po did not have even the money to buy him a coffin. It was only after his old friends got together and contributed money, that Chen Kung-po was finally laid to rest.

The life of Chu Min was also snuffed out later on the same execution stand. His wife, who went to meet him for the final parting, quietly told him:

"Your coffin is all ready. Please die at ease." This quiet and courageous devotion of Chu's wife and the composure of Chu Min himself are reminiscent of the death scene of ancient Japanese samurai. Chu Min's death poem was:

"Bright shines the morning rays betwixt the cloudy portals,
"Stilled is the pain and mirror-quiet the turmoil of a mortal
"My mind is clear, my heart knows the ecstasy of peacefulness
"As tranquilly I place my soul into the hands of death."

This fine old gentleman, during his Nanking days, was always flying kites in his yard. He was engrossed in the study of kites. However, he resigned himself to his fate. Like an escaped kite, he went to his death blown by the winds of adversity. He hid his figure for eternity in the clouds, his life string snapped by a strong gust of a cruel wind.

The death of Lin Pai-sheng was also magnificent. He and I had once seen heart to heart on a number of questions. When the Japanese Army supported the Nanking Government's opium suppression policies, this spirited young man had stood on the street corners of Nanking, had mobilized the youths of the capital city and had burned down opium dens throughout the city. Pictures of him, dead on the ground, blood pouring from his mouth were plastered prominently all over the city. But, his dead face was brimming with laughter.

The opportunists Chou Fo-hai and Chou Hsueh-chang, who constantly vacillated between Chungking and Nanking, continued their shameful conduct to the very end. Chou Fo-hai died of sickness while serving a life sentence in the Nanking Penitentiary, while the "faithful wife" of Chou Hsueh-chang filed suit against him, while he was in prison.

Chen Pi-chun's attitude was one that few men could copy. In court, she stood up and courageously said: "My cooperation with Japan was based on the Great Asia principles of Dr. Sun Yat-sen. Is not the Chinese Communist Party the vanguard of the Soviet Union and Chungking the tool of the United States? If what I did was wrong, please kill me speedily. I will go to the next world and report proudly to Dr. Sun Yat-sen and to my husband on my actions. You are quite at liberty to pass whatever sentence you desire upon me. I shall never appeal your sentence to a higher court." I heard that she was sentenced to life imprisonment in the Suchow Penitentiary. The thing that most saved her from death was her fearless attitude in court, an attitude which moved the court to silent admiration. Dr. Wang, now in the next world, should be able to rest well, after seeing his wife's magnificent comportment.

This is no time to list the merits and the sins of the Nanking Government. However, how senseless was this cruel execution of compatriots and comrades who had once participated in the same revolution! The narrow-mindedness of Chiang Kai-shek and the other leaders of the Chungking Government which did not allow them to forgive collaborators was the same narrow-mindedness that led them to oppose the Chinese Communists and to dig their own graves. The Chungking Government, which could not forgive the Nanking Government, could not possibly have taken in and absorbed the Chinese Communists. The taste of blood of internecine warfare was already to be found in the cups of toast raised to victory. The persons who once judged the Nanking Government are now in the position of being judged by the Peiping Government. Will history repeat this same pattern? Fate indeed is ironic.

Dysentery and typhus rage throughout the whole of China the year round. Cholera is also not rare, but usually breaks out between May and October.

That year cholera began appearing in the middle of April in Chung king. Neither the municipal nor the national governments seemed to take any steps to prevent the spread of the epidemics. Neither were there any isolation hospitals. Cholera spread from the city center and reached the suburbs. I often thought, "If I could only get immunization injections..." But such was an empty dream.

Despite the fact that I took particular care not to eat raw foods, I woke, with extreme pains, on the morning of May 3, the day of my father's death, and began having a running stomach. I had never experienced such before. I forced my listless body from bed and climbed the hill behind our compound and chanted sutras in requiem for my father's soul. My waist pained so much, that I thought it would come apart. I climbed down the hill and collapsed on my bed. I had to go to the toilet four or five times each hour. Soon I noticed that my stools were foamy mucous mixed with blood.

I had not eaten a grain of rice from early morning. I could not even force a drop of tea down my throat. But, I simply could not resign myself to dying from cholera before my mission had been accomplished. However, I had solemnly promised, when I submitted that letter written in my blood that I would give up my life if the Generalissimo read my report in full. Perhaps Generalissimo Chiang Kaj-shek had read and fully understood my report. If so, my life was being taken away from me, even as had pledged to the gods. In such a case, my father was probably coming from the other world to take me with him on the anniversary of his death. I hated to die of cholera, but I thought May 3 would be a fine day to end my life.

Nobody called a doctor for me; neither did any one give me any medicine. I faintly heard the worried voice of Nieh come to call me for lunch. After that I lost all consciousness. I fell into a deep stupor-like sleep; It seemed that I had gotten up by pure reflex action dozens of times because of the fierce running of my stomach, only to drop back senseless on my bed.

It was around evening, I believe, that the figures of my wife and children appeared about my bed, with worried looks on their faces.

"That's funny that they should come here all the way from Japan," and when I tried to speak to them, I realized that it was Colonel Liu's wife and her children that were at my pillow side. Colonel Liu also seemed worried, but there was nothing that they could do.

Two days passed without my eating a single bite.

"It's too early to die. I still have to ascertain the reaction to my report...."

And, whipping myself into action, I clasped my hands together and prayed to the gods and to the spirit of my father.

On the morning of May 5, the running of my stomach stopped, and I found myself regaining a little appetite. It seemed as if it were really still too early for me to die.

It was the day of the Boys' Festival. This holiday together with the Double Tenth is most feted in China. On the dinner table were, piled mountains of foods, meats, fish and tasty wines. But I couldn't reach out my hand. On this day, adults and children, men and women, officers and privates, all seemed to let themselves go and drank and ate and feasted. I could not stand the sight of the usually staid wives of the officers, quaffing their wine and playing ken, or pounding the table and blustering away. I could not bear sitting all through the feasting after, my severe illness, yet I could not leave, for it would have been the heights of bad manners. I only ate two or three "te chimaki" rice dumplings wrapped in bamboo leaves. It was the same as that eaten in Japan during the Boys' Festival. However, the Chinese place meat in their rice dumplings. Then they are wrapped in the bamboo leaves and cooked. Thus the Chinese "chimaki" tasted more delicious than anything of its kind in Japan.

It seems that I had contracted pseudo-cholera from germs contained in the Chinese bean-curd cake, which I had eaten for breakfast.

I realized that at home Boys' Festival would be celebrated with paper caps flying in the breeze against the background of new spring green. Here in this capital city of Szechwan, I missed this traditional Japanese sight.

General George C. Marshall arrived in China from the United States immediately after the war and expended every effort to affect a compromise between the Chinese Communist and the Kuomintang parties. The aim of General Marshall's efforts was to unite China into a modern democratic country, then give the unified country economic aid and, military guidance and through the stabilization of the lives of the people prevent the Bolshevization of China. However, his sincere and humanitarian mediation efforts failed to be greeted with joy either by the Chinese Communists or the Kuomintang.

The reason lay deeply rooted in the fundamental character of the Chinese Revolution. The first mistake was to view the Chinese Revolution as a modern democratic revolution.

The Chinese Revolution was a political movement inspired by the desire to overthrow the Manchu dynasty and to allow the resurgence of the Han race. Its chief aim was to establish the Sun Yat-sen Dynasty in place of the Ching Dynasty of Manchuria. Dr. Sun Yat-sen died midway through the revolution and Chiang Kai-shek and his group took over his work. True, Chiang Kai-shek did overthrow the old warlords, but the net result was the concentration of both political and military power into the hands of one person and that was Chiang Kai-shek himself.

With the Generalissimo's power as a background, his group succeeded in amassing the greater half of the wealth of the country. In the last analysis, it only meant that the

Chinese Nationalist Party and the army had taken over the place vacated by the eight Banners of Manchuria.

The Chinese Communist Party launched its counter-attack against the Kuomintang, brandishing as its slogan a rehashed version of Marxism, which they camouflaged as being democratic principles. They succeeded in their counter-attack. However this success did not signify the success of either their "ism" or their doctrines. It was a plain windfall, resulting from the hopeless corruption of the Kuomintang. In place of the Chiang dynasty, the Mao "dynasty" had risen to power. In addition, Mao Tze tung's trump card, the secret of his success, was his handling of the land question. His land reform program, sprinkled with a modicum of modernism was his greatest weapon in trapping the entire agrarian population of China. It is still too early to make a final judgment and reach a definite conclusion. We will have to watch future developments for a while. However, should we not regard the chief cause of the conflict between the Chinese Communists and the Kuomintang as one not of doctrines, but as an instinctive conflict rooted deep in the history of China?

The fact that the United States, while seeking a compromise on the one hand, aided the Kuomintang on the other, by sending warships to transport Chinese Nationalist forces into Manchuria and openly offering vast economic help and military guidance, must have convinced the Chinese Communists that the United States regarded the Chiang Kai-shek regime as the sole orthodox government of China. Such actions did not help convince the Chinese Communists of the wisdom of compromise with the Kuomintang.

The chief causes of the failure of Japan's eight years of occupation policies, were, first Japan's inability to understand the psychology of the Chinese people, second, the lowering of her moral standards which aroused the resentment of the Chinese people and, third, Japan's lack of power to save the Chinese people from famine and chaos. If the United States had really been wise and had understood the fundamental characteristics of China, she would have withdrawn her hands completely from Chinese questions following the end of her war against Japan. Then, she would have left the solution of Chinese questions to the Chinese themselves and after watching closely trends and results, she would have given economic and technical aid to the newly-arisen power in far greater quantities than the Soviet Union. At the same time, the United States should have spurred the surging racial independence movement, and should have supported the demands for the complete return to China of Outer Mongolia, Manchuria, Sinking Province and Hong Kong.

If the leaders of the United States of that time had this wisdom, she would have succeeded today in making another Tito out of the Mao Tze Tung Dynasty.

It was a mistake to think that the Chinese Nationalist Government alone was pro-American. At the same time, it was a mistake to have concluded that the Chinese Communists were only mildly pro-Soviet and nothing more.

The fact that the Chinese Communists are pro-Soviet today is recognized by all people. However, one of the big causes that made the Chinese Communists completely pro-Soviet was the lack of wisdom on the part of the United States, a fact which she should realize today.

As a conclusion, it can be said that the Chinese are eternally Chinese.

It happened on May 7. A sinister looking man, dressed in a cotton Chungshan robe and felt slippers, came unannounced to the Flower Garden of the Chang Family. He was accompanied by a major-general in uniform. The major-general turned out to be an interpreter. He did not tell me his name nor hand me his calling card. He simply sat down in front of my desk and began a random conversation. From the way in which the major general interpreted our conversation, in a fearful manner, this visitor seemed to be a fairly important person. I intuitively concluded that he was General Tai Li's successor, Major General Mao.

He knew in detail the report which I had submitted to the Generalissimo. In all probability, he had come to sound out if I had meant what I had written. In summing up the important points that I gathered from his words, his opinions were:

1. The Chungking Government did not welcome the mediation of General Marshall. The Chungking Government should not be talking about democracy at this critical time. It was time now to down the Chinese Communist Party, the anti-revolution influence, by mobilizing the entire strength of the Kuomintang.

2: Chungking's participation in a peace conference with the Chinese Communists was to gain time to move the Kuomintang forces, concentrated in the western part of China, as Far East and North as possible.

3. The Chungking Government was anxious to obtain the services of Japanese soldiers experienced in fighting the Chinese Communists. Of course, the general did not make clear so bluntly these three important points, as his own opinions to a Japanese that he had met for the first time. But, I concluded, that this was what he thought from the way in which he would blindly assent to these assertions when I expressed them as if they were my own opinions and from the way he tended to reemphasize what I had said. Although the peoples of China hoped for peace as soon as possible, both the Chinese Communist and Kuomintang parties were ready to disregard the feelings of the people and were engrossed in the establishment of their respective Chiang and Mao Dynasties.

The 4,000-year-old history of China, regardless of period, is the story of the struggle for power by various groups or individuals in complete disregard of the hopes and aspirations of the Chinese peoples. It is a history of a succession of dynasties established by ambitious men by the force of arms, men who interpreted their rise as the dictate of the gods. It is a country true to the old proverb:

"The man who has the greatest virtue becomes the Emperor-King."

However, since there is no one to judge in the final analysis who has the greatest virtue, the one who has the power, or in other words, the one that downs all the others with the sword and then designates himself as being with the greatest virtue, succeeds to the throne. Thus, the fact is that China is a country where in essence:

The man with the greatest power becomes the Emperor-King This is true today as it was at any time during her whole 4,000 years of her written history. It was natural, therefore, that the Chinese Nationalist Party, which once was on the verge of collapse, should desire to monopolize the prize of victory won after eight years of war thanks only to the support of the United States. Thus, regrettable as it was, it was also natural that the efforts of General Marshall to give half of this prize to the Chinese Communist Party, at a time when the Kuomintang was drooling over the spoils of war and their confiscation, should meet with complete failure.

First Class Private Nieh was a farmer's son, who had only graduated from public school, but when I asked him:

"For what reason must China be split into two to fight against each other?" Nieh, simply laughed and bluntly answered:

"This is because there are two people and both want the same object."

This hit the problem right on the head. The picture of the Chinese Nationalist and the Chinese Communist parties fighting for a single jewel was simply the epitome of the history of the rise and fall of dynasties over 4,000 years in China. On my desk was placed an old ink-stone, with two dragons fighting over a jewel engraved on the cover. The engraving seemed to laugh at my foolish question and to be trying to teach General Marshall a lesson!

I talked with Lu during some idle minutes after dinner (I shall write in greater detail about Lu).

"Assuming that I were in the position of Generalissimo Chiang Kai-shek, I would have retired to Fenghua immediately after China won over Japan. I would let Chang Chun take over the Kuomintang Party, make him compromise with Mao Tze-tung and give all the country north of the Yangtze to the Chinese Communists and all the country south of the Yangtze to the Kuomintang. Then I would let Mao take the Soviet Union as a patron and Chang the United States, and get both of them to worm money and gifts from their patrons as much as possible.

"Sometimes, I'd call both of them to Fenghua and let them relax and fish in the clear streams of my birthplace. Should World War Three break out, I'd make both of them cut off all ties with their patrons and let them be independent. For five years, I'd make them compete in their respective spheres against each other, to, see how much progress they would be able to make in politics and economics. Then, I would let the best man unite and rule both north and south of the Yangtze. The Generalissimo is no longer simply the leader of one party. He must be the leader of' the whole 400,000,000 people of China."

"That is exactly what the old farmers (the common people) hope for. However, the history of China shows from ancient times that the wishes and the desires of the old farmers have never been taken into consideration in the government of the country."

This pure-hearted Lu showed sadness on his face. It was indeed infrequent to find a youth with such ideas as Lu in the Military Control Bureau.

The one-month pay of a private was 9,000 yuan. If he bought ten oranges with his pay, he would have nothing left. Despite this fact, the soldiers were able to purchase shoes and to smoke cigarettes; I wondered and wondered how they made the money to buy these things. But the puzzle was soon solved. Some soldiers in connivance with merchants scalped 10 to 20 per cent on the sales and purchases of Government goods. Others cheated in doing shopping for their superiors. Others still would engage in direct theft or make money by gambling. Others would have side businesses. If you asked a soldier to do your washing or to clean your room, he would show a wry face. However, if you asked him to do some shopping for you, he would gladly go, even if it were raining cats and dogs.

There was one soldier cook in the flower garden of the Chang Family. He saved his money and bought five or six ducklings. These he fed with left-over food for about two months and then sold them to the poultry merchant. With the proceeds, plus a little more of his savings, he next bought a piglet for 10,000 yuan. In two months, the pig had grown to twice the size as when it was bought. The soldier was happy at the growth of the pig, saying that he would be able to sell it for 100,000 yuan.. But, for some unknown reason, it became sick and began to grow thinner daily. Some sort of growth appeared on the pig's chin. The soldier tied the four legs of the pig together and then with a red hot poker punctured the growth. He next poured lamp oil and salt into the wound and again burnt the surface with a hot poker. The pig squealed and howled at this rough and ready operation but failed to recover. Finally, it seemed to lose all appetite. I felt that the pig would not last another three days, when the soldier decided to kill it.

He tied the four legs of the pig together with wires, then thrust a large cooking knife into its throat. A wash basin was placed under the wound to catch the blood. While the pig still writhed in its final death agonies, the soldier put himself astride the pig and squeezed out its last drops of blood. He then placed the pig in a large tub used for bathing and poured boiling water over the carcass and plucked at the bristles until the pig was clean and pink. The soldier placed his mouth to the knife wound in the throat and began to blow. Wonder of wonders, the hide began to separate and the carcass became bloated like a drowned man's body that had been left in water for a long time. Then, the soldier skillfully split open the pig along its lower side and took off the hide cleanly. I thought that he would throw away the meat which had turned purplish, but no. Drinking strong spirits, the soldier and five or six of his friends gorged themselves in an all-night feast. The wives of the major-general and colonels all brought dishes and pans and competed in getting portions of the purplish meat. What remained, the soldier salted and put away.

When the cook was ordered to buy meat by the deputy-chief with official funds, he would sell this salted meat at the same price prevalent on the market, and finally succeeded in making up the 100,000 yuan he missed earning when the pig became sick.

For ten days after this incident, I did not eat any pork when it was placed on the dinner table. However, there was not a single Chinese who seemed to mind the purplish meat. The blood that was caught in the wash basin was drunk raw by some persons, while others put it into their soups. The bristles were gathered together and sold as material for brushes, to become eventually an important export product, while the intestines were kept to make sausages. Not a single part of the pig was thrown away, bristles, hides, bones, feet and even the tail. In Japan, the entrails, such as liver and kidney, are cheaper than meat. But, in China the entrails are more expensive than the meat. Those who engaged in such side businesses were the most honest and hardest working soldiers in the Chinese army.

China first began to think that perhaps she could defeat Japan when the Pacific War broke out. However, she never imagined that war would end so quickly. China expected that war would continue on for three to four years, even after the fall of Okinawa and made all preparations on this assumption.

The surrender on August 15, in a sense, astonished Chungking. The Japanese became confused because they had never lost in their history. But, at the same time, Chungking lost its head, since it had no experience of Victory. She was faced with great problems. How to go about the business of confiscating the vast factories operated by the Japanese in China, the extensive property and the stores of clothing and munitions of the Japanese Army? Will the Japanese army, still 1,000,000 strong, honestly throw down their arms? There was not even a single volunteer to fly into Nanking. After a long period of futile thought, Chungking finally decided to invite representatives of the Japanese army to the Chinkiang Airfield and only after verifying the intentions of the Japanese army to surrender, did the Chinese Nationalist Army enter Nanking.

The first choice of the members of the Confiscation Committee was to be dispatched to Shanghai and Nanking, the second to Peking and Tientsin, the third choice to Manchuria, the fourth choice to Taiwan and the fifth choice to South China and French Indo-China.

The higher-ups in the army and the government dispatched their underlings ahead in all directions and gathered up enough wealth at one stroke to last several generations ahead. The Chinese Communists, who had watched with fingers in their mouths, were therefore in no mood to become friendly with the Chinese Nationalists. The Chinese Reds stood by, determined at the first opportunity to take all this wealth away from the Chinese Nationalists with the force of their arms.

The broadcast by Generalissimo Chiang Kai-shek: "Reward hatred with virtue," made the Japanese cry: "We have also been defeated morally,"

However, this proclamation was no more than a sedative, even as was the appeal made to the rebel soldiers during the February 26 Incident [Ni Ni-Roku Jiken, coup d'état attempt, Japan, February 26, 1936]: "It is not too late as yet...."

The generalissimo's proclamation was made at an ominous time when 1,000,000 Japanese soldiers still remained armed, with rifles in their hands.

Of Chungking's population of one million, roughly 400,000 were government workers and their families. It was not an easy task to move this vast army of officials and their families to Nanking from Szechwan Province with its poor transportation facilities. Four to five hundred people were transported daily by air from Chungking's two airfields, but air transportation was limited to high officials and dignitaries and members of their families, and the majority of minor government officials had to rely on river transportation. The greater part of the river barges, made of wood, were half rotten, their bottoms ready to fall out at any time. Thus, there were not a few who sank to a watery grave in the Yangtze River on their way to Nanking. As the civil war increased in intensity, land, air and river transportation facilities were practically all mobilized for the movement of armed forces and government officials were forced to stand by, waiting with fingers in their mouths and appealing unsuccessfully to heaven. Many had sold all their household goods and through bribes several times larger than actual costs had managed to buy tickets for the boat trip. However, day by day their departure was postponed. There were, therefore, not a few who committed suicide after spending all their travel expenses in much-needed food. It was two years later that all Government officials were finally moved to Nanking. Once they got there, they were again faced with the necessity of dispersing either back to Chungking or to Canton.

Generalissimo Chiang Kai-shek left for Nanking on May 2. His official residence was to have been that of Wang Ching-wei's. However, when he got there, he said: "Too extravagant," and refused to accept the mansion. He made his home in the official residence of the principal of the Wangpoo Military Academy. It was a small and simple residence, but there were none to follow the example set by the Generalissimo.

He neither smoked or drank, not even tea, and kept himself above corruption in the hope that others would follow his example. However, after the death of General Tai Li, there was no one among the people around him to offer him unwelcome advice. Not even a small portion of a single report on the corruption and degradation arising from the taking over of Japanese property and the return of the capital to Nanking ever reached his ears. In this way, the Generalissimo became a man on clouds, far removed from this earth. Below him in the lower world hundreds of devils ran rampage during the day while the "evil spirits of mountains and river" filled every nook of the capital city. And, the chief leaders of this evil band were the men closest to the Generalissimo, Kung, Chen, Sung and Sung, who took over half the confiscated industries as their own private property. The fact that the Chinese Communists began demanding: "Overthrow bureaucratic capitalism and the four great cliques of wealth!" pointed to the undeniable existence of corruption in high places." The remarkable thing was that great factories, directly operated by the Kuomintang appeared in Tientsin, Tsingtao and in Shanghai and the profits from these establishments poured into the pockets of party leaders and enrichened their private lives. If General Tai Li had been living, there would never have been such ugly and seamy examples of corruption.

The most pitiful among the Kuomintang officials were those that had chosen to go to Manchuria to handle the confiscation of Japanese property and investments there. In order to take over highly specialized industrial facilities, all technical personnel dispatched to Manchuria were forced to sell their property even that accumulated through generations by their ancestors. When they got there, all they found were the chimneys and the pillars, with all the machinery dismantled and taken elsewhere. They were soon involved in the civil war between the Chinese Communists and the Kuomintang and finally fell into the clutches of the Chinese Communists.

It could not be said that there were none who looked ahead. One old officer, who had risen from the ranks to become a lieutenant-colonel and an aide to a unit commander told me:

"In two years time, Chungking will again become the capital," and this soldier went ahead and bought up homes, which had fallen in price to less than a tenth of their value, and household goods being thrown on the market for almost nothing.

The whole city of Chungking from the highest to the lowest levels was engrossed in the problem of how to move even a day earlier and ahead of others.

It was an interesting study for me to read with great care from the first to the last page, the ten newspapers published in Chungking and to observe statistically the movement of politics, economy and military matters. Gazing at a single map of the world on the war, I quietly thought of next world developments.

The organ of the Chinese Communist Party, the "Hsinhua Jihpao," was delivered to me daily. However, the people hated seeing me read this paper. It was natural that they did not wish their Japanese guest to read bitter and sharp denunciations of themselves by the Chinese Communists. I bribed First Class Private Nieh to bring me papers when all the rest of the people had gone to sleep and then read them far into the night until I could almost make out the other side of the pages.

Despite the fact that the unit with which I lived was the Anti-Communist Propaganda Department, the only reference material on the Chinese Communist Party was a work, used for long years by the Japanese Army in North China. This was utilized as the only and the highest authority on Chinese Communism. I translated this document, written in the Japanese language, into poor Chinese and made the men study. Despite the fact that the Communists were the same Chinese, the ability of the Chinese Nationalists to study and evaluate scientifically and systematically their rivals was surprising low. Their guest became the teacher on such matters as the "Three-Three System" and the Land Reform Program of the Chinese Communists and I taught them for over 100 days.

In this typical example of Chinese officialdom, where both private rely failed and official lives were led together, what astonished me most was the way the wives of the officers spent their entire day from morning till victorious night chattering away and singing, whenever they had time, in high than one pitched voices anti-Japanese songs. The last words of the song:

"We won't lose. We won't lose," could be heard held out thinly for a long time. It was obviously a national song, composed during the war against Japan, in which the

Japanese military clique was cursed, racial raining enmity stressed and the will to fight to victory emphasized.

The four children that I had cared for and loved as if they were decided my own from the very beginning of my stay here said without batting an eye, just as they were about to leave for Nanking:

"The Japanese soldiers are bad. They killed the children of China."

One day one of the children now in his second grade in public school came to me bringing with him geography textbook. There was a page which contained all the flags of the world. I asked this child which flag he liked the most. He first picked out the Chinese Nationalist flag, then Burma and the Stars and Stripes, then the Union Jack, then the French tricolor, and after a long pause he picked out the Hammer and Sickle and then the German flag. No matter how long I waited he would not point to the Japanese flag. I finally asked him what he thought of this, the Rising Sun flag. He immediately answered with a brazen face:

"I hate it the most."

"I'm Japanese. Do you like me or don't you? "

The child seemed at a loss what to say, and then he gave a very clever answer:

"You are a foreigner. Foreigners are not bad."

It was truly worthy of a son of Chinese.

These were the final parting words of a young child that I had looked after for three months. This one comment seemed to indicate how deeply the feelings of resistance to Japan, hatred for Japan, fear of Japan and disdain toward Japan, had been instilled into even Chinese women and children. Verily, it seemed that these feelings had been worked into them to the very marrow of their bones.

It was simplicity itself for the Japanese to jump to the conclusion that the racial feelings of hatred toward the Japanese, instilled through half a century of mistaken policies toward China, could be swung around immediately in a 180 degree turn by Generalissimo Chang Kai-shek's proclamation to "reward hatred with virtue." Of course, I had only seen a very small part of the whole of China as revealed in the families of right wing members of the Kuomintang Party. However, the hatred of the Chinese for the Japanese was not to be mitigated so simply and quickly as imagined by the easily moved and easily tiring Japanese people.

There was a youth named Lu who returned to Chungking from Chengtu in the first week of June. At the beginning, he was introduced to me as a student attending a university in Chengtu. However, I eventually learned that he was captain attached to the Military Control Bureau. Lu was a bright and likeable fellow and I felt he was too good a youngster to leave in such a department as ours. He became my teacher of Chinese. Every day he taught me the language using newspaper editorials as teaching material. I was blessed with the opportunity of hearing from his mouth on frequent occasions fragmentary references to the situation within the Chungking Government.

The majority of the intellectual youths in China believed that the greatest statesman in Japan's past was Prince Saionji [Kinmochi]. I was surprised that the principles of racial equality, voiced by Prince Saionji at the Versailles Peace Conference as part of Japan's diplomatic policy, had left such a strong impression among the oppressed peoples.

[Prince Saionji Kinmochi, born December 7, 1849, died November 24, 1940. Twice Prime Minister of Japan.]

A large number of American landing boats were provided to help in the work of completing the return of the capital to Nanking. At the beginning of June, some were also allotted to our agency with orders for the entire group to move together to Nanking. For this purpose, preparations were begun by the officers and men. However, Major-General Mao had since given me no instructions. It seemed that I had been forgotten.

On June 28, I climbed the hill behind our compound and worshipping the crimson sun appearing over the eastern horizon, I chanted a sutra. When I had finished a flash occurred to me. Good, I would make my last report. By noon, I had completed a report something like the following:

"Your Excellency Mao Jen-huang:

"I have no words to express my gratitude to you for your excessive kindnesses to me at a time when you have been so busily occupied with official duties. Since my arrival in Chungking, a hundred days have passed, during which my work has been limited only to reading newspapers. As a result, I have come to realize that Japan-China collaboration, which I had pondered over for a long time has completely failed and that it was a mistaken idea from the very start.

"To think that a defeated Japan could cooperate with a victorious China and thereby atone for its past crimes was nothing more than one-sided fancy on my part. When I view your great ally, the United States, with her far greater potentials and her excellent techniques, according you great help in the development of your industries, the construction of factories, the establishment of welfare facilities and in the training of your military personnel and thereby solidifying the foundations for eternal US-China collaboration, I realize my past mistake and have decided to surrender myself to the American military authorities and make my way to the gallows.

"However, if by chance I should ever speak about the steps as they actually occurred in my escape from. Bangkok, it would cause trouble to General Hsieh and a large number of other Chinese comrades under him. Thus, I sincerely pray that you will make it seem superficially that I was arrested while in hiding in Hanoi by the Military Control Bureau; brought to Chungking and in Chungking handed over by the Bureau to the American military authorities.

"However, I would finally ask of you that you fully understand deep in your heart the will of His Majesty of Japan, as outlined in the report which I submitted to Generalissimo Chiang Kai-shek on March 31.

Today, when all my long-standing hopes of Japan-China collaboration have been completely squashed, I can no longer, as a soldier of Japan, continue to bear your warm-hearted treatment. It is more painful than being executed by a firing squad.

"I pray that you will hand me as quickly as possible over to the American Headquarters so that I may be executed."

"Will you kindly transmit my best regards to Lieutenant-General Cheng and Lieutenant-General Hsieh."

Absolutely alone, and with empty hands and fists, I completed my last missile, risking my life once again. Whether it would be wind or rain, I left everything up to fate and that night. For the first time in a long while I slept soundly. Deputy commander Li, evidently flustered by my letter, hurriedly took my third report and submitted it to Major-General Mao.

The next morning he came back with this answer:

"Major-General Mao said that he would try to send you by plane to Nanking as soon as possible and asked you to wait a few days."

I resigned myself to waiting another month, since "several days" usually extends into months in China. However, that afternoon at 2 p.m., one officer who spoke Japanese, came from the Military Control Bureau and handed me this order:

"You will leave by plane tomorrow for Nanking. Therefore, hurry and go to Chungking. All the members under Major-General Meng stationed in the Flower Garden of the Chang Family will also leave for Nanking within the day."

This was like a bolt from the blue. It took me only an hour to get ready, since my personal belongings consisted only of one wicker suitcase and one bed roll. However, the several families that had lived within the compound for several years were thrown into turmoil at the unexpected order. It was as if a hornet's nest had been overturned. The plodding Chinese people usually think and do things slowly. This, I was surprised at their quickness in getting down to preparations. Not a single person voiced complaint. They were happy in their busy preparations for the return to the capital city of Nanking. A dream which they had dreamt for eight long years had finally materialized, my final report, which I had sent off in anger after growing impatient of my long wait, had served to bring happy results in its wake to these people also.

First Class Private Nieh who had looked after me for so long, saw me off with tears in his eyes. It seemed that a man's sincerity is candidly and honestly accepted by soldiers and uneducated farmer folk. It was not that I had given him a lot of money. For that matter, I was not in a position to make such gifts. I had only treated him kindly as I would a younger brother, unable to suppress the sympathy that I felt for this innocent youth and his aged father and his mother, now about to be drawn into the vortices of civil war. I quietened this Chinese soldier, who seemed on the verge of breaking down and crying, by telling him that I would return. Then together with my luggage, I got on the truck they had sent out. The figure of the First Class Private Nieh, who saw me off, was soon swallowed up in the cloud of dust rising behind our speeding car.

About 15 of us, sitting astride suitcases and tubs, were shaken mercilessly. Several times, we were almost thrown off. However, we managed to arrive safely in Chungking in the evening.

Our quarters that night was a corner of an army barrack. We were forced to lay out our blankets on a dirt floor. This might not have been too great a hardship for the men to bear, but I wondered how the wives and the children would fare.

Eventually, two young officers came to call for me from the Military Control Bureau headquarters and said: "Dr. Shih (me) alone please come and stay at the headquarters."

The two officers did not even glance toward Major-General Meng, but took up my suitcase between them and led me away. The headquarters turned out to be an imposing four story building looking out on the Liangkiang River. In front, there was still displayed a signboard signifying that this had been the offices of the US-China Collaboration Society., This was the supreme headquarters of all wartime information and underground activities, the home base of General Tai Li. The two officers vacated their room on the third floor and let me stay there.

The luggage allowed on the plane was limited to five kilograms. Once again I was stripped to bare essentials. The majority of my belongings I entrusted to Lu, asking him to send them on to Nanking for me.

One of the two officers who had come to take me to the headquarters was a major, called Wang Chih-lung. He had been dispatched to North Burma as a chief of an intelligence unit and had gone underground in the town of Lashio in order to probe the military secrets of the Japanese Army. I did not dream that this young officer could be a hero. He had disguised himself as a Burmese and had worked as a coolie in the Lashio railway station, gathering information on the Japanese Army. He knew in comparative detail the distribution of the units of our army crops.

Yesterday's enemy became today's friend. In the quarters where I stayed were some who were leaving for Manchuria and others who had once studied in Japan. It was a merry gathering.

After supper, I lay my tired body on my bed, but I could not sleep all night for the bed bugs and mosquitoes. I do not know how many times woke to kill hundreds of bedbugs, but the insects far outnumbered my futile efforts. The person next to me slept on soundly, obviously immune to these attacks, and continued to snore stentoriously through the night. I felt again the deep-rooted and strong power to live peculiar to the Chinese race alone. Or it might have been that the insects all flocked, to attack a new guest out of sheer curiosity or to sample the different taste of a new arrival.

Early on the morning of July 1, five of us got on a jeep and headed for the airport.

This time, it was not the airfield on the river bed, but one out in the suburbs known as the Pai-Shih-I Airfield. On our way, we passed by a village of mountain caves where the Generalissimo's air raid shelter was located. This underground residence had served eight years an important role and was now quietly resting.

The road that connected this village of mountain caves with Chungking, several miles distance, was solidly paved. I heard that the road was completed in less than a year under the supervision of American engineers. The air-raid residence of the Generalissimo lay in the side of this mountain covered by woods. So carefully had it been constructed that even the heaviest bombs of Japan could not have touched it. In all probability, even an atom bomb would not have had much effect.

It was past ten when we reached the airport. The runways were built to enable their use by B-29's. However, the buildings attached to the airfield were all temporary structures. In the case of Japanese airfields in Manchuria and at Nanking, more money was spent on accessory facilities than on the runways themselves. In China, the opposite was true. In all probability, this was due to American supervision. We were scheduled to leave at one in the afternoon and had to wait three hours in the steamy hot waiting room. This deflated my high spirits.

Eventually, one large aircraft from Hankow landed on the airfield. Among the passengers were military men and merchants, who carried assorted merchandise with them. They were utilizing air travel for business. Once they made a round trip, they could make a profit several ten times more than the air freight. Despite the fact that travel to

Nanking was so congested that people had committed suicide at being unable to get a seat on the plane, here high government officials and smugglers were using up air space and making one round trip a week.

This time my name was changed to the alias Wut Chieh-nan. It was perhaps because my other aliases, Shih Cheng-hsin and Shih Kuang-yu, had already been overused. It was no longer any, trouble for me to pass under an alias. All I had to be careful was to be sure I, caught my name, when it was called.

The plane was a large-size aircraft capable of carrying a truck. There were thus more than 50 passengers aboard. The summer heat was over 100 degrees Fahrenheit and it was like a steaming hell within the duralumin body of the plane broiling under the sun. Perspiration flowed like a waterfall and wet my summer suit through and through. I was soon as soaked as if I had fallen into a river and had to wring put several times the towel with which I wiped myself.

Once off the ground, we felt suddenly cool. I could not but feel deep respect for the way in which the American pilot handled the craft with deliberate calm. Leaving behind 100 days of memory in Chungking, felt an attachment to the mountains and rivers slipping by under me.

"Good-bye Chungking. I pray that you will never have to meet again the fate of becoming a wartime capital." This was the prayer that could not help offer for these peaceful hills and streams.

Over Hankow, our plane dropped altitude slightly. But, without landing, it continued eastward along the winding Yangtze. The same scenery, which I had seen two years ago when I was chased out of Hankow to the battlefields of Burma, was again unfolded under my eyes, though seen under different circumstances and from a different plane gazed at the placid flow of the great Yangtze, the wind-filled white sails, and the white walled mansion's scattered in the woods-all making up a familiar scenery.

And all the while the plane sped eastward I prayed in my heart that we would be able to throw our mutual hatreds of the past into this great river and to progress together toward the union of the whole of Asia. I also prayed that we would be able through a high morality and through the rigorous application of this morality to change the feelings of fear, repulsion and disdain for Japan, instilled into the very marrows of the bones of the Chinese. From the bottom of my heart, I prayed that the day would come soon when we would be able to link arm-in-arm and march toward a common goal.

Chapter 5 Nanking, After the Return of the Government

Viewing the Chungshan Mausoleum on our left, we landed on the air-field outside the walled city of Nanking. On the airfield were lined over 100 American-made fighter planes and large transports. It was 6 p.m., July 1 [1946]. Chiang Chen-chun and another youth from the Military Control Bureau came to the airport in a jeep to meet me. The four of us, Major Wang, who had accompanied me from Chungking, the two officers who had come to meet me and myself, got on the jeep, and entered the walled city of Nanking.

The deep green of the Tzu-chin-shan (the Purple Mountain) seemed to set off the proud white edifice of the Chungshan Mausoleum. The walls of the city were plastered with posters, gone mad over the victorious return of the capital to Nanking. It seemed that the markets were busier and more prosperous than usual. The Japanese stores along Chungshan East Road had been returned to their former owners and were decorated in festive array as if to compete with one another, while the faces of the people walking the streets seemed filled with bright hopes.

On the jeep, Chiang showed me my appointment papers. On it was written: "Shih Cheng-hsin, Official of the Second Section of the National Defense Department. Minister: Pai Chung-hsi." And the paper was stamped by a huge chop. It seemed that I had been turned over from the Military Control Bureau to the National Defense Department. In all probability, this was the first time that an officer of a defeated country like Japan formally entered as an official into the National Defense Department of the Chinese Republic.

We passed through a gate guarded by a stern looking sentinel. The wartime Military Administration Department, the Military Operations Department, the Military Control Bureau and other organizations seemed to have been unified into a new Department of Defense, which was just starting to the open shop in the former Central Military Academy on Wangpoo Road. The Second Section, to which I had been assigned, was concerned with information. This was not my specialty, but it was probably impossible to place me, a foreigner, in the Third Section (operations). When I thought about the future of Asia, I felt that it was not absolutely without significance for a sinful man like myself, who was no longer needed by Japan, to repay a part of Japan's debt to a neighboring and friendly country.

Major-General Wang, with whom I shook hands for the first time, had once studied in Japan. He was a member of the 42nd graduating class of the Military Academy. His face showed warmth and he wore a smart uniform. My first impressions of him were very good. The Department of Defense had been born at the advice of the United States Military Advisory Group and had been formed by unifying the Military Administration Department, the Military Operations Department, the Military Control Bureau and all kinds of other military agencies. The newly-formed Department of Defense had not yet started its work. All that this vast new household was occupied in doing was negotiations, bargaining and maneuvering over personnel matters in an effort to get as many family members and relatives into the Department as possible and to land as lucrative a post as possible.

I went to my designated room and finished a simple meal. On the entrance was hung a sign:

"Comprehensive Study Group."

However, there was only one desk, one chair and one bed. There was no mosquito netting. I lay down on the bed, but I could not keep away the fierce attacks of a vast host of mosquitoes. Time and time again, I woke to swat blood-bloated mosquitoes all over my body, leaving red stains spattered on the white wall. A non-commissioned officer was allotted to me as my orderly. He spoke a word or two of Japanese, having been employed once as a bell boy in the Japanese Officers' Club in Peking. Wherever I went, either to wash my face or to the toilet, he would follow me persistently. At the entrance to our building stood a sentinel armed with a Mauser pistol, eyeing me with suspicion.

The latrine was worthy of a military academy. It was built so as to accommodate several hundred men at a time. Below in a deep trench continually flowed water, like a valley stream. The door to each cell was open at the top, hiding only the lower half of a person's body. The idea was not to allow any soldier to abuse himself or engage in activities.

The official residence of the Generalissimo was situated within a separate compound. It was not even 500 yards from where I was located inside the toilet. The next morning around eight o'clock, a vast crowd of officers gathered on in the academy grounds. It was soon filled with jeeps. A light rain began to fall and many of the officers appeared, carrying umbrellas. The Generalissimo was scheduled to give an address of instructions the preparations for a decisive fight with the Chinese Communists.

Soon, the Generalissimo appeared, carrying a walking stick and wearing a mantle. He ascended the central rostrum and spoke in a high pitched voice. I could not hear what he said. I watched the whole proceedings from my second-story window, since I had received no notification to attend. After the meeting, Major-General Wang appeared, bringing with him a young officer. This was Lieutenant-Colonel Chung. Chung had once studied in Japan and he was assigned temporarily to look after my needs. A native of Canton, he had a very good grasp of the Japanese language. Led by this youth, we set out from the Department of Defense on a jeep. I interpreted in good faith the fact that no instructions had been given me concerning my work as a result of the Department's lack of preparations to start work as yet. Led by Chung, I watched with nostalgia familiar sights along the streets of Nanking as we drove in the direction of the former Supreme Headquarters of the Japanese Army in China.

The former official residence of Chief of Staff Itagaki, I found, was Japanese Army in China. Shusei Club by Kung Hsiang-hsi; and the former official residence occupied by the Governor of Yunnan Province, General Lung Yun; the Prince Mikasa by General Hsu Ying-chang. Each gate sported new name plates of the new occupants.

The quarters where I was finally taken to were located in one former billets for petty Japanese government officials, situated below the horse exercise grounds attached to Prince Mikasa's former residence. My quarters were a single room in a long tenement house. This proved to be Chung's private home. The tenement house was the billets Chinese officers of the rank of major and lieutenant-colonel. There was hardly anyone who lived alone. Several families occupied one room sleeping in confused piles–the families of friends and relatives arriving in close succession from Chungking. Lieutenant Colonel Chung had the foresight to rent another room in an adjoining tenement house. Chung had brought in from somewhere a wooden bed stand, six feet long and three feet wide. It was made out of odds and ends picked up here and there. He also brought in a red Japanese Army blanket, still retaining the tab, marking the year of manufacture as 1897. The sole other piece of bedding was a thin pancake mattress, black with dirt.

When I had slept on this bed for two or three days, a red stain oozed out of the planks and colored the whole of my mattress. I realized that the pieces of wood were undoubtedly taken from vats used by the Japanese Army to hold soya sauce or miso paste. I could not but feel that my life was a polite form of prisoner hood. There was, however, one piece of gorgeous furniture, completely out of keeping with the bare room–a piano. I thought that I had seen it somewhere before. I looked at it carefully and saw the mark of the Shusei Club burnt into the underside. Somebody had looted the piano and was secretly hiding it here.

A lieutenant-colonel's monthly salary was 150,000 yuan. The rent for the two rooms was 100,000 yuan a month. His beautiful wife was working in some commercial firm, bringing home another 250,000 yuan a month. With this total income of 400,000 yuan, Chung had also to feed an aged mother and one housekeeper. The "amah" was a farmer's wife from Supei, whose home had been burned down by the Chinese Communists. She had escaped with her life, leaving behind two children. She was not paid a cent of salary, content only to be fed. However, she managed to earn a sizeable sum by making a profit on the family shopping.

Food, the same as that eaten by the Chinese soldier, was brought to me noon and evenings from the unit, to which Chung belonged. It was usually brought in a bucket and sometimes in a dirty and grime-streaked wash basin. The food was always reddish rice and bean sprouts cooked in salt. This continued for roughly a month. I had the "amah" buy for me daily 1,000 yuan worth of subsidiary foods (1,000 yuan of that time would be equivalent to 10 yen today in Japan or around three United States cents). Every day dozens of peddlers came around. One egg cost 200 yuan, one water melon 2,000 yuan and one toasted rice cake 20 yuan. As long as one had the money, one need never starve.

What troubled me most as usual was the toilet, or more properly, the lack of a toilet. Every family seemed to meet their needs with a single chamber pot. Adults and children would be answering the call of nature without any reservation in this chamber pot placed in the combination guest-room-living-room-dining room. This was an unbearable burden and at nights I would slip out into the neighboring fields and hide myself from the eyes of people. Every morning the "amah" of each household would take these chamber pots out and empty them into a vacant lot in the vicinity. This lot and its environs thus became one big toilet, smelling to high heaven. The chamber pots were washed in a nearby pond. The same water in the pond was being used to wash rice. It was also used to wash dinner dishes and also for washing the face and rinsing out the mouth.

An old woman came around daily selling newspapers. I could not buy papers at their marked price. I always had to pay a 50 per cent premium. And I read every item printed.

Later a newly married Captain and his wife moved into the next room. The wife was a comparative beauty, but she spat snot-green phlegm on the floor and then rubbed it dry under her felt slippers. Each time I saw this act, I was seized with a desire to vomit. The monthly salary of a captain was 130,000 yuan. However, after paying 50,000 yuan rent for one room, the captain could not hope to live on the remainder. The stern reality of the life of a Chinese Government employee was the necessity of doing evil in order to eat and keep alive.

Practically without exception, the Chinese Government employee loaned out a lump sum to merchants at black market interest rates. They would lend a million yuan at 15 per cent interest monthly. With the resultant 150,000, plus their salary, they somehow managed to make ends meet. This million yuan, in the first place, was made by taking a cut on property confiscated from the Japanese Army and Japanese residents in China and selling the stolen portion on the black market.

Four or five days after my arrival, five or six tough-looking soldiers came to get the piano in my room. They took it away to a certain major-general attached to the Department of Defense.

There were practically no youths engaged in reading books. Ninety percent of the lower and middle rung officials spent their nights eating, drinking, gambling and whoring. In the next room, the people stayed up every night until 3 a.m., whiling away their time in foolish and noisy amusement, never-rising until nearly noon.

The wave of new arrivals from Chungking pushed into this small tenement house. Two to three families or from seven to eight people, slept folded on top of each other it seemed, in rooms no larger than 12 feet square. On the other hand, the higher officials in both Government and military circles monopolized two to three large western-style mansions where second and third wives were kept. Perhaps because the age-old tradition of China for the weak to be beset by the strong, there were none to complain or to become indignant. In fact, the lives of the high officials were viewed with longing by less fortunate men who stimulated their imagination with a dream of a similar life. Their chief ambition was to rise as soon as possible to a comparable position. But, usually these hopes constituted the hidden side of the life of Chinese government official.

I spent roughly one month in this atmosphere and in these surroundings. On August 4, Chung told me unexpectedly to change my lodgings. My new quarter was the kitchen of the former Japanese Supreme Headquarters. The room that was given me was one used by the kitchen duty sergeant-major.

Lieutenant-Colonel Tung, who took Lieutenant-Colonel Chung's place as my mentor and caretaker, was a gentle youth, although he knew no Japanese. My new quarters, superficially, was not very pleasant. But the environs were much better. The great vats that had been used in the past to wash rice now became a bath tub for me, and I found it more convenient to do my washing here. There was a captain, a staff officer, who always came to get water from the tap in our kitchen. While his wife lounged listlessly on her bed, her husband was forced to carry water for her in his uniform.

This kitchen was part of the barracks of Lieutenant-Colonel Chung's unit. One day a Chinese soldier came to me and in a very friendly manner told me that there was Japanese in the unit. I felt my heart leap with joy and that night had this youth called to my room. He was the first. Japanese I met after my arrival in China.

The youth went by the pseudonym, Taichi Fujita, and was a native of Fukui Prefecture. He had studied at the Port Arthur Technological University and had served on the French Indo-China border as regimental cryptogram expert. Through an unhappy mistake he had been taken prisoner and had been sent to Chungking two years before the end of the war. There he had been asked under duress to decipher secret code messages of the Japanese Army. For three years he had managed to dig up one excuse after another and 'had' continued to refuse. Now, given the treatment equivalent of a sub-lieutenant, as a member of Lieutenant-Colonel Chung's unit, he was engaged in deciphering the secret code messages of the Chinese Communist Party. At the beginning, he was all self-abasement at having been taken prisoner and was not at ease with me. However, I treated him as I would a younger brother, comforting him in his miserable self-torment from the bottom of my heart, until he came to trust me implicitly. He was only a First Class Private, but his first idea was to make himself out as a lieutenant, sell himself into a favorable job in Chungking, then cross over to the United States and there try and make good. It was a youthful dream. However, the reality in China was a cold resentment and a deep revulsion against the Japanese. He told me that when a code book, covered with blood, had been recovered from a Japanese plane that had been shot down, and it had been brought to him with the strong injunction to decipher it, he had thought of preventing it being unraveled even at the cost of his own life.

Many thousand Japanese prisoners of war had been sent to Chungking. One third of them had died of malnutrition Men who showed any signs of resistance were shot down in cold blood on the spot, or starved to death, their feet chained together in prisons, located in isolated mountain caves. Some managed to escape and reach the vicinity of Ichang, only to be caught and thrown alive into the Yangtze River.

After the war, the Japanese Prime Minister was easily moved by the Generalissimo's "reward hatred with virtue" broadcast and when prisoners were sent back from Chungking, he issued a statement thanking China for its humanitarian treatment of Japanese. The Prime Minister's statement was reported in Chinese papers, which coldly criticized it as "the crafty speech of Japanese premier." There was, I felt, reason for this criticism on the part of the Chinese.

Japanese diplomats, who thought that Japanese-Chinese relation, would immediately improve after the end of the war, freely expressed to the press cheap and unmeant flattery of China. All the while, youths captured and sent to Chungking were storing up frightful observations of what they saw and of how they were treated. They also found out what Wataru Shikaji and Kazuo Aoyama did in Chungking and what treatment they were accorded.

The Chinese openly showed that they did not like me associating with Fujita. However, for two months after our first meeting, I lived with Fujita, talked with and strolled together with him. My surroundings were such that I was comforted even to hear the Japanese language spoken. One day the kitchen duty soldier neglected bringing me my food. Fujita asked for the food. To this, the soldier insulted Fujita and said:

"You are Japanese. You take him the food."

Fujita lost his temper and struck the Chinese soldier. The incident created a tense atmosphere which might well have developed grave consequences. Fujita had seen time and time again in Chungking Japanese soldiers being shot on the spot for hitting Chinese soldiers. But, righteous indignation had so seized Fujita that he had forgotten himself and had struck the Chinese. This was sufficient to indicate the atmosphere in which we lived at that time.

I went for strolls with this lad around Lake Hsuan Wu, and on 14 numerous shopping trips. I felt my love for Japanese, even the worst type of Japanese, was far greater than for a foreigner of the best type. This was particularly so in the case of such a pure-hearted youth like Fujita.

Through Fujita's introduction I met a Sergeant-Major Mori (an alias). He had been wounded in the front line in Central China and had been taken prisoner. His prisoner's life had already totaled three years. In contrast to Fujita, Mori had become one bundle of complexes.

One day, while visiting Lieutenant-Colonel Chung's home, I accidentally met Interpreter Ikeda. Formerly employed as an interpreter with the Japanese Embassy in Nanking, he was detained in the services of the Chinese Nationalist Government at the end of the war. He lived with his family, composed of his wife and one child, in one room rented in a Chinese home.

I felt great pity for their 13-year-old child who could not attend school. But at the same time, I felt envy at seeing the warm atmosphere: of their home life, with one family altogether under one roof. There was one Japanese girl staying with the Ikeda family. She, I heard, had been the mistress of a certain Japanese officer. Ikeda often had to go on business trips to Hsuchow, leaving his wife and child at home. Fearing that there might be possible raping and looting by Chinese ruffians, during his absence, I often visited his home to see that all was well. I was surprised at my yearning and love for fellow Japanese. While a handful of Japanese sought each other's company in far away China, the Japanese at home in Japan were constantly fighting and opposing each other, one group engaged in stool-pigeon activities against the other. I could not understand why such had to be.

It was the last week in August. Clammy with sweat from the steamy heat of the day, I was washing down my body when Lieutenant-Colonel Tung came over to give me a message. He said that the higher-ups of the Second Section of the Department of Defense were coming to see me. In my room next to the kitchen, I sat waiting on my bed (no chairs). The group immediately arrived on jeeps.

The group was composed of six men, all nuclear leaders, headed by Major-General. Kung, acting head of the section during the absence of Lieutenant-General Chieh in Peking. They seemed surprised at the bareness of my room, with its single, broken-down bed and without even a desk or a chair.

The majority of the visiting group at one time or another had studied in Japan and for roughly one hour I spoke on the future Japanese-Chinese collaboration. However, they did not seem to have any interest in this. All they were interested in was the collection of information and intelligence.

During our conversation the name of Lieutenant-General Tsuchida (an alias) cropped up. He had been Chief of Staff at the end of the war in Shanghai and since he was an expert on Soviet Russia, I concluded that he had been detained and was in the employ of the Chinese Nationalists. I asked about [Colonel Takushiro] Hattori, and learned that he had been sent back to Japan by plane in May at the request of the American authorities. I asked whether he had been designated a war criminal. If so, I said, I could not stay idle like this in China. I should present myself and take over the whole responsibility for any of the crimes attributed to him. Major-General Wang, however, told me that Hattori had been called back to collate data and information. At this, I calmed myself.

According to Staff Officer Tung, the Department of Defense had been deeply impressed by my report on observations of a possible Third World War, a report submitted early in August. My guests left me whiskey and tobacco as gifts. Whatever the case, I guessed that there would be work for me to do in the near future.

It was soon after this that Colonel Chin was appointed to take the place of Lieutenant-Colonel Tung as my mentor and companion. A graduate of Waseda University, Chin spoke fluent Japanese. Despite the fact that Chin was evidently a highly prized officer in the Military Control Bureau, he was an open-minded and extremely pleasant youth. I was able to speak openly and bluntly to him and we argued for hours over the futures of our native countries and the corruption of China. His father had also studied in Japan and was at that time President of the Transportation Bank in Chenkiang.

Colonel Chin felt no attraction for money–a rare characteristic for a Chinese. However, he was extremely sloppy. In the mornings, he woke around nine. He also wore his clothes for a month without washing them. His bed was never made and was always mussed up. He spent every spare moment reading novels. His duty seemed to have been liaison and surveillance. However, he never followed me. He stated bluntly:

"My duty is to keep surveillance over you and maintain liaison between you and the Defense Department. However, I respect your character. Please go out by yourself, whenever you so feel, and stroll about."

I realized that behind these words was his determination to shoulder all responsibilities should anything happen. He was as different from Lieutenant-Colonel Chung as was the mud and the clouds. I was surprised to find such a person, though rare, among the chronically suspicious Chinese people.

We lived together for roughly two months and I never met such a pleasant man. That winter, Colonel Chin married. He opened for his wife a small tobacco shop, with a street front of only six feet, on a corner on Shenshi Road. He boasted that it was the smallest store in China, but the profits from that store came to twice the salary of colonel. I realized why in such a country a good man never entered the military field. The majority wanted to take off their uniforms in some way or the other and return to civilian life. However, once you became a member of the Military Control Bureau, it was no easy task to get out despite the fact that since his father was an affluent banker without any money cares, Colonel Chin hated having to sponge on his father.

The building in which our kitchen was situated had been confiscated by a certain high official in the Defense Department, whose wife often came around on an inspection tour.

An old captain, risen from the ranks, was this high official's aide and was custodian of this building. The wife was a rude woman, who did not even acknowledge the polite bow that I made. She came into my room without knocking and stared at me with contempt in her eyes as if to say:

"You damn Japanese prisoner."

I sat on the bed and glared back at her.

Another aide was a thin-faced lieutenant. He was the epitome of corruption. Thanks to the kindness of Ikeda, I had managed to borrow a desk from the Liaison Office of the Supreme Japanese Headquarters. The aide, as soon as he saw this desk, ordered one of his soldiers, to paste a label on it marking it as belonging to the government. Outraged at this, I stoutly refused to let the captain carry out his intentions.

"This is a desk I borrowed from Ikeda and it does not belong to this unit."

"What in the hell are you saying. Japanese can own no private property. Everything belongs to China."

"Take it away, you fool!" I shouted, completely forgetting myself for my anger.

The captain also was excited and with trembling lips called the guards from the window. Four or five soldiers, with fixed bayonets, rushed into the room. The atmosphere was tense and a spark would have set off an explosion, when Colonel Chin came in and saved the day.

Colonel Chung's unit occupied the former site of the monitoring branch of the Japanese Army and was engaged in intercepting the messages of the Chinese Communists.

In the vicinity were the former billets of minor Japanese officials. The Second Section of the Department of Defense designated one of the rooms in these tenement houses as my quarters. It still had its straw mats. However, the wolves of Chung's unit would not obediently carry out the orders of the Second Section. They immediately occupied the room, before I could make a step with dirty shoes they tramped over the "tatami" (straw mats) and brought in beds to accommodate four or five men. It was only late in August that I was finally able to claim the quarters as my home, after a herculaneum house-cleaning task, for the soldiers had spit everywhere and had trampled all over with dirty shoes.

It was a sunny eight-mat room (12 feet square) and I immediately began my work. However, the wily skunks could not resign themselves to seeing this fine room occupied all alone by Japanese. They viewed me with jealousy and with hateful stares. However, I had lost all feelings of sympathy toward them by this time, and did not mind the feelings of these people.

A young and innocent-looking 16 or 17-year-old soldier was the orderly for this area in general. Every day he would bring me my food. One day, I awoke to find missing the pair of shoes that I had placed inside the shoe box the night before. I realized that it had been stolen only by someone from nearby. I was unable to go outside, even to wash my face. I asked Fujita to go to town and buy me a pair. He returned with ones that set me back 40,000 yuan. At the same time my shoes were stolen, several million yuan was also found missing from Colonel Chin's suitcase next door. There was no doubt that the culprit was someone inside our unit. An investigation revealed that the innocent-looking orderly was the criminal The aide with whom I had quarreled over the desk threw the youngster into detention in a nearby shed and locked the door from the outside.

Night and day I could hear the pitiful crying of this boy. I asked the captain through Colonel Chin to release him but the aide refused obstinately. Perhaps he had forgotten that he himself engaged in still greater thefts, or perhaps he wanted to make out as if he were a saint. He insisted that evil-doers must be strictly punished and left the boy for four or five days as if he had forgotten all about him.

Later the youngster's father was called out. The boy was brought out and forced to sit on the ground, his arms tied behind his back. Then in front of the whole unit, the officer whipped the boy, until the father paid up several hundred thousand yuan as indemnity. Only then was the boy released.

However, not a yuan of this money taken from the father was given to the actual persons who had suffered from the theft. It all went into the captain's pockets. The young lad, on leaving the place with his father, stared with stark hatred at the aide.

Late in September, five major-generals, Wang, Huang, Ma, Lin and Chin, all wearing their uniforms visited my eight mat rooms in the former billets of petty Japanese officials. They were all men who had studied in Japan and none had as yet reached the age of 40. However, unless I addressed them as " Your Excellencies," they showed displeasure on their faces. The orderlies, the sub-lieutenants and the lieutenants seemed surprised at this sudden invasion of my room by general officials. I realized that someone from this group was going to work with me.

General Huang stuttered slightly, but looked just like Japanese. This was perhaps because he had studied in Japan from childhood, but more probably because his mother was Japanese. After our conversation, was treated to my first real dinner with the group in Colonel Chin's home.

The next day, Major General Huang came alone. He told me that a Third Research Branch, composed mainly of Japanese, had been organized and that it would begin soon on the work of gathering and compiling information and intelligence. Major-General Huang was the chief in charge, Major-General Wang the deputy-chief in charge, while Lieutenant-General Chieh was nominal head of the branch.

My first assignment was the compilation of "Reference Material for Operations in Cold Zones." Since the 400,000 soldiers from South China, now occupying Manchuria, would be experiencing their first cold winter, the wish expressed that I hurry so that the instructions would be in time for the coming winter. Recalling the experiences of my days as staff officer with the Kwantung Army, I wrote out in one night roughly forty pages of instructions and the next morning had the work delivered to the home of Major-General Wang. This was my maiden work in my new position in the Department of Defense. The Chinese were astonished at the speed in which I delivered my first assignment. Hounded by Generalissimo Chiang and Chief of Staff Chen Cheng, my work was translated word for word, without a single revision, immediately printed in the name of the Defense Department and issued to front line troops.

Since I had written the pamphlet, laying prime stress on detailed precautions to be taken in meeting the cold Manchurian winter in a manner understandable to the middle and lower echelon officers and men, I felt that I was able to save many soldiers from frost-bites and freezing limbs. 3,000 copies of this pamphlet were flown to Manchuria and were in the hands of the various units by the end of September. After that, one assignment after another was given me, on plans for the collection of information and on measures to counteract espionage activities. All these assignments I completed at a night's sitting. Soon later I received a message saying that Deputy Chief of Staff Liu would like to meet me.

Accompanied by Major-General Wang and Colonel Chin I set off to meet Lieutenant-General Liu at his home, the former residence of Chief Public Information Officer Mishina on Chih Pei (Red Wall) Road. I found the General a man over 50 years of age, tall of figure and oval of face. His first words of greeting constituted a question regarding the health of to my family. This query moved me greatly. I realized that even among the Chinese, there were still generals full of human feelings and capable of understanding the sentiments of another human. Next, he said:

"It is indeed regrettable that Japan, which had prepared for years for defense against the Soviet Union, should have mistaken its direction and advanced south."

I told him that it was doubtful whether Japan and Germany, attacking from the east and the west, could have forced the Soviet Union to surrender. However, if the Japanese Army, I continued, had fought against the Soviet Union to its heart's content and then had lost, the Japanese would not have felt regret at defeat. In addition, he listened intently to my opinions on future Japan-China collaboration. Thus, I felt that I must not let this opportunity slip by and touched upon the question of Japanese war criminals I stressed that war crimes should never be placed on the shoulders of designated men but should be borne by all of us officers. I said:

"In order to atone for even a part of this common sin of ours, I am willing to give up my life for the sake of cooperating with China. Please, therefore, strive to lighten the crime of all the war criminals."

I pointed out that a certain commander was my father-in-law, a certain colonel my dear friend and a certain other person my brother-in-law said that I could not stay placidly in the Defense Department watching these people lose their lives on the execution stand. I said that if it were necessary I would gladly give up my life and in return, I hoped that he would try and forgive the sins, of these people. The instant he heard my heart-felt pleas, he showed slight pain on his face. At any, rate, he replied that he would try his hardest and we parted after a firm handshake.

Deputy Chief of Staff Liu had studied in the Infantry School and the War University in Japan and had made a thorough study of my country. However, despite the fact that he was friendly toward Japan, he was during the war the man responsible for military operations against Japan. General Liu was greatly loved by the nuclear officers of Chungking and was a general most deeply trusted by the Generalissimo. There was no mistaking the fact that he was regarded as one of the treasures of the Defense Department. When the tide of war pressed hard against the Chungking Government and it faced imminent collapse, he was chosen to head a peace mission to negotiate with the Chinese Communists. Together with Lieutenant-General Chang Chih, he flew to Peking but failed to return to Nanking. Newspaper reports stated that the two generals had joined the side of the Chinese Communists. However, none knows whether they were forced to join the Communists or whether their actions were voluntary and planned beforehand.

With his composure and with his far-sightedness, he might have known sooner than anybody else that the Chinese Communists would eventually govern the whole of China for a fairly long period. Perhaps, out of this realization, he might have gone over to the side of the Chinese Communists, without a worry over possible slander or a thought for reward, and only for the sake of saving the 400,000,000 people of China from bloody civil war.

Leaving aside for the moment all criticisms of his actions, I do not want to believe that he was a man who would have sold his principles for the sake of personal gain.

During my 100 days in Chungking, I carefully studied every day magazines and newspapers, staring all the while at a map of the world, and the net product of all this thinking and observation was a work entitled: "Observations on World War Three." After my arrival, in Nanking, I gathered further material and in a corner of Lieutenant- Colonel Chung's room, I completed my manuscript, a total of about 100 pages. I drew a map of the possible strategy of both the United States and Soviet armies and also touched upon factors in a war in East Asia, as part of the over-all picture.

It was August 2 when I submitted my report in poor Chinese.

On September 30, an automobile came to call for me to take me to the Defense Department. In the large lecture hall, I found my manuscript beautifully rewritten and posted in entirety on the walls. Roughly 50 higher officials, above the position of section chiefs, were gathered to hear me lecture directly. The officers were of all rank from full generals to full colonels. I was overwhelmed at the gathering of so many of the elite. At the beginning, Major General Wang Liang interpreted but part of the way through, he became stuck and Colonel Chen took over.

This was the gist of the lecture that I gave:

1. Preamble:

This study is a common sense evaluation of a possible World War Three, based mainly on systematic observations of public material in newspapers and magazines over a period of roughly four months since my arrival in China and on my own past experiences. It is not altogether free of criticisms of containing paper gossip. However, would be happy if it should be found to contain suggestions that might be of value to you.

2. Observations on the date of World War Three:

World War Three will occur sometimes between 1950 and 1955. The reason for this is that the Soviet Union will complete its fourth Five Year Plan during this period and United States aid to West Europe will bear fruit about this time. The Soviet Union will also succeed in making the atom bomb itself.

3. The Nature of World War Three:

World War Three will be a long-term affair. The reason for this is that the Soviet Union will have completed its measures to cope with atom bomb attacks by the United States, with Soviet industries dispersed all over their country, and important strategic resources stockpiled in huge quantities. In addition, the Soviet Union will be favored by General Winter and with the inborn resilience of the Soviet soldier. On the other hand, America's industrial strength during this period will far outpace that of the Soviet Union. However, its industries will be concentrated in large cities, and in the event that the Soviet Union succeeds in manufacturing a considerable number of atom bombs, the industries of the United States will receive a mortal blow at the very outbreak of war.

However, it will be impossible for the Soviet Union, no matter what the situation, to land on the homeland of the United States. Neither will she ever be able to make the United States surrender. In such a case, this war will last from four to five years and both sides will be grievously wounded.

4. The position of the Far East.

The fate of both the United States and the Soviet Union will be decided in Europe. The Far East will be a secondary battlefield and victory or defeat in the Far East will not influence the general trend of war.

The armed forces which the Soviet Union will use in the Far East will be less than 30 divisions at the most. This strength is limited by the capacity of the Trans-Siberian railway to keep this army supplied. This limitation also arises from the fact that the Soviet's ability to supply her own needs in the Far East is insufficient.

The above is based on the premise that the Chinese Nationalist Government maintains its hold on Manchuria, and that the United States is able to procure on the Chinese continent, land, sea and air bases for its attacks on the Soviet Union.

However, in all probability by then, the Soviet Union will come to support the Chinese Communists and seek to obtain control of the whole of China. If not, she would plan control of all that part of China north of the Yangtze River, maintaining a defense line linking Outer Mongolia, North China, Port Arthur and Dairen and North Korea and creating thereby a vast buffer zone to protect her own territory.

When the Soviet Union succeeds in this, she will limit the troops to be used in the Far East to a minimum.

5. The place of China:

China should immediately solve its Chinese Communist problem. The objective should be, accede politically to the requests of the Chinese Communists as magnanimously as possible and militarily to prevent the defection of troops.

More important still, the Chinese Nationalist Government must purge itself politically and maintain the trust of the people. It should immediately and sincerely tackle the land problem with an eye to a quick solution and should accomplish a social revolution to stabilize popular living. Through this, China should strive to prevent the vast agrarian population from running over to the Chinese Communist camp.

I cannot but voice the bitter counsel to you that the political corruption of the Nationalist Government is the hotbed of Chinese Communist expansion.

It took me three hours to explain just the gist of this comprehensive and vast study. Despite the fact that not a single minute of intermission was given, the listeners busily jotted down notes. When I finished, Vice Minister Chieh Chung gave an address.

When I look back today, my lecture was nothing more than plain common sense. However, the China of 1946 had not as yet awakened to reality and was still, drunk with victory. There was hardly anybody that gave detailed thought to the possibilities of a Third World War.

This work might have only been a dream, but because it was a dream, I poured heart and soul into its completion. It was the fruit of four months of study. I had no way of judging that or how great an impression my lecture made on my audience. However, the map that I used to outline the probable strategy of the two countries and the report presented seemed to have been kept with care. Until the very end, the report did not come back to my hands.

Lieutenant-General Tsuchida (pseudonym), who had been detained in Shanghai for service with the Chinese Nationalist Government, came to Nanking around the end of July. I was not allowed for some time to meet him, but early in September had a chance to become acquainted for the first time in his gorgeous official residence on Taping Avenue. The place seemed like paradise to me in comparison with my room in the dirty tenement house for minor officials with its phlegm and spit scattered all over the place. I had never worked together with General Tsuchida in the past and so at the beginning I did not seem to be able to get close to him.

But, as I talked with him, I found him extremely interesting.

On the morning of October 18, Major General Wang Liang came to call me on a jeep. He told me that the Third Research Group had opened up a new shop on Chien-yeh Road. I felt an overwhelming sadness at having to part with Fujita, with whom I had lived together for nearly two months. I also felt sadness at having to part with Colonel Chin, with whom I had spoken to as man to man. Our new offices turned out to be the residence where a certain gendarmerie colonel, chief of the Special Mission of the Japanese Army in Nanking, had kept his Chinese mistress. Here again was exposed to the eyes of the Chinese a facet of the corruption of the Japanese Army of the past. Next to this building was the Central Political School, but the surrounding were crowded with small and dusty houses and it was by no means an ideal environment.

The Chinese staff to which I was introduced was composed of Staff Officers Chang and Chao, both of whom had studied in Japan, five or six interpreters, one typist and roughly 20 clerks. It was a fairly large family.

There was one Japanese on this staff. To my surprise, this man was Captain Samura (pseudonym), who had been involved in a forced landing, together with Captain Okino of the Japanese Navy, during a trip from Hankow to Nanking. Samura had been wounded in the head, captured by the Chinese and taken to Chungking. A graduate of a higher normal school, he was a third grade judo expert. He talked to me in detail of both the dark and the bright side of life in Chungking, as seen during his three years as a prisoner. It was valuable data and helped clear in my mind the realities of such wolf-in-lamb's-clothing expressions as reward hatred with virtue and the "Emperor Concept" of the Kuomintang.

In the indoctrination program for Japanese prisoners of war, the overthrow of the Emperor was clearly indicated, signed and sealed. After the war, it was only the simple Japanese that greeted with joyous tears the broadcast:

"The fate of the Japanese Emperor system should be decided by the Japanese people themselves." The intellectual class of China believed implicitly that the Japanese Emperor would meet at the hand of the Japanese people the same fate as the German Kaiser and the Romanoff family of Imperial Russia.

Samura had tried several times to commit suicide in order to wipe clean the blot of having fallen prisoner, but each time he had failed. He protected Captain Okino by sacrificing himself and was flown from Chungking to the United States and back again to Chungking. From Chungking he had come to Nanking. His experiences and the observations that he made during these trying journeys were of extreme value,

The man in charge of the Third Research Group, Major-General Huang, was a cultivated man, who wrote poems in the Japanese language. However, he was a Chinese who was poor at Chinese. He also liked oil paintings, and every day while gazing out of the corners of his eyes at the typist, he would paint away with his brushes. He was a relative of General Ho Ying-ching. However, in a Defense Department whose key positions were all held by members of the Cheng faction, his background was a liability. He was a likeable fellow.

The room that was allotted to me was a small six-mat chamber that had once been used by the maid. It had only one small window facing west. Even during the daylight hours, it was dark and the rendezvous of mosquitoes. I was forced to sleep under a mosquito net until the end of November. It seemed that this room had once been occupied by a Chinese who had died of tuberculosis, and since nobody wanted the room, it had been given to me. However, I resolved that I would not complain about food, clothing or shelter. When I thought of the number of Japanese families whose homes had been burnt during the war, and who could not have even such a room as I, I knew that I was not in a position to be pampered.

From morning to night I sat at my desk and tried to rely on all that was in my old brains. I had no reference material, yet I was asked to write a basic manual for training in modern warfare. All I could rely on were what had been left unburned by the Japanese Army and ordinary publications. The ideas in these book and documents had to be studied and coordinated. Luckily I had spent the greater part of my military life on the battlefield. Thus, my experiences in fighting soldiers of various countries became the storehouse of my knowledge. Every day I wrote from 20 to 30 pages. This was translated into Chinese by five or six translators, typed on stencils by two typists and mimeographed by several clerks. Yet, they failed to keep up with me and the Chinese submitted the manual, one or two months after I had completed it.

Thus, all my opinions on immediate problems lost their timeliness and only the basic material was of any use.

The main part of my work was to compile reference material on strategy and tactics, and on the use of strategic topography in the disposition to troops. In particular, I was asked to compile a treatise of the strategic use of topography of the outlying parts of China.

The fact that China had to ask a foreigner to compile such fundamental military reference material was indicative of the general inefficiency and powerlessness of the whole Defense Department.

Early in December, young Major Kanda (a pseudonym) accompanied by his wife arrived in Nanking from the Kwantung Army as a detainee in the employ of the Chinese Nationalist Government. He was weak with tuberculosis. He had been in Manchuria at the end of the war and all his worldly belongings had been stolen from him. Despite the fact that it was already winter in Nanking, he arrived in a ragged summer uniform. The bone-piercing cold had accelerated the progress of his sickness. His wife, however, was a fine and strong-willed woman, very beautiful to boot. She had been a teacher in Mutankiang in Manchuria when the war ended. She was a person that had managed to escape with her life after having gone through a number of amazing experiences. While nursing her sick husband, she busied herself cooking miso soup and making Japanese pickles for the few Japanese gathered there. Her kindnesses helped lighten my solitude and encouraged me greatly in such gloomy circumstances.

She told us hideous memories of her first encounter with the Soviet Army. [She is referring to the August 9, 1945, invasion of Manchuria by the Soviet Red Army. The Kwantung Army surrendered to the Soviets on August 20, 1945. The surrender lead to Soviet forces committed widespread looting, rape, and indiscriminate killings of civilians.] Russian soldiers had attacked the girls high school where she had taught and had raped several dozen girls there. The girls eventually decided that death was better than this humiliation and all committed suicide in a body, drinking potassium cyanide. She also revealed that after she had been moved to a concentration camp, she saw a newly-married bride raped in turn by several Soviet soldiers in front of her husband. The bride then changed to the ceremonial clothes she was wed, covered herself with gasoline, struck a match and then burned herself to death.

Such acts, at which even beasts would turn aside their faces, were carried out in broad daylight in all parts of Manchuria, and in the sight of many people. Those who resisted were immediately shot to death. Marxism views men as machines and is not surprised at the possession of women as common property. Perhaps, this doctrine lay at the root of the actions of Soviet soldiers.

The true characteristics of the Soviet soldier were to be seen in their view that it was no crime to rape women in broad daylight and in front of people and in their imperviousness to any sense of shame at such actions.

The Chinese Nationalists saw that Major Kanda could never work for long with them because of his poor health, and the Defense Department also agreed to his return. On December 23, as the year drew to a close, I said farewell to the young couple. We had a simple lunch together and I saw them off to the gate with tears in my eyes at the forlorn sight they made as they left.

Kanda was a native of the town: of Matsumoto in Nagano Prefecture. I wonder where he and his wife are today and how they are faring. Mrs. Kanda sent me a letter, dated December 28, and written on the pier in Shanghai. It came to me through the Chinese mails, one month later. One meeting and one parting–the old Buddhist saying:

"You meet to part."

Samura also petitioned for his return to Japan. However, approval was not forthcoming immediately. As long as the Chinese Nationalists felt that a man was useful to them, they detained and used him. I could not but suppress a feeling of displeasure at this attitude of disregarding the feeling and the hopes of the individual.

When Samura was finally allowed to go home, it was March of the next year.

The code designation of the Third Research Group was "The Bamboo Shelter." It was a name reminiscent of a smart Japanese-style restaurant. It was fully worthy of Major-General Huang, who had chosen it.

The Chinese who were gathered in the "Bamboo Shelter," from major-generals to service soldiers, were practically all very bad-mannered. During meals, a certain colonel would blow his nose on the floor with his hands, not to speak of spitting out phlegm. When it came to eating, even this colonel would .become a completely different man, the very color of his eyes changing, as he crammed down his food. Thus, practically all the subsidiary dishes on the table would be completely gone by the time he was ready for his second bowl of rice.

"Chih Kuo La Ma," (have you eaten?) has become a common every day form of greeting in China. You begin to understand why this is so, after you have lived for a time in the country. The greatest task of the day seems to be to fill your stomach fuller and faster than the man next to you.

Everywhere, I encountered in the words of the officers and soldiers, in the streets and in the shops, the remnants of a feeling of hatred and contempt for the Japanese. But, if this one Japanese had escaped or had disappeared from the "Bamboo Shelter," all would have lost employment. Thus, there was none here who complained directly in my face, or expressed any anti-Japanese sentiment.

Staff Officer Nieh was newly appointed to our group. This was the first time that he had ever consorted with Japanese. During our conversation at our first meeting, he referred time and time again to Japan as "Erhpen," whereas "Jihpen" was the usual term. I wondered why he did this and after some questions, I discovered that "Erhpen was a widely used term of derogation during the war.

During the several I months that I spent with the Chinese, eating and sleeping with them, I discovered that they were very weak in military efficiency, but very good at amusing themselves and in making money.

In China, this was a common saying:

"If a man has served as intendance officer or accountant for three years, execute him immediately without any investigation." I found that this was very true. I noticed one major general set up another ghost of a major-general, and then receive two man's salary, stamping the ghost major-general's receipts with a counterfeit chop. On other occasions, I noticed how the salaries of the entire staff were received in advance, by the paymaster, but was held up and loaned to merchants at very high daily interest and then paid out to the staff at the very end of the month. It was a common practice among soldiers and officers to connive with merchants and cheat on the prices of goods purchased with official funds for official use.

Although the Chinese did little work, they were not fools when it came to women. There was a new typist who had not only a beautiful face but a beautiful form. Her name was Wu Teh-ying.

The faces of the major-generals, colonels, lieutenant-colonels and all the men were turned away from their work and concentrated on watching this young typist. In answer to her request for autographs, everyone from Major-General Wang down was soon absorbed in writing. With beautiful expressions and lovely phrases, they tried to win her interest. Finally, she came around to the Japanese. In clumsy calligraphs, I wrote her the following cynical poem:

"The Bamboo Shelter greets a branch of blooming plum, which multiplies ten million times the volume of the work accomplished."

Soon after, a young girl, Yang Yao-lan, was added to the staff as draftsman. She did not use any makeup and her clothes were simple, yet she was endowed with a natural beauty that outshone all the other girls. Suddenly, she became the center of attraction. A certain staff officer accompanied her to and fro from work and tried to make a conquest, another staff officer sent her love letters, while a third asked for her hand in marriage, only to be rejected. All the while, young Wu grew jealous and more jealous and put more and more makeup on her face, in a futile attempt to win back the attention of the men folk. Nobody seemed to be doing any work. However, in their attention to women, they were absolutely without peers.

One day, I became engrossed in my work and forgot about supper. I was ten minutes late in going to the dining hall. When I got there, the soldiers on kitchen duty were wolfing down what food had been left. I was thus forced to go outside. I went out alone, had a bowl of Chinese noodles, dropped in at a barber's and came home in a leisurely manner.

"He's come back! He's come back!"

There was shouting and commotion as I entered.

Major-General Huang came in with a pallid face and indignantly told me:

"Mr. Tsuji, why did you go out?"

"I was so engrossed in work I forgot all about supper. I couldn't very well go to sleep without something to eat...."

It was forbidden me to leave this office alone. Nominally, this was for my, own protection. However, it was to maintain strict surveillance of my action.

At any rate, Major-General Huang apologized and went home.

Later, Mrs. Kanda, I heard, had said: "You can't get anything on Mr. Tsuji. You simply can't eat him, whether you boil him or fry him."

I knew that her comment hit the mark. My meat was all sinew and my body full of shrapnel and bullets, which would have broken the teeth of anyone trying to put one over on me and "eat" me.

One day, one of the orderlies brought home one small baby pigeon that he had caught. He hoped to fry and eat it. I felt so sorry for the bird, I bought it for 2,000 yuan. Then, I made a little nest for it in my room and fed it. One week later, when it became used to me, I, let it go. However, morning and night, it would return without fail and ask for food. It was perhaps a reaction to my hateful surroundings that I felt such an attachment to this small bird.

Major-General. Huang had once served as commander of a carrier pigeon unit. He took advantage of his acquaintances and brought over a pair of pigeons to be kept in the Bamboo Shelter. Soon, the female laid an egg and brooded over it in turns with the male. Several days, later, after the female had been exercised, it was returned to the box and the male taken out. However, a stray cat tried to catch it and in fright it flew away. The female kept warming the egg for four or five days, without eating anything, as if lonely at the departure of its mate. Several days later, it too ran away somewhere, leaving behind two unhatched eggs in the box. I wondered if mother pigeons had a stronger attachment to their mates than toward their young. I took the two eggs, abandoned by the mother, and carried them around in my pocket trying to keep them warm in the hope that they would hatch. However, nothing seemed to come of this, and I dug a hole in the back yard and carefully buried the two eggs.

The Chief of the General Affairs Division brought over a new pair of native pigeons. They had black spots on their heads and on their tails and seemed like hybrids between a crow and a pigeon. They were not pretty at all or loveable.

Major-General Wang was transferred to the Chinese Mission in Tokyo and Major Generals Yeh and Niu were appointed in his stead. Both were youngsters of around 36 or 37, and although they were ten years younger than myself, they would show displeasure if I failed to address them as "Your Excellencies."

Major-General Yeh's father was one of the elders of the Chinese Nationalist Party and belonged to the pro-Europe-America faction. Major-General Niu belonged to the faction composed of men who had studied in Japan. Both were influential and up and coming nuclear members of the Defense Department. Niu looked after the clerical side of the Bamboo Shelter and Yeh the actual intelligence work. As the old Chinese proverb states: "A snake with two heads cannot undertake a long travel," this combination was not to last long.

As a result of the changes in the heads of our group, our staff doubled in size, while key men were all changed. People having ties, either by blood or by virtue of a common hometown, came seeking work in droves. It was the custom in China for the lower echelons to be changed whenever the heads changed. It was not a question of hiring people to do a certain work, but one of making a job for certain people. This special characteristic of the Chinese officialdom cannot possibly be changed quickly even under the Chinese Communist regime.

The Third Research Group was a miserable intelligence unit. However, the, post of group chief was held concurrently by Lieutenant-General Cheng Chieh-min. Following the death of General Tai Li all confidence and hopes were placed on General Cheng as a possible successor and the consensus was that he was highly trusted by the Generalissimo. I had long expected him to come around on a tour of inspection, but it was not until six months after our agency was established, that he appeared in the last week of April. I realized that the appointment of such a busy and important man to the concurrent post of head of our agency was not because the agency was regarded as being that valuable, but because it was easy to obtain a large budget for such a type of work.

He was also concurrently head of the Military Mediation Department in Peiping, an on-the-spot mediation organ for disputes between the Chinese Nationalists and the Chinese Communists. I had read in the papers how he had escorted with great deference officers and soldiers of the Communist Party that had been killed and wounded in a certain clash with the Chinese Nationalists. When I read the item, I said to myself that here was a rare man of character. Thus, I had placed no uncertain expectations upon him. In addition, since he was a fellow countryman of General Hsieh, as well as being a bosom friend, I had felt an additional sense of friendship toward him. However, on his first inspection tour, he gave an address of instructions to the entire personnel, and asked us to expend our entire efforts in the compilation of a manual on the strategic topography of Siberia. When I heard this, I felt a sharp pang of disappointment in the man.

It was the heights of foolishness to be still dreaming of fighting the Soviet Union on Siberian territory. On the other hand, if he had planned to sell this information to the United States in exchange for some remuneration, such a work would be miserably inadequate.

At a time when a half year was being spent vainly in translating and printing the manual on the strategic topography of Siberia, published by the Japanese Staff Headquarters in 1924, the war situation was hourly becoming more critical. Yet, there were no works on the stratetopography of Manchuria–a work needed at that time more than anything else. I could not but be struck by the waste of six months of valuable time and effort in a work of no consequence.

After he had finished the address, he gave me the opportunity of meeting him in private. His first words were:

"The reason why Japan failed was because she lost the trust of the natives of Manchuria. It was because she was engrossed solely in preparations for war and did not think of ameliorating the livelihood of the people." How different were these words from those of Deputy Chief of Staff Liu.

The Military Control Bureau was abolished hand in hand with the establishment of the Defense Department. The greater part of the Bureau was absorbed into the Second Section. However, a part of the crack elements of the Bureau were organized into an extra-Departmental organ, known as the Secrecy Maintenance Bureau, whose existence was known only to a few. Then, in great secrecy, the majority of these men went underground. The chief of this bureau was General Chieh, this was another concurrent post. Whenever commodity prices jumped in Shanghai, he was reported as having gone there.

The loss of General Tai Li to the Generalissimo was a great blow. The caliber of the men who were chosen as Tai Li's successors amply emphasized this great loss. The men left around the Generalissimo were such that they could not halt the onrushing disintegration of the Chinese Nationalist Government and Army.

One of my daily worries was how my fatherless children were faring. I often wondered how my wife could keep the home fires burning in a home that did not exist. I did not know even where they were and how they had managed to live through the past several years. They did not have a single cent in savings and had not a single yen of income. I often wondered if they had not all starved to death already. Or perhaps, the whole family had ended their lives in mass suicide. One of the reasons, I was able to forget my family and sacrifice my past 30 years in official work for the sake of the Emperor and the sake of my country, was because of the guarantees given me by the state. Lulled by this sense of security provided by the state, I had failed to provide adequately for the future of my family. Time and time again, I had chided myself for my lack of will power and my attachment to my family. However, this was a human failing, a human weakness and a human sadness for one's thought to turn constantly toward his family. In August of the year before, the Chungking Government had guaranteed me liaison between myself and my people in Japan. However, since then I had not heard anything further, neither from the Chinese nor from my family. Yet, I had not seen any bad dreams and I comforted myself with the thought that they must still be living somewhere in Japan.

I had no idea when I would be able to return to Japan. I was also a criminal who knew not when or where he might be called upon to give up his life. I decided to write down the steps which a father had taken in 40 years of life and leave this record to my children so that they would not have to repeat the same mistakes which the father had made. I began writing from my childhood memories, and completed five volumes by the end of January 1947. The five volumes covered *"My First Aspirations," "My First Taste of War," "Bitter Battles," "Good Battles"* and *" Evil Battles."* I spent 70 odd days in this autobiography of more than 400,000 words. It was a book of confessions and at the same time a book of precepts for my offspring. I wrapped and sealed the document and sent it to my family in the care of Staff Officer Chao who was leaving for Tokyo in the company of Major-General Wang. In the 400,000 words were enclosed a spirit of martyrdom for my country and my concept of humanity. I read over this document written in blood and tears and prayed: "My children proceed with courage and with justice."

My whole heart went with this document.

At the beginning of March, I was lying in my semi-dark room that had once belonged to the maid, sick with a high fever–the result of overwork and fatigue. Suddenly, Staff Officer Liu called on me and said:

"Congratulations. A letter has come for you from your home."

I thought that I was seeing a hallucination as I started at the letter which had already been opened and censored. It was written in pencil in childish characters, but I immediately recognized my eldest son's handwriting.

"Please forgive these long years of my unfilial neglect. We are all healthy and well. Eiko is going to Shimizutani. I have quit school and through the kind efforts of Mr. Kawamichi am working in the Kimura Food Store. Mother is doing dress-making...."

Oh, my children were all alive. The gods and Buddha had protected the children of this sinful father and the wife of a sinful husband. My family had somehow managed to keep alive, without a single one dying. My wife and my children had suffered and persevered with desperate effort. My wife was working during the day in a clothing factory, treading a sewing machine. At nights she was attending a dress-making school. She had gritted her teeth, had suffered all manner of shame and had brought up my five children.

My son had quit high school midway and had found work in a bakery shop. My heart went out to my eldest son, who was helping his younger sisters through school. I wonder how my eldest daughter felt as she studied at the alma mater of her own mother–a girls' school that had welcomed her with warm sympathy. My wife, it seemed, did not write a letter, being absent at a mill. However, father-in-law's letter hinted at the difficult life my family was leading.

Whatever the case, they were all alive. They had all been allowed to live. I was the husband of an innocent wife and the father of innocent children at whom people pointed their fingers and whispered in contempt. I could only thank the gods and apologize to my wife. I wanted to say: "Please bear a little while longer." Yet, I did not know whether

I would be able to meet them again. In the first half of my life, during which I had forgotten my family and my children, had taken too great an advantage of the state, I had been too self-conceited in my own position. I forgot my temperature of 39 degrees Centigrade (102 degrees Fahrenheit), raised the upper part of my body and worshipped in the direction of the east. Hot tears dropped unchecked upon the letter from my child.

Immediately after my arrival in Nanking, the war crimes trials were begun. The first to be sentenced and executed was Lieutenant-General [Takashi] Sakai. While he was living, I did not have any connections with him either in private or public life. However, his last moments were too cruel for words.

The general was dragged around in an open truck through their streets of a Nanking sweltering in the oven of a mid-August day, with the burning sun turning the city into a hell, and was exposed to the voices of hatred from some half a million people. Then, he was dragged to the execution stand [September 30, 1946]. Tens of thousands of spectators went mad and shouted and applauded with insane glee, as he fell in serenity, bearing on his shoulders the crimes of a million soldiers. His last words were:

"And I dying shall be the stepping stone to Japan-China unity."

His body was exposed for several days on the execution stand. His clothing was stripped from him by unthinking, and inflamed mobs. The crows were allowed to peck away at his eyes.

Greatly enlarged pictures of the dead general, shot in the back of his head and blood flowing from his mouth, were plastered on every street corner of the city.

This was Act I of the drama: "Reward Hatred with Virtue."

Lieutenant-General [Hisao] Tani was brought to Nanking, together, with Lieutenant-General Isogai, as the commander responsible for the "Nanking Massacre." His military trial, held in the Li-Shih (former Japanese Officers' Club), lasted for over a month. He realized that he would not be able to escape execution and night and evening he chanted sutras. In the spring of 1947 [April 26, 1947], he was shot to death on the same execution stand, already spattered with the blood of Lieutenant-General Sakai. The picture of the general, vomiting blood, was displayed on all street corners and carried prominently in all papers. His last words:

"Though my body perish on the execution stand, my spirit shall return to my motherland," was carried with derisive comments in all the newspapers.

This was Act II of the play, "Reward Hatred with Virtue."

That day, Lieutenant-General Hou, who had become acting-head of our group for Lieutenant-General Chieh, inspected the Bamboo Shelter. However, I pleaded sickness and refused to meet him. Major-General Huang came to call me, asking me to make an appearance. Without saying a word, I showed them the death picture of Lieutenant-General Tani, cut out of a newspaper. Only Major-General Niu understood my feelings and said:

"War is not the crime of a single individual. This picture is too cruel. I will speak to my superiors and try and do something to put halt to this."

I pasted this picture on the wall of my room and with a brush wrote in large letters:

"Reward hatred with virtue?"

Every morning and evening I chanted sutras before the picture and said requiem for the soul of the dead general. Then, for roughly two months after that, there was not a single Chinese who dared enter my room. It was only when I left that home on Chien Yeh Road, that they tore the picture off the wall.

Lieutenant-Colonel Gunkichi Tanaka, Major Noda and Major Mukai were brought to Nanking late the same year as perpetrators of the Nanking Massacre. The three men had been detained once in Sugamo Prison in Tokyo and had been released after investigation on the grounds of insufficient evidence. Since Lieutenant-Colonel Tanaka was the model of a novel by Minetaro Yamanaka, Majors Noda and Mukai were featured in a news story in a certain newspaper as having killed one hundred enemy soldiers with their swords, their pleas of not guilty were unheard and they were executed on the basis of evidence found only in novels and newspapers.

[Minetaro Yamanaka, born December 15, 1885, died April 28, 1966. The novel referred to is *Tekichu Odan Sanbyakuri* (750 Miles Through Enemy Lines), a military adventure story based on the exploits of Lieutenant Yoshitsugu Tatekawa during the Russo-Japanese War. This novel was serialized in *Shonen Kurabu* in 1931 and later published in book format.]

[On December 12, 1937, the *Osaka Mainichi Shimbun* newspaper published an article entitled, "Hundred-Man Killing Contest." The article described a contest between then Second Lieutenants Toshiaki Mukai and Tsuyoshi Noda to see who could kill 100 men the fastest using a sword. The victims were Chinese prisoners of war. The newspaper quoted Toshiaki Mukai as saying, "Without realising, we both surpassed 100 people. It was quite pleasant."]

[Because it could not be determined which soldier killed 100 men first and so won the contest, they decided to hold a second contest to kill 150 men with a sword. The second contest was held December 11, 1937. The *Nichi Nichi Shimbun* newspaper reported the results of the second contest were Mukai, 106 men killed, Noda, 105 men killed.]

[A few contemporary historians have suggested the story was partially fabricated. According to Wikipedia, Noda made the following statement about the contest: "I killed only four or five with sword in the real combat.... After we captured an enemy trench, we'd tell them, 'Ni Lai Lai.' The Chinese soldiers were stupid enough to come out the trench toward us one after another. We'd line them up and cut them down from one end to the other."–https://en.wikipedia.org/wiki/Hundred_man_killing_contest]

It cannot be claimed that company and platoon commanders, through their long years in the front lines, were absolutely without sin. However, when evidence is taken from old newspapers and novels in sentencing a man to death, then nothing can be said. The year was drawing to its close when the three men were placed on the execution stand. Everything was white hoar frost. They calmly smoked their last cigarettes, shouted in unison "banzai" for the Emperor and calmly went to their death with smiles on their faces [January 28, 1948]. The whole three expressed in their final words practically the same sentiment:

"If the hatred of the Chinese people can be resolved through our death, willingly we die."

These words were reported in the Chinese newspapers. Lieutenant-Colonel Tanaka wrote at the end of his last statement:

"A timeless loyal subject of His Majesty, Gunkichi Tanaka."

His end was typical of the man.

Thus ended Act III of "Reward Hatred with Virture."

The final moments of Commander Hisaichi Tanaka, who was executed in Canton, were particularly gruesome: While driven around the city in an open truck, he was severely wounded by stones thrown at him by inflamed mobs. When he fell, blood frothing from his mouth, his dead, body was immediately buried under a heap of flying stones [March 27, 1947].

This was Act IV of "Reward Hatred with Virture."

Yoshiko Kawashima was executed in Peking. The only daughter of Prince Su of the Ching Dynasty, she had been brought up as the adopted daughter of Naniwa Kawashima and was the heroine of a chequered career.

Spring had barely come to Peiping on that morning of March 25, when Yoshiko was shaken out of her slumber even before dawn and dragged outside in her night clothes. She realized that her hour had come and quietly asked the guards, stamping for cold on the frost-covered courtyard, to be allowed to change into her day clothes. She was refused. The only thing that was permitted her was to write her last letter. In beautiful calligraphs, she wrote:

> *"To Naniwa Kawashima: Thank you for your long years of loving care. Yoshiko will be executed in Peking at dawn on March 25. I shall come again, reborn as another person. Then I shall be able to work for the sake of China and for the sake of Japan. Please report before the grave of my father (Prince Su) the sad realities of China today. Also remember me to the people of Japan. I shall continue to pray in the next world for the long life of my foster-father. Good-by.*
>
> *From Yoshiko."*

She asked the guards why she was not allowed to complete her final toilet. But, she seemed to have resigned herself completely. Smilingly, she joked with the guards:

"Please cremate the 'picture of my dear pet monkey with me, so that I can take it with me to Hades.'"

She stepped calmly to the execution stand, turned toward the east and made a deep obeisance and met her death with deep tranquility. The newspapers carried side by side the picture of her smiling face taken just before her execution and the picture of a lifeless corpse, bleeding from the mouth and laid out on a door. One guard who attended her execution said:

"I have carried out many executions. But, I never saw anyone who did not fear death like Yoshiko."

There were around 150 women prisoners in the Peking Penitentiary. Without anyone making the suggestion, a spontaneous movement to save Yoshiko arose. In an unprecedented action, all the women prisoners signed a petition asking leniency for Yoshiko.

The Chinese Nationalists feared the consequences of their execution of Yoshiko, the last remaining seed of the Ching Dynasty, and held the execution in secrecy. However, one American newspaperman was allowed to attend. At this, all the newspapers of China banded together and protested against the Government. The press howled that the banning of Chinese newspapermen from the execution, while allowing an American, was a national humiliation. (All the accounts given above are based on facts published in the Chinese press).

Why did they have to kill such a weak and defenseless woman?

[Yoshiko Kawashima was born Aisin Gioro Xianyu on May 24, 1907, the 14th daughter of Prince Su (Shanqi), a member of the Qing imperial family. She was adopted at age eight by Naniwa Kawashima, a Japanese espionage agent. Renamed Yoshiko Kawashima, she was raised in Japan and received a Japanese education. During the Japanese occupation of Manchuria by the Kwantung Army, she worked as a spy. She was arrested by Chinese authorities in Beijing on November 11, 1945. She was detained at Hebei Model Prison and charged with treason. During her trial, Kawashima claimed Japanese citizenship, arguing that she could not betray China if she were no longer a Chinese citizen. However, the court rejected this defense, noting that she had never formally renounced her Chinese citizenship. On October 20, 1947, she was convicted of treason. She was executed by a single gunshot to the back of the head on March 25, 1948, in the courtyard of Beijing Prison No. 1. Her body was publicly displayed after the execution, serving as a warning to others. A Japanese monk collected her remains, which were cremated and later interred at Shorinji temple in Matsumoto, Nagano Prefecture.]

Such was the cruelty of this execution, that I could not but think that Chiang Kai-shek, the chief actor in the Chinese Revolution which over-threw the Manchu rulers and gave rise to the Han peoples, had condoned the killing of this woman, the last of the Ching Dynasty, in order to put the finishing touches to the initial task that he had embarked upon.

It was natural for the Kuomintang, without the virtue of magnanimity, to tread of its own accord the path to self-destruction.

It became evident, hand in hand with the increasing scope of the war against the Chinese Communists, that the Chinese Nationalist Government was planning to hire former Japanese Army intelligence officers. With only Lieutenant-General Tsuchida, detained in Shanghai for service with the Kuomintang, and myself available, the Kuomintang did not have sufficient personnel to gather vital intelligence, although we had gathered together basic material to guide the Kuomintang in waging a modern war. However, I was not consulted in any way as to whom should be brought from Japan.

One May 1, I was again prey to an attack of acute malaria as a result of long accumulated fatigue. With a fever raging over 104 degrees Fahrenheit, I lay suffering once again in my small maid's room. There was nobody to nurse me. It was indeed a miserable experience to become sick all by oneself in China. The young girl Wu brought me an ice pillow from her home, for which I was deeply grateful. She had seen fit to extend a modicum of sympathy to this ragged old man of a defeated Japan. My fever refused to come down, so I greeted once again May 3, the anniversary of my father's death. On this day, last year, I had been on the threshold of death with false cholera in Chungking: This year I was sick again on the same day with acute malaria. I wondered at the significance of this quirk of fate.

Lieutenant-General Tsuchida was on an official trip to Shanghai and his gorgeous residence was unoccupied. The Chinese evidently realized that I needed a good long rest and transferred me to the third floor of Tsuchida's home. From the window of my new room, I could see a magpie feeding four young nestlings in a nest built high on a branch of a Chinese parasol tree. I watched with fascination, the way in which the mother bird popped food into the mouths of its young in set order. And as I watched, my thoughts turned to my children waiting at home like these fledgling birds for their father.

It was this day that Colonel Okawa (pseudonym) arrived in Nanking from Japan. Colonel Okawa had been my teacher in the past and a senior from the same prefecture as myself. I wanted to meet him badly and from my sick-bed I asked Major-General Yeh a number of times to have Colonel Okawa come and see me. However, each time General Yeh would put me off, telling me that Okawa had gone to Hsuchow or that he was in Shanghai. Yeh refused absolutely to let me meet my friend. (All these excuses I later found were complete lies).

Lieutenant-General Tsuchida, with his characteristic stubbornness, was not amenable to the wishes of his Chinese master. It was a fact that he was politely being kept at arm's distance. The Chinese feared greatly that the Japanese Army officers in their employ would form a common front around him. For this reason, their policy was apparently to keep each of the Japanese isolated, using them separately, without allowing any liaison between them.

It was such atmosphere that greeted the arrival of Major-General Yamada and Colonel Matsuyama (both pseudonyms).

The Chinese Mission in Tokyo had promised these men treatment as state guests, when they were persuaded to accepted work with the Chinese Government. However, they were shipped aboard a vessel of only 500 tons and finally brought to China after two weeks of perilous journeying, more dead than alive. I had just moved to our new offices on Kiangsu Road after my illness, when these two men, both my respected seniors in the Army, arrived under Major-General Huang's escort direct from the station. Filled with nostalgia, I gripped their hands in a warm handshake that was filled with all kinds of emotions and it was with an effort that I halted the tears flowing from my eyes. I had expected some word through them from my family, but there were no messages. They told me that Colonel Okawa had brought something for me, but I had not been allowed to meet him. The reason given was that our, respective assignment were in different fields.

Tsuchida, Matsuyama, Yamada and myself asked to be allowed to work jointly. However, the Chinese were intent on keeping Tsuchida apart and would not listen to our suggestion. The night the two new arrivals came, we all had a simple dinner at the newly-built mansion of Major-General Yeh.

Our new offices on Kiangsu Road had been prepared with the view of enabling the whole four of us to work together and it was a Splendid office, forming no comparison whatever with the Bamboo Shelter.

The day Yamada and Matsuyama arrived, we talked until three in the morning and I listened with intent to the first reports on what was happening in Japan after the surrender. You could not argue over Matsuyama's background and upbringing. Concealed in a peaceful and friendly exterior burned fiercely the flame of a strong sense of justice. My new friends who had been persuaded to come to China with the bait of state guest treatment had even prepared formal clothes in which they expected to meet Generalissimo Chiang Kai-shek. However, not even Lieutenant-General Hou showed his face, let alone the Generalissimo. They were first surprised on their arrival at the lack of sincerity on the part of the Chinese. Then, they must have felt no small degree of unpleasantness to find Major-General Huang, more than 110 years their junior, push them around as if they were subordinates.

Major-General Wang Liang, who had brought the two from Tokyo, arrived in Nanking after stopping over for four or five days in Shanghai. He brought me a letter from my wife, the first in four years. Even during the heart-breaking period immediately after the end of the war, they had managed to keep from starving, aided and abetted by such comrades as Kimura, Nagai and Kodama (author of "I was Defeated"). My wife had placed the two youngest children in an orphanage, sent my eldest son out to work and had herself found a job in a factory. In this way she had managed to keep the wolf away from the door, but I found tear stains imprinted on the pages of her letter. In closing, she had written:

"If the worst comes to the worst, please die in such a manner as to make people say that it was just like Tsuji. Please die in such a manner that the children will not be ashamed of you."

The fact that I had gone underground, without committing suicide after the war, was an action sufficient to spur my wife to tell me how to die. I realized that my wife felt that my words and my deeds did not match. People had laughed at her for my having gone into hiding out of a fear for my life. It was not easy to guess that my wife felt shame in front of the eyes of others. I had fully resolved to expect such a reaction from certain people, when I had decided to start on my long travel in disguise. But, it was painful to think that people were cursing me as a coward. I wondered how long I would have to bear with this.

It was as if my very vitals were being torn into shreds....

Who could understand my pain and my suffering?

However, this letter had been opened, translated and. censored by officials of the Defense Department, before it had been had been handed to me.

The elderly Wang Chan-chun was an old friend of mine. In 1932, this old man had accompanied me as an interpreter on a trip through remote Sinking Province. In Lanchow, we had both been thrown into prison and had stood at the threshold of death. Since then for ten odd years we had been friends tied by bonds transcending all racial differences. His son, Wang Ho-cheng, had graduated from the Kyoto Medical University and had successfully won his doctorate. Now he was head of a certain hospital in Peking and had become a famous doctor. One day I chanced upon his name in the newspapers and immediately sent him a letter asking him to transmit my best wishes to his father.

Almost immediately, an answer, almost wild with joy, arrived from the father. Wang had concluded that I was detained as a war crimes suspect. He had resolved to allay the suspicions held against a friend at the risk of his own life, take the stand for me in the courtroom and if the worst came to the worst, to be executed with me. With this in mind he had sold all his private possessions and had flown all the way to Nanking. His elder brother-in-law, Shao Li-tzu, was one of the most outstanding elders of the Kuomintang Party at that time. Wang, however, did not even drop off at his brother's home but had come directly to my address on Kiangsu Road. He wore a dirty suit and torn shoes and looked no better than a beggar. This old man, who had neither fame nor fortune, had spent his life writing both verse and prose. He had not bothered to rely on his brother-in-law, even in moments of great poverty.

He stayed for several days at the home of Major-General Wang and spent the greater part of his time conversing with me. In a loud voice that could be heard in the vicinity and the neighboring rooms, he denounced the corruption of the Koumintang and predicted its inevitable disintegration. But, he was not one to praise the Chinese Communists, saying:

"In China there are only black pieces of glass and no transparent pieces."

Born a Manchurian Bannerman, he had become a major-general at the age of 16. During the Chang Tso-lin regime, he was chief of the Fengtien (Mudken) Arsenal, and was one of the greatest experts in China on land mines. Unable to stand idly by while the transportation network of China was being severed in all parts of the country, as a result of the intensive Chinese Communist tactics of disrupting communications, he had come to offer to the Kuomintang Government his technical skill and experience in mine detection. Through this service, he had planned to plead for leniency for my war crimes. He had also heard that there were still some 100,000 Japanese people being utilized in Manchuria and he had come to ask me to join him in flying to Manchuria, to parachute down with him, and to try to save the Japanese soldiers and civilians. I saw that he was fully determined to throw away his aged frame, if necessary, in carrying out this venture. However, neither Lieutenant-General Hou nor Major-General Yeh, both members of the pro Europe-America faction would listen to this. Although old Wang's plans were dashed to the ground, he seemed completely relieved to find his old friend safe and sound, aimlessly crossed over to Formosa with the determination to spend the last remaining days of his old age composing poetry. Knowing that he was completely without funds, I borrowed a month's salary in advance and gave this to him to help him out with his travel expenses. On parting, the old man left me a sandalwood fan, on which he wrote a poem by Wen Tien-hsiang. Perhaps he is lolling away today in the hot springs of Peitou (near Taipei), composing verse would like very much to meet him again. The scent from his sandalwood fan seems still to breathe the friendship of this old man for me.

Despite the fact that both General Yamada and Colonel Matsukawa had risked their lives time and time again to come to Nanking all the way from Tokyo, the higher Chungking officials, with the exception of Major General Yeh, did not seek to meet them. In addition, although our work was a joint affair, headed by Lieutenant General Tsuchida, there was every indication of attempts to block the meeting of us four Japanese. Major-General Yeh left to carry out negotiations with the Chinese Communists concerning a local truce along the Hwangho River and was absent during May and June, He had left no instructions with us concerning our work and these two months passed fruitlessly. We were not allowed to go out and were kept like birds in a cage. It was no wonder that we should come to hate the treatment they accorded us. On top of this, the intendance officer, who lived with us constantly embezzled official funds to enrich his own pockets. He would steal from our salaries. To add insult to injury, he would come into the dining room clad only in a pair of BVD's. He was indeed an uncouth and rude individual.

On July 9, the new arrivals finally decided upon their future course of action. The whole four of us expressed to the Chinese our determination to go back to Japan. Of the verbal promises made to the two generals in Tokyo not a single one had been kept. The free liaison between Japanese officers, necessary to keep up our work, was obstructed. Our treatment was equivalent to detention. In addition, placed under a dishonest intendance officer, we were being cheated out of our full salaries. Thus, we concluded, unless this situation was radically remedied, we could not continue to stay in China. Daily, the surrounding atmosphere deteriorated, making it harder for us to remain. Major-General Yeh, upon his return to Nanking late in July, finally gave us an answer, transmitted to him from his superiors.

"The Third Research Group will be dissolved. Tsuchida, Yamadá and Matsuyama will return to Japan. However, since the British authorities are still looking for Tsuji, he had better stay alone for the time being."

The other three were satisfied with this answer. However, I alone could not understand why I should be left behind. My initial high hopes of paving the way for eventual Japan-China collaboration had been completely deflated. I could not bear to stay longer in China, if I were merely to be asked to sell my obsolete knowledge as a soldier of a defeated country.

"I will stay behind if you think that I am necessary for Japan-China collaboration. However, if you are not permitting my return to Japan, out of a desire to protect me, I absolutely refuse to stay behind."

Major-General Yeh asked me to wait two or three days and then departed. Later, he came back with the answer:

"We would like you to stay behind and help us in our work."

"If that's the case, it can't be helped. The big assignments that have started are still incomplete and in the sense of fulfilling my promise to finish this work, I shall postpone my return to Japan for a few months."

With this, I alone agreed to stay in China.

The blame for sending back my friends, who had come all the way from Japan without leaving behind a single contribution, lay on both sides. The Chinese were to blame for not according a certain amount of respect and deference to these generals, soldiers that they may have been of a defeated country. On the other hand, we were to blame for the fact that' we did not strive to bear suffering, persevere and hold ourselves in complete humility.

On August 12, Lieutenant-General Tsuchida quietly left Nanking, not seen off by anyone. General Yamada and Colonel Matsuyama were to leave on August 25. Matsuyama left behind several verses on his parting. One Chinese verse read:

"Look back and see the thousand peaks, like sheep's backs, that I've crossed."
"Many a thorny path I've trod to climb precipitous cliffs."
"Yet, I'm not old, but full of life and vim."
"I point with a laugh to the crests above the white clouds."

The other verses in Japanese were as follows:

1. "Two are the paths, in number and in kind, that parting go...
"Wild will the mists be and heavy the rains."

2. "The path that you go single and alone"
"Through the deep valley winds"
"With dew pearl heavy on the early autumn vines."

3. "Hearten yourself for the peak is not far"
"Though the thorns tear deep into your feet."

4. "Brave is the figure of a parting friend,"
" Who through the cool breezes goes."
"Sad is his song as he thinks of his home."
"In the night, under the Milky Way."

On the night of August 15, Major-General Tsao, acting on behalf of Lieutenant-General Hou, together with seven or eight others, including Major Generals Yeh and Huang, invited us to a modest farewell banquet. Although it was merely a matter of formality, I could perceive a token of friendship in the gesture. After the dinner, we went to the Nanking Station. There in a hot steamy carriage, I said farewell. Matsuyama's hand seemed to tremble as he shook my hands. I asked him to take back to my child the sandalwood fan given me by Wang Chang-chun.

I returned alone to my quarters and spent my first night of self-imposed solitude in loneliness. My friends who had come all the way from Japan had returned without having done anything. I. moved from my garret quarters on the third floor to the room that had been occupied by Matsuyama. In my new room I found a pot of morning glories which Matsuyama had carefully looked after. Many days passed during which I found my only comfort in looking at these soulless flowers.

The material that the Kanda couple had brought with them from Manchuria was extremely voluminous. The majority of this material had already been made, public in various publications. However, if the data were coordinated an outline of the strategic topography of Manchuria could be compiled.

Although the Chinese Nationalists had dispatched a force of 500,000 crack troops to South Manchuria for decisive battle with the Chinese Communists, there were no preparations made to supply maps or data on strategic topography. The Japanese Army had also prepared over a number of years for war with the Soviet Union. However, perhaps because Japan had no intention of fighting the Soviet Union on Manchurian territory, there was no outline of Manchuria's topography to fill strategic and tactical needs. In addition, there was no person in the Chinese Nationalist Government capable of undertaking the compilation of such an outline.

Requested, by Major-General Yeh to complete such a work, I made him promise me the following:

1. To provide me with sufficient personnel necessary to keep up with the speed of my writing, making the entire process of compilation, translation and printing a conveyor-belt operation.

2. To allow me six month's time.

I tackled a mountain of material contained in over 500 odd volumes and began on my work on March 25. At the end of April, I had completed my first volume, "General Outline;" at the end of June the second volume, "South Manchuria;" at the end of July the third volume, "North-west Manchuria" and on September 15 the fourth and last volume, "Northeast Manchuria." I had kept my promise to a day. The complete work totaled 2,500 sheet of manuscript paper included 120 appended maps and charts and aggregated a total of over one million words. In this-way, I brought to completion a big task.

Despite the fact that the war situation was becoming more critical dally, the higher ups in the Defense Department diverted a half of our personnel and energies to the compilation of the strategic topographical outline for Siberia. Finally, they even disregarded their initial promise to me and it was only at the end of December that my entire work was completely printed. In the meantime, General Chen Cheng had been appointed commander-in-chief of the Chinese forces in Manchuria and he kept cabling daily to us to speed up our work. However, when the material was sent to the front lines, it was already late in January of the next year. But then, the war situation in Manchuria had so deteriorated as to rule out all possibility of a recoup. In this way, the greater part of my work, in which I had poured forth all my energies, failed to be of any use.

The Generalissimo granted me a letter of appreciation and a monetary gift of 5,000,000 yuan (Y5,000 in Japanese money, $13.90 in US dollars). However, somebody embezzled the money and it failed to reach my hands.

Whenever the war situation turned in disfavor of the Chinese Communists, the Nationalists would come to ask for an evaluation of the enemy's future movements. My predictions practically always hit the mark. This was because of simply extreme cold judgment, made from the standpoint of a third person, and because of my thorough knowledge of the complete corruption of the Chinese Nationalist Army and of its military weaknesses. However, my unreserved predictions of Chinese Communist victories were never told to the Generalissimo by the fearful intermediary staff officers. Thus, my evaluations of the military situation did not serve any purpose in actuality. Despite the gradual approach of a military crisis, I saw not a single man among the backbone group of the Defense Department applying himself with any seriousness to the study or the guidance of operations. The Generalissimo's approval was necessary even before a single regiment could be moved. With the exception of the Generalissimo, all the rest of the generals and officers seemed to be looking at the war as if it were the affairs of other people. They spent their days in dancing and in feasting.

All alone, I sat quietly and systematically studied various newspapers, pondering over the future course of the war and predicting in an impartial manner the victory or the defeat of the Chinese Communist and Nationalist forces. I had neither friends to talk to, nor the means to amuse myself. The only thing I had with me was my sense of responsibility for the defeat of Japan. I wondered whether I could expect from such a party as the Kuomintang the possibilities of long-lasting collaboration between Japan and China.

I compared the past when I had observed China from the standpoint of a victor and the present when I was reevaluating China from the standpoint of the vanquished. And I could not wonder at the difference in my outlook. Perplexed in my judgment of trends and puzzled in my evaluations, I finally decided to catch only the realities of the situation and to formulate a conclusion later.

Japan's policies toward China in the past were simply based on immediate interests and on a subjective view of China. For this reason, Japan met with failure. How foolish Japan's policies had been in the past. She had obstructed the course of the Chinese Revolution at China, she had placed fond expectations in the restoration of the dying Ching Dynasty, she had viewed with jealousy the currency reforms sponsored by Great Britain and she had tried to force her notorious "21 Demands on China," in an attempt to fish in troubled waters.

The fact that the Communists and the Nationalists were fighting in grim battle was a purely domestic political quarrel. Those who were doomed to die were digging their own graves, while those who were rising to the fore possessed a certain progressive vigor and energy. If my counsel had been heard, and if I had been utilized with greater confidence, I, with my trait of being impressed more deeply than others by a show of appreciation of my efforts, would have probably risked my life to help the Generalissimo. However, the Generalissimo was a man worshipped a top a cloud and was himself, always covered by dark mists. Thus, there was nothing that I could do but look on from the sidelines as a spectator.

It was immediately after the departure of Generals Tsuchida and Yamada and Colonel Matsuyama that I was finally allowed to meet Colonel Okawa. The reason why the Chinese did not give me permission earlier was their fear that General Tsuchida's obstinacy would be catching. Colonel Okawa was an old man, who had taken care of me from my student days in the Military Academy. With an attitude full of sympathy and true friendship, he comforted me in my loneliness at being left in China by myself. He also told me in detail of the struggles of my wife and children placed in the circumstances of "hearing the war cry of the enemy on all sides." The Colonel, together with several of his former subordinates, was being employed, under detention, in another agency. However, one of the key members of this unit turned out to be Colonel Chin, who had shown me such unlimited kindness and consideration in the past. Colonel Chin, I learnt, was also treating this old friend of mine with the same thoughtfulness. What a coincidence this was. Colonel Okawa and I visited each other once a week and comforted each other in our respective solitude. From then on, we would get together and make our clumsy Japanese dishes–miso soup, and Japanese stewed vegetables flavored with soy sauce, then eat these dishes and also rice steeped in tea with pickles. We cooked these dishes, with the feeling of returning to naked malehood. The joint meals were the greatest pleasures during the next half year.

His Excellency Lieutenant-General [Rensuke] Isogai was an old acquaintance and my most respected senior in the Japanese Army. As commander of the Seventh Infantry Regiment in Kanazawa, he had first guided me when I was still a lieutenant. From that time for more than 20 years he had educated me both directly and indirectly and after the death of my own father I had regarded General Isogai as my own parent. He was brought to Nanking together with Lieutenant-General Tani as a war criminal. From the time I first saw a newspaper picture of the General, showing him seated on the grass in the yard of the Defense Department, I had kept close watch for anything concerned with him, feeling that if they were to kill mercilessly this father of mine, I myself could never continue living. Deputy Chief of Staff Liu had promised to do all he could for General Isogai. However, it was to be doubted that he could hold out against the pressure of public opinion and the persistence of the British. The general's final trial was opened one year after his arrival in Nanking. General Isogai was charged with having sold opium in Manchuria, of having allowed his subordinates during the Hsuchow Battle to engage in acts of barbarism and of allowing many Chinese to starve during the mass dispersal of Chinese Residents from Hong Kong, when he was governor of that British colony. I expected the worst for General Isogai, in the same way as I did for Lieutenant-General Tani, shouldered with the responsibility for the Nanking Massacre.

I produced evidence showing that each of the charges were absolutely without foundation and requested that I be allowed to stand in court in his defense, if necessary. During the three days of the trial, I purified my body and chanted the Lotus Sutra morning and night and prayed for the General. A sentence of life imprisonment was handed down with all solemnity [July 22, 1947]. At least his life had been spared. However, at his age, life imprisonment was nothing more than a death sentence. [General Isogai was released in 1952 and died June 6, 1967.]

Time and time again I asked to be allowed to send him something daily. Finally in October this was allowed, thanks to the kindness of Major-General Tsao.

I emptied my skimpy purse and succeeded in sending him my first parcel, containing blankets, shirts, a winter overcoat, butter, sugar and other items. The General sent me a letter of thanks in his own familiar handwriting. There were Chinese present when I received the letter, but I broke down and cried. I wondered how greatly he must be suffering. Did I have to cooperate with China while my father was in chains? I wanted to meet the General personally. At every opportunity I asked to be allowed to see him. It was on March 15 of the next year that they finally agreed to a visit on one condition: "This will be the first and last time."

I was taken to the second-floor waiting room of the jail in the compound of the Defense Department and waited several minutes. When I saw the general, thin and grey, come in wearing a dirty padded coat and trousers such as worn by Chinese soldiers, I could not speak for a while so pitiful did he seem. The General also had not expected me. The both of us stared at each for a time, tears flowing down our cheeks and this was the sight of my father in his old age. Shouldered with the grave responsibility of defeat, I, who should naturally have been sent to the gallows, had escaped and gone underground. In my stead this father of my mine, this aged general, was now suffering in chains. I felt as if my heart would break.

"For the past year, I haven't had any fatty food. The tea and the sugar that you gave me have helped greatly. I've chewed them together a little bit at a time every day and have managed to make both last till today...."

This General, who had never once complained about food, clothing or shelter, was now pitifully emaciated. His hands were crooked and his waist bent. At this rate, I realized he would not last another year.

The Chinese had told me:

"The food provided war criminals are the same as that eaten by officers of the Chinese Nationalist Army There is no need for you to send in things to him."

What I wanted to ask was whether Chinese officers never had any fat during a whole year, or whether Chinese officers lived the whole year through on one bowl of reddish rice two times a day, with only a mess of vegetable, slush cooked in salt and a bowl of salty soup as subsidiary foods. Whatever the regulations might have stipulated, the reality was quite different. It was stark and undeniable. The intendance officer in charge of the feeding of war criminals took his usual percentage; the chief warden also got his share. By the time the prisoners got their food it was the minimum needed to keep them from starvation. I, held down a torment of seething emotions and talked with him for roughly an hour, and then took leave. I left a small comfort bag, ashamed at not being able to do more. As I left the prison, I noticed several children of the prison warden playing in the yard. Their cheeks were as red as apples. I, who usually felt deeply attracted to children, on that occasion alone, felt hatred toward them. As soon as I returned, I wrote a letter to the Chief of Staff:

"...Life imprisonment is only a nominal sentence. It is nothing more than a graduated death sentence resulting from malnutrition.... If his treatment is not improved, please also throw me in prison...."

Immediately, General Isogai was removed to the Shanghai Penitentiary. In Shanghai, the entire budget for the food of prisoners was handed over directly to the Japanese, who had organized an autonomous self sufficient system. They reared their own pigs and raised their own chicken. They also make confection and their own bean cured cake. I, who had prayed day and night that the General would be transferred to this prison, was able to see a picture of the handcuffed General, taken at the Nanking Railway Station, in a newspaper.

However, when I quietly looked back over the past, I realized that all this was the sin of Japan's policies toward China, stretched over a long period of years. It was a sin which should have been borne by the state, 80 per cent of it, in particular, by Japanese army men.

One night, an unknown Chinese youth, under the cover of darkness, paid me an unexpected visit.

He said:

"I am a subordinate of Dr. Chang who once belonged to the East Asia Federation. I am also your comrade. During the war, I was dispatched from Chungking to North China and worked in the Chinese underground. I entered the North China Political Affairs Committee in disguise and served as secretary to Wang Ko-min."

The youth spoke Japanese like a native. After a wary glance around he took a letter from an inner pocket. From the handwriting, I realized immediately that it was from my friend Dr. Chang with whom I had been on intimate terms through the East Asia Federation. During the war, Dr. Chang had been the head of the whole underground movement in North China and I had come to know him through Miao Pin. Since then, for roughly ten years I had kept in contact with him, sometimes talking with him in Peiping, at other times in Nanking. I also had the opportunity of obtaining his release, when he was arrested by the Japanese gendarmerie. Because I knew he had channels to Chungking, I protected him and tried through him to use these channels to bring about peace between Japan and China. An influential leader of the Kuomintang, he was one of the few top-flight men who could speak unreservedly to Generalissimo Chiang Kai-shek. Now we had changed positions. I opened the letter from this dear friend:

"My hopes regarding the future of Japan and China are the same today as they were in the past. I can have no greater happiness than to know that you have been well and alive. I expect you to become without fail a tie between the two countries in a period of new collaboration. You have experienced personally, I believe, the corruption of the Kuomintang. I am ashamed to have a foreign comrade see this corruption. However, I have not lost all hope."

"My conscience will not permit me to enter this crucible of corruption. At the present time, I have no job, no master and suffer even for my daily bread. No matter if I should starve, I will not be fed by these corrupt elements."

"Over half of our comrades in the East Asia Federation have entered the Chinese Communist Party and are in the midst of efforts to lead the Chinese Communist party along racial and Oriental lines. Despite the fact that you are endowed with far-seeing sight, you are now running counter to the direction of the Chinese Revolution. You are helping the dying Kuomintang and are headed for a double suicide with the Nationalists. This I cannot stand by and watch in silence. I pray that you will be able to return home as soon as possible and I wait the day when you will be able to cooperate with the new China from the bottom of your heart."

To My Comrade Shin Sheng-hsin.

In the eyes of a Chinese, the Chinese Communist and. Nationalist parties are the same as the Democratic and Liberal parties in Japan. Any shift from the one to other is viewed in the same way as the Japanese view the transference of allegiance by Japanese Diet members from the Democratic to the Liberal Party and vice versa, or the splitting up of parties and their coming together in new groups in Japan. The reason why the officers and the soldiers of the Chinese Nationalist Party, which fought so bravely against Japan for eight whole years, should so easily defect and cross over into the Chinese Communist camp or the reason why they should ally themselves so easily to be disarmed arises from just this sort of feeling.

It might be the American White House's way to regard the fight between White China and Red China as a conflict between two worlds. However, this viewpoint met with failure despite the efforts of General Marshall. The question of China cannot be saved unless left to the good sense of the Chinese themselves. The old Japanese proverb:

"Even a dog does not join in a quarrel between husband and wife," should hold true here. Those who interfere will fail for sure.

I realized anew the basic principle underlying this truth. I had come to Chunking at the risk of my life and had gambled my very existence in trying to move the Generalissimo. However, I was already too late when I had come to the conclusion that I could help the Kuomintang in its work of post-war reconstruction and the stabilization of East Asia. It was absolutely impossible with my two hands to engage in the Sisyphean task of emptying the sea dry with a sieve. I had gradually begun to think that the great central pillar, holding high the Chinese Nationalist structure, was already eaten through and through by the white termites of corruption and stood ready to topple at the first gust of a strong wind. Dr. Chang's secret letter not only contained much food for thought but served also to substantiate the accuracy my premonitions."

According to the words of this youth, Dr Chang was having a hard time making ends meet, but he did not want to obtain help from any Chinese. Dr. Chang, the youth continued, had said that a foreigner like Tsuji would understand his position well. However, my circumstances had gone from bad to worse. I had bought a ticket for Wang Chang-chung to Formosa by borrowing a month's salary in advance. And I had spent whatever I had left to send gifts to General Isogai in prison. However, I knew that I must return the confidence placed in me by Dr. Chang. I left to myself only the barest essentials and sold everything in my possession. With this, I raised a total sum of 30,000,000 yuan (roughly 30,000 Japanese yen and about $85 US). This I sent through secret, routes to Dr. Chang in Peiping.

Colonel Chen of the Chinese Mission in Tokyo returned to Nanking in September for liaison purposes. According to his report, the hunt for me by the Japanese police was still being carried out with thoroughness. Later, when Lieutenant General Lin Hsun-nan returned from Tokyo, he brought with him a letter for me from Hattori. It was the first token of Hattori's friendship for me since our parting three years previously. In all my 20 odd years of army life I never felt as grateful for a trust worthy senior as Hattori. If any other man had given me warning, I might not have listened. However, I felt that I should follow Hattori's advice unconditionally. In sum, he advised me that objective factors still would not be in favor of my return to Japan at that time. Thanking him in my heart for his advice, I resolved to stay another half year in China. Everything called for me waiting until my "time." This advice was against that given me by my friend in Peiping, but in essence it was the very same thing.

The valuable data, "The Industrial and Military Potentials of the Soviet Union," complied by the Kwantung Army over a period of roughly five years, by mobilizing all the authorities on Soviet Russia in Manchuria, was buried underground by an employee of the South Manchuria Railway Company in a certain place in Fengtien (Mukden), immediately after the end of the war. [Mukden was the site of the false flag event staged by Lieutenant Suemori Kawamoto of the 29th Japanese Infantry to justify an invasion of Manchuria.] This was dug up by American Army officers and was about to be shipped by air freight to Washington, when on the night before the shipment was to be made, soldiers belonging to the Blue Shirt Society stole this data and sent it by a Chinese plane to Nanking. The compilation was contained in one thousand several hundred volumes and was precious data, the like of which was not to be found in any other country of the world. However, to the Defense Department of that time, this material was like pearls to swine. There was not a single person who even tried to make something out of it.

"Alright, I'll tackle this. It will be a work, greater but not less than my work on the strategic topography of Manchuria. At the same time it will be valuable research for me.

Major-General Yeh was very pleased at this. From the end of October, I immersed myself in a pile of half-rotten material. And forgetting all things, I became completely engrossed night and day in this work. I had been under the impression that I had applied myself diligently during the past ten old years in a study of the Soviet Union in preparation for war. However, I found out that this study had been partial and incomplete, limited only to tactics and to political trends. I felt deeply the mistake of having completely neglected a study of the Soviet Union's industrial strength, forming as it does the basis of all things. This was also the same mistake that was made in the Pacific War.

I was able to study penetratingly the historical steps taken in the construction of a planned economy undertaken with ruthless cruelty for over 20 odd years. I searched carefully for the reasons which enabled the Soviet Union to defeat Hitler. Finally, I was able to find in detail and in figures the material war potentials which the Soviet Union had built up for a new war. This study was more rewarding than any that I had ever undertaken in the past. By accident, I was able to find data backing up the gist of the lecture that I had given in the great auditorium of the National Defense Department two years ago.

I gathered up the results of my six months' of study in a work entitled, "An Evaluation of the Material War potentials of the Soviet Union," and submitted it to Generalissimo Chiang Kai-shek. It was a huge volume, numbering over 1,000 pages and totaling roughly 400,000 words.

The gist of the conclusions reached were:

1. The industrial war potential of the. Soviet Union will reach from 40 to 50 percent of that of the United States at the end of 1950.

2. The point in which Soviet war potential is superior to that of America lies in the fact that Soviet industry has been systematically scattered over the whole country. In addition, each of these scattered industrial areas, have been built up into self-sufficient blocs, which have been, in turn, correlated into a larger whole, with the relation between the lack of certain materials and the possession of certain materials adjusted.

3. Soviet Russia's stockpile of strategic material is astonishingly great. Even if, by chance, all of her above ground industrial facilities and functions are completely wiped out by atom bombs dropped by the United States, the Soviet Union has stored deep underground all the materials necessary to last another three to four years.

4. Another point in which the Soviet Union is superior to the United States is her vast manpower resources and the huge stock piles of food to feed this manpower.

5. The greatest weakness of the Soviet Union is her oil resources, both in volume and in distribution. The Soviet Union is striving to overcome this weakness by large stores of petroleum. (Seventy percent of the entire oil production of the Soviet Union still continues to be situated lop-sidedly in the Baku area).

6. The ratio of the importance of the area east of Baikal to the whole of the Soviet Union is only from five to ten percent. However, if necessary, it is possible for the Soviet

Union to enable the whole present population into the Far East to be self-sufficient and in addition, sustain in some way from 500,000 to a million troops.

Thus, the military importance of the area east of the Baikal is to be regarded as less than 10 percent of the rest of Russia and its loss or its consolidation will have little relation to the fate of the entire war. However, if Manchuria is placed completely under Soviet control and its heavy industries revived, the Far East would come to have a far greater importance.

The decisive war will clearly be fought in West Europe. The next important battlefields will be the Middle and the Near East oil fields.

7. The industries for the manufacture of the atom bomb will utilize the electric power generated in the Angara and Baikal areas. Raw material for the atom bomb, uranium, is available in Kuzbas (Kuznetz Basin), Central Asia, and Czechoslovakia. Factories for the manufacture of the atom bomb seem to be under construction in the Kuzbas, or the Angara-Baikal and Ural areas. The Soviet Union will succeed in supplying itself with the atom bomb by 1950 at the latest.

8. The point to be specially noted with regard to the Soviet Union's munitions factories is her great capacity to manufacture aircraft. Her annual production is close to that or greater than that of the United States. In quality, the Soviet small aircraft is superior to that of the United States, while in larger aircraft the Soviet Union is inferior.

In both tanks and fire weapons, the Soviet Union might be superior to the United States but not inferior. The Soviet productive capacity of tractors for farm use exceeds that of the United States and in the event of war, the Soviet Union can easily mobilize speedily a large mechanized force.

9. At this rate, together with the completion of the Fifth Five Year Plan, the Soviet Union's industrial potential will in 1955 catch up to that of the United States and in certain fields will forge ahead of the United States.

10. Victory or defeat in World War Three will not be decided by aircraft alone. It would necessitate the occupation by American land forces of the entire Soviet territory. Thus war will become a long drawn-out affair, exceeding over five years and in the worst event, both the United States and the Soviet Union will be grievously wounded.

11. The forces which the Soviet Union will be able to mobilize on the basis of industrial strength as outlined above will be:

Army:		
Sharpshooters	roughly	500 divisions
Cavalry	"	50 divisions
Tanks	"	25,000
	(annual production roughly 50,000)	
Air Force:		
Front line planes		15,000 machines
	(annual production 7,500 planes)	
Total Mobilizable Force:		
	roughly	24,000,000 men

The biggest cause of Hitler's defeat was his under-estimation of Soviet Russia's war potential.

The Soviet Union will be able to increase by 1950 her industrial power to at least 50 per cent of that immediately before World War Two.

Leaders of the United States should ponder carefully over the fact that even if by 1950 the U.S. should have 500 atom bombs, it will not be easy to crush the Soviet Union. The reason why you were able to a wipe out the industries of Japan was because these factories could be accurately pinpointed as targets above the earth. However, on the other side of the Iron Curtain, it is absolutely difficult to aim at Soviet Industries which are dispersed over a vast area, scattered and camouflaged. In fact, it is well nigh impossible.

In contrast to this, if the Soviet Union were permitted to stockpile 50 atom bombs, she would be able to deal a grave blow to the heavy industries of the United States at the very outset of war, concentrated as these industries are in all the larger cities.

As long as the Soviet navy is weaker than that of the United States, it will be impossible for the Soviet Union to affect a landing on the American mainland. On the other hand, if the United States enters deep into Soviet territory, she will meet up with all kinds of hardships.

Thus, World War Three, when seen from behind stage, will become a great battle of attrition, lasting for over five years, whose outcome cannot easily be forecast.

Today, the Chinese Communists have completely won domination of the 400,000,000 people of China. If they should take this entire population over to the side of Stalin in the next war, victory for the United States will become that much more difficult. At any rate, it will be a decisive battle between giants.

This was the conclusion reached through my work of one half year of study.

Manchuria greeted its second winter of Nationalist control. Chief of Staff, General Chen Cheng, had been transferred to Manchuria to take over the command of the Chinese Nationalist forces there. He boasted that he would occupy and maintain by armed force the whole of Manchuria. However, the war developed daily in his disfavor. Studying the general situation, I advised boldly and bluntly the policy of a northern and southern dynasty in opposition, saying:

"Abandon Manchuria, hold on for a time to North China, but withdraw all the rest of the Nationalist forces to the south bank of the Yangtze and consolidate your positions there." However, this heartfelt advice was not accepted. Rather it was greeted with resentment. All the men in positions of responsibilities from the Second Section chief down did not even try to transmit my suggestions to the Generalissimo, perhaps afraid of the gravity of the situation. They were simply absorbed in immediate war Operations and perpetrated the disgraceful scene of discussing the local withdrawal from Kirin and Changchun in Manchuria, without reaching a decision. I felt deeply the lack of capable men in the Defense Department. I realized that there was no way left of saving the Chinese Nationalist Government. There was no meaning left in my continuing longer to be fed by the Defense Department.

Awaiting the completion of "An Evaluation of the Material War Potential of the Soviet Union," I presented formally in February my resignation. The Chinese used this appeal and that plea to try and get me to stay. However, this time there was nothing left to make me change my mind.

"The British are still looking for you. If you return home you will surely be caught and sent to the gallows. Bear for a while and wait."

This was always the final plea that they brought out.

"I am very grateful for your thoughtful consideration of my future. Since I was appointed to serve in the Defense Department, I have with sincerity produced material and data only to see them, time and time again, put aside unused. Even at that, if I were not wanted in Japan as a war criminal, I could bear this and stay. However, I cannot remain a day longer in China if my purpose were to escape pursuit. Whether I am arrested or not, I leave to fate. In my native home, I have an old mother close to 80 years of age, sick in bed. If I, just because I am afraid for my own life, am unable to be at my mother's side at her death, knowing beforehand as I do now that she no longer knows whether she will live the morrow, then I would never be able to continue my life as one who failed to carry out his filial duties. If I should be arrested upon my return home, I shall make no pleas in my own defense but shall go the gallows with a smile on my lips. This is the only natural way in which I can atone for the sin of defeat."

This was the gist of an appeal which I made directly to the Chief of Staff. It seems that they could not oppose this final pleas of mine. Finally, at the end of April, permission came for my return home.

The answer was:

"You will be allowed a two-month vacation. We would like you to come back once again and help us out."

I doubted whether they really desired the return of an old and tough colonel of a defeated army, judging from their past attitude.

Finally, on May 13, when I presented myself at the Defense Department to pay my farewell respect, the Second Section Chief deeply thanked me for my services during the past two years and asked me to come back again as soon as possible. At this farewell greeting, I should perhaps from a standpoint of diplomacy have accepted the invitation and have left it at that. However, my conscience would not permit me to do so. I, thus, replied:

"I have never in the past lied to a Chinese. On this last parting with you, I would like you to permit me to say the truth. During, the past two years, I banked my whole life in my work and in my advice to you. However, the net result is that nothing that I did was of any value to you. The reason why I resolved on returning to Japan was because I realized that the whole trend toward disintegration could no longer be prevented, no matter how I tried. I want to tell you that I have neither hopes nor intentions of coming again to China. I wanted, at least, to go home without telling a lie until the very end."

Lieutenant-General Hou's face clouded in deep thought. I parted company at Major-General Huang's home with Wang, Major-General Yeh and the others with whom I had worked together for two years despite the frequent quarrels. I completed all preparations on May 14. Lieutenant-General Isogai was now in Shanghai, but Lieutenant-General Hiroshi Takahashi was still in the Nanking Penitentiary. I sent him a final parcel, at which he wrote an answer that choked me with tears.

There was a major-general, Tsaof Tai-chung by name, a native of Manchuria. He had studied in Japan and had all the three attributes necessary for a good soldier, brains, brawn, and guts. Because he was native of Manchuria, all were afraid of him and none could make full use of his abilities. During the war, while Major-General Sakurai was engaged in Honan Province in the work of persuading Chungking troops to defect to the Nanking camp, Tsao mixed in with the surrendering troops in disguise. Then, catching the Japanese forces in an unguarded moment, he incited the troops to mutiny and leading the greater majority of the defecting troops, calmly marched all the way to Hsian. He still retained the make of a sword cut inflicted ion him by a Japanese officer at this time. He was the principal of the Central Training Corps and was a confidant and a trusted aide of the Corps Commander, Lieutenant-General Huang Shieh. I had fought against Lieutenant General Huang in North Burma, where he had commanded the Chinese forces and my memories of him as a ferocious soldier were still fresh in my mind.

Tsao came over on the eve of my departure for Japan and asked me to meet General Huang. I readily agreed and the meeting was arranged to take place in Tsao's home. General Huang, who arrived clad in a Chungshan robe, I found was a man at his prime, being around 43 or 44. I was two or three years older than the General but was only a colonel in rank. He had already graduated from the post of an army commander and was engaged as a proxy of the Generalissimo at that time in the re-education of middle class officers. In addition, General Huang was one of the most trusted of the Generalissimo's soldiers. Our conversation flowered with recollections of our battles in Burma, on the battlefield at Lungleng and Mangshih and I noticed that his features had gradually reddened with excitement. We praised each other for having fought well and we promised to be comrades in the future. He had come, at the secret orders of the Generalissimo, to listen that night to my final unrestricted and uninhibited opinions. I realized that this was one opportunity in a thousand. I finally found myself talking until three in the morning. The gist of my conversation was:

1. There is no person close to the Generalissimo who is not afraid to offer blunt advice and bitter counsel. Premier Tojo failed because of this shortcoming. If the Generalissimo allows this situation to continue, he will also meet up with failure. The upper and middle strata of the Defense Department should be cleaned out en masse and staff officers, not afraid of giving straight-from-the-shoulder advice, should be given their way in overcoming the present crisis.

2. The war is decisively disadvantageous to the Chinese Nationalists. The successful capture of Juichin more than a dozen years ago still seems to be burnt into the Generalissimo's brain. However, the actual fighting prowess of the Chinese Communist Army today is much superior to that of the Chinese Nationalists. The Generalissimo should evaluate impartially and detachedly the respective war strength of both the Chinese Communist and Nationalist side and should make a prompt and resolute decision. There is only one way to arise from death and find life again. Manchuria should be abandoned immediately. North China should be evacuated and the whole army be withdrawn to the Yangtze, where a line of should be consolidated. The provinces of Fukien, Kwangtung, Kwangsi, Hunan, Kiangsi, Yunnan and Taiwan should be built up immediately into bases to continue the war over a long period. The capital should be immediately moved to Canton. If the present situation continues, Manchuria cannot be held till the end of this year (1948) and Nanking and Shanghai will fall in the spring of next year (1949).

3. The post war diplomacy of China has been extremely weak. China still continues its old policy of playing off one foreign nation against another, thereby losing the trust of her allies. China should immediately change her attitude, abandon her right of veto in the United Nations Security Council and together with the United States and Great Britain should conclude peace with Japan. With the help of Japanese technical aid and American capital, the Generalissimo should push the development of South China.

4. It is the heights of foolishness to remove factories from Japan as reparations. At the present time, precious machine-tools lie rusting on the piers at Shanghai and are degenerating in scrap. This negligence should be immediately halted and the machinery sent to important factories. In addition, the industrial facilities of Japan should be allowed to begin operations, with China providing the raw material and receiving in return the finished products.

5. The political corruption of the Kuomintang Government should be thoroughly purged. Before punishing the small malefactors, several-high officials, engaged in large-scale larceny and swindling, should be executed, thereby putting a fresh concept of life into the hearts of the people. Political corruption has thrown China's economy into chaos, alienated the hearts of the people from the Government and has caused the army to lose its fighting spirit. The basis of anti-Communist administration is to carry out land reform policies and win over the agrarian masses. That which nurtures the growth of the Communist Party is the corrupt elements within the Kuomintang itself.

If these five platforms are carried out, the Kuomintang will be able to retain control over half of China's vast territory, bring about a radical change in the international situation and maintain a foothold on which it would be able to build up and rise again. If these measures are not carried out, the Kuomintang will not be able to last another year.

Lieutenant-General Hung condensed the stenographic report of this conversation into roughly 30 penciled pages and asked me for my signature at the end. We three, Huang, Tsao and myself jointly signed the document. This, my final advice, was to be submitted directly to the Generalissimo, without being rewritten in clean form early next morning. Lieutenant-General Huang expressed complete accord with what I had talked about into the early hours of the morning, and said:

"If only I had met you one year earlier."

This was a feeling entertained mutually by the whole three of us laying aside the possibility of what would have happened if we had met one year ago, I felt that if the Kuomintang Government had put into effect the measures that I advocated even at that late time, it would have held today the areas south of the Yangtze.

At dawn of May 15, I returned to Kiangsu Road. "A parting bird leaves no dirt," and I finished clearing up my entire surroundings. To First Class Private Yu Chih-lin, who had looked after me for one year, I gave all the spare money that I had. A farmer's son from Supei, his home had been burned down during the war and his father killed. Together with his mother, he had fled to Nanking. A pure-hearted and a kind youth, I had treated him as best I could, shared what little I had received. He cried and asked to be taken to Japan. I consoled the boy, telling him that I would come back for sure once again.

Finally, I bade a sad farewell to Colonel Okawa and his men. We prayed for each other's health. Then it headed for the Nanking Railway Station. I shook hands with Liu, Ma, Huang and other Chinese friends. It was probably my last parting from a Nanking that would not least another year. If fortunately I should ever return, a red flag would be flying over the ancient walls of this city. I felt an indescribable feeling well within me as I thought that the fate of these friends would be decided within a year. Major-General Ma came all the way to the station to bid me good-bye, bringing with him sugar as a farewell gift. He tendered to me to the very end his deep friendship, staying until the train pulled out. Private First Class Yu Chih-lin broke down and cried in loud voice right in the middle of the platform and ran after the moving train, shouting:

"Tsai Lai Ya, Tsai Lai Ya." ("Please come back again please, come back again.")

It seemed that the sincerity of one Japanese had penetrated most deeply into the heart of an uneducated soldier. Eighty percent of the 400,000,000 population of China, I thought, is composed of just such people.

In Shanghai, Staff Officer Liu took me to the home of Wang Wen-cheng. The reason for this lay in the fact that Liu's family shared the home, living on the third floor. Wang's real name was Wang Tzu-hui. During the war, he had crossed over in disguise to the

Peace Camp, with the understanding of the Chungking Government and had become Minister of Industry of the Provisional Government. In this position, Wang had worked to bring about peace between Japan and China without success. After the end of the war, he was saved from being branded a traitor and in 1946 had crossed over to Japan to work for Japan-China collaboration. However, because of complex rivalries with the Kuomintang he had been recalled.

On the second floor of this home, lay Commander-in-Chief General Neiji Okamura of the Japanese Expeditionary Forces in China, sick with tuberculosis. Lieutenant-General Matsui was living in a Japanese home, nearby and often came to see General Okamura. I had no particular desire to meet the two. However, I felt that I should pay as a former army officer, at least my final respects to them. I wondered how troubled Commander Okamura and General Matsui must be to be living in ease outside of prison walls, while so many of their subordinates had been executed as war criminals and so many others still lay in chains. However, contrary to expectations, both seemed untroubled and full of bright spirits.

For Wang Wen-cheng to protect former Japanese officers after the war and continue his relations with them was not an easy thing to do when the safety of his own self was at stake. However, not only Wang, but his wife and his children greeted me with joy and looked after me with loving care. For this I had nothing but deep gratitude.

All of my personal belongings were contained in one suitcase and one roll of bedding. However; I felt that as a father with children, should at least take some candy home to them. I went to the Yingan Kungsu (Yingan Stores). While shopping I engaged in idle talk with the Chinese clerk:

"When will the Communist Party come to Shanghai?"

"Perhaps, within one year."

"Which side do you prefer?"

"It's just the same. Whoever comes, it's 'mei fa tzu' (can't be helped) for us. The master of the house will be the only one to be shot."

Chinese Communist Party Fifth Columnists had for some time already begun their underground work, lulling the people into a false sense of security. Despite the propaganda of the Chinese Nationalists, who tried to paint the Communists as devils and fiends, the popular masses, sick and tired of the corruption of the Kuomintang, were waiting with a mingled feeling of curiosity and interest the arrival of the new guests. Of course, the general citizenry of Shanghai, for instance, knew nothing of Marxism and their anticipations in the coming of the Chinese Communists had no connection with "isms" or policies.

After I had completed what little shopping I had to do, I spent the rest of my money purchasing gifts to be given to those held in the Shanghai Penitentiary. These parcels, I entrusted to Staff Officer Liu for delivery. I wanted, if possible, to visit my seniors and comrades again, but that was not permitted. Neither did I have the time to spare. My ship, I was told, was to set sail the next morning. I went to the concentration camp for Japanese and a completed my procedures for returning to Japan, registering myself as "Professor [Kiuchi Katauji of Archeology], Peking University."

The majority of the Japanese in China had already gone home. Those who remained were a handful of technicians who had been left behind to serve with the Chinese Nationalists. The Japanese in the concentration camp, to my surprise and disgust, were engaged in slander and stool pigeoning on each other. So disgusting was this spectacle that the inmates were even despised by the Chinese. Despite the fact that they had come face-to-face with a great racial tragedy, they had not the slightest desire to protect each other. They only wished to kick down the next man and seek life for themselves alone.

I wondered if this was the national habit of the Japanese people. I wondered if the Japanese were such a stupid race.

On May 16, I bid farewell to the family; of Wang Wen-cheng and to Major-General Wang Pei-cheng. Then, together with Staff Officer Lia I went to the pier. Around 150 Japanese civilians and 50 or 60 war crimes suspects were lined up under a burning sun, waiting to pass through customs inspection. This, I learnt was the last repatriation ship from China. Because I had been busy during the last few days, I had not attached luggage tags to my belongings. A Japanese representative accompanying the inspection group shouted at me in a rage:

"What in the hell do you think you're doing. Don't be a nitwit and hurry and put your name tags on."

These words, the last words said to me on leaving the continent, and from a Japanese to boot, came from a representative of the Japanese Residents' Association of Shanghai.

"I'm sorry. I arrived late in Shanghai last night and did not have time to get around to it," I apologized. Wiping the sweat from my brow, I managed somehow to suppress the flames of anger rising in my breast. Then, as if in abject apology, I bowed low. A young high school girl, who had come to see friends off, had been standing by, watching this scene. Pitying me in my plight, she immediately came to my aid. She tore a towel into strips and with needle and thread, she sewed tags on my luggage. Then, she borrowed ink and brush from neighboring Japanese and deeply wrote my name on the pieces of cloth. And, presto, I had my extemporaneous name tags.

The customs inspection was extremely strict. Everything of value was confiscated by the customs officers and the gendarmerie. I felt pity at the sight of the Japanese, standing by helpless at this outrage, unable to cry out in protest, no matter how indignant they may have felt.

I feared that if my belongings were searched in such a thorough fashion, the records of my whole three years of travel incognito would be exposed. I wracked my brains for some way in which I could avoid this inspection. Staff Officer Liu seemed to read my mind and arranged with the officials to have myself alone passed without inspection. I felt that my ugly features did not deserve such an exemption from regulations.

I had landed on this very pier in February 1932, during the first Shanghai Incident, as a company commander [of 7th Infantry Regiment]. This was also the same pier from which I had left on my triumphant voyage home. It was not a final parting for me from Shanghai, which had been intimately bound up with the vicissitudes and changes in the secret history of 17 of my years of the first half of my life. I was now leaving this same pier, upon which I had trodden in the past, gripping my sword with soaring spirits. But, in what different circumstances! Now, I was a timid and pitiful-looking individual, seeking to avoid the eyes of people.

The customs inspection took almost a half day. From luggage accumulated by blood and sweat had been confiscated a mountain of clothing, bedding, cloth and food. These were piled into trucks and taken away.

Just before dusk, the vessel carrying a load of emotion-filled people quietly pulled up its anchor. The ropes were let loose and slowly it edged away from the continent.

The white handkerchief which Staff Officer Liu kept waving, remained for a long while in my mind.... Shanghai, good-bye!

Chapter 6 Aspects of a Changing China

I had travelled a long way in disguise in my underground escape, crossing one barrier of death after another, with a heart that refused to let me die. And during this long sojourn, what had come to my sight was the sad image of a China torn by internecine war and bloodshed.

The Chinese Nationalist Army, which had shown such resilient strength during its eight years of resistance against Japan, had crumbled swiftly in face of the Communist Army. Why had the Nationalists been so easily defeated? The world asked this question. Yet, it was a fact, a fact which I myself could scarcely believe.

"China cannot be conquered by force," is the conclusion that comes from a study of her turbulent history.

However, this conclusion does not mean that the Government of Chiang Kai-shek could not be conquered or be made to disintegrate. The "July 7 Incident" was not a welcome event. However, Japan must be given the credit for having made Chiang Kai-shek into the hero that united the whole of China, at a time when the task of unification remained still incomplete. If at that time, the China Incident did not occur, China would have continued to be torn until today by civil war between the Kuomintang and the Chinese Communist parties.

If the General Staff Headquarters of Japan had had plans and prepared rations for an all-out war when the July 7 Incident broke out, and Japan had engaged in a thorough war against Chiang Kai-shek, Chungking would probably have fallen within two year's time and the Chiang Regime, or the Kuomintang Government would have collapsed. At the same time, Japan would have been drawn into long-term guerrilla warfare with her enemy, the newly-risen force of the Chinese Communists. With the same low moral level as exposed in the past, Japan would never have been able to win the cooperation of the 400,000,000 people of China and would have come face to face eventually with the same fate as today. The net result would have been to bring about the unification of China under the Chinese Communists.

The fact that Chiang Kai-shek was able to continue eight years of resistance against Japan was not because he was strong but because Japan's strategy was mistaken. The person who knows best the weakness of the Kuomintang was none other than Chiang Kai-shek and it was he who expended every effort to bring about war between Japan and the United States in order to escape from inevitable defeat.

Immediately before the outbreak of the Pacific War, when the plane carrying Admiral Osumi of the Japanese Navy crashed in the vicinity of Canton, the Chinese captured a most important document, a document containing complete data on the fighting strength of the entire Japanese Navy. This was immediately flown to Chungking by the Military Control Bureau.

The Chief of the Military Operations Department (Chief of Staff of the Chinese Nationalist Army) at that time attempted immediately to reveal this precious data to Chungking's ally, the United States. Chiang Kai-shek was greatly angered and said:

"If this data is shown the United States, she would fear the fighting strength of the Japanese Navy and would hesitate in her decision to fight Japan. China has no hope of continuing its existence except through making Japan fight the United States."

The whole 4,000-year history of China has been one continuous policy of playing off one foreign power against another. We should realize fully that this had indeed become a national trait of the Chinese people.

This incident was told to me with pride, as an example of "the farsightedness of the supreme leader," by a responsible high official who was directly connected with the affair.

The moralists and the intelligentsia of Japan who shed tears of adoration at the Generalissimo's instructions to "reward hatred with virtue" and who humiliated themselves, saying, "We were also defeated morally by China," should further reflect with prudence, and view frankly the unshakeable realities of China's past.

Generalissimo Chiang Kai-shek and the Kuomintang Party which he heads have as their prime credo the philosophy and the principle, that there is no country outside of China: Japan does not count; neither, for that matter, the United States. The Generalissimo is definitely not a believer in the principle of "Asia is one." It seems that he is neither an internationalist.

During China's war of resistance against Japan, there were three occasions on which the Chinese Nationalist Government stood on the brink of collapse.

The first time was immediately after the fall of Hankow. If, at that time, the Japanese Army had pushed on without taking a breath straight to Chungking, a situation would have arisen in which Chiang Kai-shek could have been disregarded, even without waiting for Premier Fumimaro Konoe's notorious statement to this effect.

The second time was immediately after the outbreak of the Pacific War, when the Burma Road was cut off. If at that time, the Japanese Supreme Command in China had resolutely undertaken a drive to Chungking, the Chiang Kai-shek Government would have had no other alternative than to surrender.

The third time was during the Kweilin Operations, the year before the end of the Pacific War. When part of a Japanese army corps appeared at Tushihkou in Kweichow Province, Chungking was caught in a state of complete vacuum and was head over heels in preparations to move the capital to Chengtu. Of course, at that time, the general war situation was disadvantageous to Japan and it might have been difficult to crush Chiang Kai-shek's will to continue fighting. However, it was a fact that a Government, worn out by war and sick and tired of fighting, had come to the verge of collapse.

Even after the fall of Okinawa, the Chungking Government believed that war would continue for another three years. Thus, when Japan surrendered after two atom bombs were dropped on her, China was thunderstruck.

The Kuomintang had neither preparations for victory, plans for taking over Japanese assets nor measures to cope with the Chinese Communists.

There was a 180 degree swing from tension to relaxation. Everything was in turmoil and already the portents of decline were evident.

It was Japan's so-called continental policies that enabled Chiang Kai-shek to bring about so handily the unification of China and it was Japan's unexpected early surrender that was one of the causes of the Kuomintang's surprisingly early collapse.

Japan today suffers through her lack of experience of defeat. China through a lack of experience as victor failed to consolidate her victory and invited the tragic fate of eventual collapse.

Despite the fact that Premier Konoe declared in a solemn statement that Japan would have nothing to do with Chiang Kai-shek, various peace negotiations were carried on through various channels by various peoples. There were efforts made by the China Section of the Japanese Army, by the Black Dragon Society under Mitsuru Toyama, and by people connected with Steward of the United States. In addition, anyone who claimed to be a China expert, every Tom, Dick and Harry, had some secret tales concerned with peace negotiations with the Chungking Government: In contrast to the shameful contest among the Japanese to claim the honor of successfully negotiating peace first with the Kuomintang, the Chinese side unified all peace negotiation efforts through the one channel, that controlled by General Tai Li Nay, nobody outside of Tai Li could have handled such negotiations. If one did so, he would have had to resolve himself to being killed by pistol-carrying members of the Blue Shirt Society. When I met the man in charge of peace negotiations on the Chungking end, this person said:

"You were at least three years late in coming. If it were before the fall of Saipan, China would have given up Manchuria to Japan, by covering its eyes, in return for the restoration of at least Formosa, as à matter of face. Then, there could have been the possibility of also dragging in the United States to make peace with Japan, along the lines of opposition to Soviet Russia and Communism."

But, it was already, the day after the fair. There was no use crying over spilt milk.

Despite the fact that I was held in detention like any other prisoner of war, the cold winds of the outside world blew upon me. The changing moods and feelings of military bureaucrats and the words and actions of their families and employees all of whom I daily met impressed deeply upon me the shifting aspects of the world and citizenry about me. In particular, the newspapers, which were my only comfort, told me bluntly of the raw realities of a China undergoing swift transformation. The Chinese citizen that I observed was seen only through the small window of my solitary room. Yet, when woven into the general fabric of my experiences of the past, these threads of observation helped form a picture of the Chinese citizen as a whole. Let me tell of few of the changes saw and the conclusions which I reached.

Even the Chinese Communists could not deny that the reputation of Generalissimo Chiang Kai-shek among the whole 400,000,000 people of China was at its zenith immediately after the war. Every street in Nanking, every home, even the walls of toilets, were covered with pictures and photographs of the Generalissimo. In reaction, this adoration of the Generalissimo naturally also took the shape of denunciations of the Wang Ching-wei Government and calumny against the Japanese Army. This reaction of hatred toward all past masters filled the streets of all Chinese cities and towns in the areas under Japanese occupation. Then came the military bureaucrats of the Kuomintang Government. In dirty Chung-shan robes, they fell upon Nanking like a swarm of locusts. They requisitioned homes and under the pretext of hunting for collaborators began a campaign of extortion. Gradually, day after day, they alienated the feelings of the people. Then, when civil war started, the voices of adoration of the Chiang Kai-shek Government changed to those of denunciations.

The student movement has always formed the spearhead of revolution in China. Viewing the corruption of high government officials after the end of the war, their livelihood threatened by ever rising prices and with salaries completely pegged at an unchanging level, the students, no longer able to eke out a living, suddenly rose.

At first, their objective was to gain an increase in the food allowance of Government-subsidized students. Soon the students began to oppose the civil war, then they began to demand the withdrawal of American forces. The student movement had developed into a full-scale political movement. In Peiping a clash occurred between the Kuomintang garrison forces and demonstrating students. Several students were killed and wounded. The study body of the entire nation went on strike in protest and demonstrations flared up everywhere. In Shanghai students of the Tungchi University struck Mayor Wu Kuo-chen. The incident developed into a student riot.

Just at this time, Great Britain began the compulsory evacuation of Chinese living within the Kowloon Leased Territory. The student movement suddenly expanded into an anti-British movement. Resentment and hatred swept through the student ranks of the entire nation, leading eventually to the setting a fire, by students of the British Consulate in Shamien. The strategy of the Chinese Communist Party was to create incidents between the Chinese and foreign countries, in particular, Great Britain and the United States and to place the Kuomintang in a desperate position. This trend became more evident with the passing of days.

The CC Corps and the Military Control Bureau began the arrest of Chinese Communist Party underground members in all parts of the country in an effort to counteract this strategy. However, this only served to incite the general masses still further.

The Kuomintang had underestimated the student movement, thinking that students armed even without a single pistol, could hardly do anything. It was a grave mistake. The Communists, who supported and exploited the students, gained great successes.

Thus, Chinese parents, with children still students, were alienated from the Kuomintang and came to have sympathy toward the Chinese Communists. At the same time, the words, “Wang Puppet Regime,” and “Traitor Wang Ching-wei,” soon became “Wang Government” and “Dr. Wang Ching-wei.” And gradually the pressure against and the punishment of members of families of collaborationist were relaxed.

Immediately after the return of the capital to Nanking, I had gone into a barbershop and had been recognized as Japanese. The barber deliberately struck my head with his clippers again and again while trimming my hair. One year later, this barber was speaking to me with real friendship. Two years later, this barber would bring his mouth to my ears and whisper in Japanese:

“It was better when the Japanese Army was here.”

In my early days in Nanking, I was not a very welcome customer when I went out shopping in the markets. Two years later, it had all changed. Merchants smiled and tried to please me. Sore out of sympathy would even give me a discount of ten per cent. Pedicab drivers, who demanded exorbitant rates whenever they saw a once Japanese, would say:

“You’ve got no money, so I’ll make it cheap.”

The citizens of Nanking, who had waited for eight long years for the return of the capital to their city, saw that the corruption of the officials of the Nationalist Government was much greater than that of officials of the Wang Ching-wei Government and swung over to the side of the Chinese Communists, at the same time as yearning for the “good old days,” when the Japanese Army occupied the country.

Servants, long employed by Japanese families, often could be overheard talking to each: "I wish the Japanese soldiers would come back once again...."

Inflation followed on the heels of war in all belligerent countries throughout the war. But, in no victor country and not even among the vanquished, was inflation as acute as in China. This is how the commodity price index soared in China during her eight years of war against Japan and after.

Taking the index of the year 1937, just before the outbreak of the China Incident, as one, prices had climbed to 635 eight years later. The Japanese Army which saw this spectacular rise believed that the Chung-king Government would face economic collapse.

One year after the war the index had soared to over 6,000; two years later to 60,000; three years later to 1,000,000 and four years later to over 15,000,000. The reason for the defeat of the Kuomintang lay in such an economic collapse, which influenced military operations. Military defeats further accelerated economic deterioration, completing the vicious cycle that brought the Chungking Government to its present straits.

Immediately before the China Incident, 10,000 yuan would have bought a beautiful three-story western-style mansion with land included. On the day of victory, 10,000 yuan would only buy one head of cattle, one year later one pig, two years later one chicken, three years later one egg and four years later a single sheet of thin writing paper.

Victorious China dropped to a condition far worse than that of vanquished Japan.

During the period from the V-J Day until civil war became full-scale hostilities, 80 per cent of the "racial industries" which arose during the war in Szechwan Province went into bankruptcy, while the factories in Shanghai taken over from the Japanese sent no smoke up their chimneys, with the exception of a few textile mills. The chief reason for this was the dumping of American goods in China. The flood of cheap and yet good-quality foreign goods drove out domestic products, and forced one industry after another and one family artisan after another into unemployment and bankruptcy.

The Sino-American treaty of commerce and trade, signed against such a background, gave complete equality to China and the Kuomintang loudly publicized that this was China's first equal treaty since the opening up of the country to trade with the west.

As a result of this treaty, luxury goods and non-urgent commodities flowed like a tidal wave into China and the Government's dollar holdings, totaling some 750,000,000 dollars, dwindled to nothing overnight. Then came the flow of gold from China, which accelerated China's inflation. Finally, the Government was forced to limit imports and adopt a policy of subsidizing exports to spur outgoing trade. This measure met with the opposition of foreign firms and was abandoned immediately. In the meanwhile, civil war had disrupted communications, preventing the smooth interflow of commodities and adding impetus to the mounting inflation. The capital of the battlefield areas flowed into Shanghai, spurring still higher the cost of living and commodities. Economic collapse now threatened national livelihood and political corruption fed the rising tide of Communist popularity. Following this trend step by step with impartial eyes, a discriminating American observer stated that one half of the responsibility for the bolshevization of China should be borne by the United States.

The first thing that government officials did on reaching their desks each morning was to look through the columns of newspapers containing commodity price lists. The look in their eyes would be intense, the most intense eyes for the whole day. As soon as they get their pay, they set out immediately for the markets and, leaving only a very small amount, changed their money into commodities. They bought cotton cloth, tobacco and canned goods. The high Government leaders spent their day hunting around for American dollars and gold bullion.

Commodity prices did not go up with a day as a standard. They rose hourly or even every ten minutes.

At one store, I thought of buying a blanket for 1,000,000 yuan. However, I decided to look around. After dropping in at two or three other stores, I came back. The price had gone up to 1,100,000 yuan. I complained, but the storekeeper placidly replied:

"That was the price two hours ago.

The commodities which the petty officials hoarded would be sold later a little at a time, depending on the movement of the prices. With the resulting margin, they would buy their pitiful daily necessities in order to eat that day. The only way to prevent the value of money from changing, in face of the minutely rising prices, was to change money into commodities.

The collapse of the "racial industries" completely transformed the character of large towns and cities. These cities changed from productive centers to great consumer-areas. Production became limited to agrarian areas alone in China. When the agricultural districts, attracted by the Chinese Communist program of land reforms, fell under Communist control, the Kuomintang became only the guardian of great consumer areas.

During this period, devaluation of the currency was carried out a number of times and the Government planned again and again to issue large denomination notes. At that time, the largest denomination was the 10,000 yuan note. The Finance Ministry announced in the paper that the Government would absolutely not issue 50,000 yuan notes and attempted to allay the existing chaos in financial circles. Less than ten days after this statement was made, the Government issued 100,000 yuan notes. Shanghai financiers, hard-headed and daring as they were, could not close their mouths opened in a moment of stunned amazement.

At the end of 1948, the Government issued a proclamation designating all gold and silver and foreign currency as Government property. It strictly banned all private possession of this form of wealth. Persons found having even a single American dollar in their possession were shot to death on the spot. The eldest son of the Generalissimo, Chiang Ching-kuo, was placed in charge of carrying out this Government order, but so drastic and so strict was his enforcement of the edict that the whole of Shanghai was thrown into a terror. The flow of commodities into Shanghai, particularly food, halted altogether and the citizens were driven into dire straits. The Chinese Communists took advantage of this situation to spread further their propaganda. In addition, the defeat of Chinese Nationalist forces in all parts of the country put skids under the deteriorating reputation of the Chinese Nationalist Government. In addition, various provincial governments rebelling against the monetary controls of the Central Government began issuing currency of their own, based on silver, as a self-defense measure. As a result, the Government began to lose control of outlying districts. The resultant loss of authority by the Chinese Nationalist Government spurred the defection of various Chinese Nationalist units to the side of the Chinese Communists.

The masses of China, used to repeated revolutions, changes in their rulers and economic collapses, anticipated the Government's changing policies by resorting to self-defensive measures, and skillfully swam through the rough waves of national disintegration. Despite this fact there were also many tragic incidents too numerous to recount here. In one country village an old women had scraped and saved for 50 odd years some 6,000 yuan in silver currency. This she had placed in the walls of her mud hut and plastered it over with mud. The money was saved to keep her alive in her old age. Some Government officials heard of her wealth and on the grounds of the Government ban on the possession of hard currency tore down the walls of her home and took away the silver money. In return, they gave her 6,000 yuan in inflated notes of the Kuomintang Government. The shocked woman stood staring at the fistful of notes, representing not even enough money to buy a package of cigarettes. That night she threw herself into the Yangtze River.

General Marshall spent one year in China expending every effort to bring about a reconciliation between the Chinese Nationalists and the Chinese Communists. But all his endeavors were in vain. On the eve of his departure from Nanking, he publicly stated that the corruption and the bigotry within the Kuomintang were beyond salvation. Lieutenant-General Albert Wedemeyer, who succeeded General Marshall, frankly exposed at the final afternoon tea held in his honor, in the presence of Generalissimo Chiang Kai-shek and all party and military leaders of the Kuomintang, the political corruption of the Chinese Nationalist Government. General Wedemeyer even went as far as to say:

"It is the Kuomintang that is helping the Chinese Communist Party."

His was the attitude of an inspector making his final criticism and his judgments were absolutely correct. Yet, why did he not go a step further and study the basic reasons for the rottenness of the Kuomintang Government?

No one can guarantee that an American, transplanted to China without a single personal belonging and forced to live life among the natives of China, would not become as corrupt as the Chinese people themselves.

The reasons why the politics of China becomes corrupt are first modern industries, capable of absorbing youthful intellectuals, have not been developed and second, and more fundamental, the population is too great in relation to the land and its resources.

It is rare to find a people in the entire world so oppressed as the farming population of China and it is rare to see elsewhere on this globe land so exploited as that of the vast areas of China. No one can talk about the politics of China, without having seen poor farmers digging up even the roots of grass for fuel on completely denuded mountains.

If a man cannot live unless he attempts the impossible, his whole life work is concentrated on how to accomplish the impossible with the least effort. This effort to live is concentrated on the lowest scale in the mutual protection of a family unit. This is next expanded to included relatives, then members of one clan and is eventually broadened into a union of people from the same village, town or province. The operation of the life of the Chinese, in its political, economic and military phases, not only in China itself but even overseas, is based on these blood and birthplace relations. This peculiarity must be viewed as resulting from the history of a people who never during their long existence enjoyed a state authority which they could trust to afford protection of their lives and property. Thus, Chinese society is based on vertical units, so to speak, with people from the top to the bottom knit together by a unity of relations aimed at mutual aid and mutual protection. Thus, if the head of a certain Government department is transferred, then the section chiefs and the section members right down to the messenger boys change. If a Government employee is once fired he has neither unemployment insurance or a pension to tide him over until his next job. Thus, it is the failing of any human being, not necessarily of the Chinese alone, to want to save sufficient money while employed to support his family the rest of his life or at least for two or three years until better opportunities arrive.

Year after year, the several tens of thousands of high school and university graduates gravitate to the large metropolises seeking help from blood relatives and from senior home towners. Thus, in China, a job does not look around for a suitable worker. It has become a custom of the Chinese officialdom to make a job or create a new desk for a new arrival thus, even as "a vessel piloted by too many people runs atop a mountain," Chinese officialdom becomes extremely inefficient. Nay, this is not quite correct. If a man works efficiently at one work that comes up, and clears it up immediately, he has nothing to do until a new job appears. Since he hates to be regarded as a loafer, it would be common sense for him to let one bit of work last as long as possible or until the next assignment comes up.

The novel, "Kuan-i Hsien-hsing Chi" (A Record of a Government Office), depicts and exposes bluntly the corruption of the Chinese official world during the last days of the Ching Dynasty. However, the Chinese themselves said that the official world of the Kuomintang Government was more corrupt than that of the Ching Dynasty. There was in the corruption of the preceding era a certain order and self-discipline. The bribes for a gate keeper, a secretary or the head of an agency were set figures. Thus, people lightheartedly gave these bribes with the feeling that they were paying a certain form of tax. However, with the advent of the Kuomintang to power, even this discipline and order disappeared. The Chinese people complained that no matter how much you gave Kuomintang officials in bribes, they were never satisfied. Perhaps, this is to be viewed as a result of the increased difficulties of making a living in modern China compared with the days of the Ching Dynasty.

The most important point in the revolutionary doctrine of Sun Yat-sen was democracy and the nuclear principle of this was the "equality of land rights." If the Kuomintang had brought into being the ideal of "Kengche Yuchitien" (the cultivator owns the land he tills) it would never have crumbled so, quickly in face of the Chinese Communists.

Everyone in the Kuomintang, from the elders to the minor officials, became land owners or connived with land owners. For the Kuomintang to have tackled the land reform question would have been to strangle itself. Such would have ended in bathos. The fact that the Chinese Communists are bravely carrying out their land reform program gives us grounds to believe that they have not as yet become large land owners.

The Chinese Communists, after all, are Chinese and are human beings. In order to live, they fought with hands and teeth to overthrow the Kuomintang. However, they themselves take over the position vacated by the Kuomintang, it is natural to expect that they will expose their human traits, become Chinese and Shanghailanders, no matter how great a hero and leader may appear, and no matter how he may set an example, it will be impossible for him to purge Chinese politics of its inherent corruption. The great Chinese statesman will be one who will be able to lead Chinese politics along right lines, fully accepting its weaknesses and its tendencies toward corruption.

If the politics of China is to be radically purged, the only way of doing this would be to redistribute the land and the resources of the world in proportion to population and thus stabilize the livelihood of all members of the human race.

Might not Japan today be also treading the same path of corruption of as trod by China's officialdom?

Practically all Japanese people, nay the entire people of the world, regard Sun Yat-sen as the person most responsible for bringing about the Chinese Revolution. This is not necessarily a correct view. He was undoubtedly one of the leaders of the Chinese

Revolution. However, every time he stumbled, he went on trips to foreign countries to escape danger and only returned when the situation had improved. He was an individual closely resembling Japan's Sanzo Nozaka.

The men who were really responsible for the success of the Chinese Revolution were practically all killed before the Revolution was brought to fruition. Sun Yat-sen luckily managed to live on to receive the crowning glory of being regarded as the Father of the Chinese Revolution.

It was the same thing that took place in the Meiji Restoration of Japan. Large numbers of patriots who strove to restore the Emperor died without seeing any hope of success. The men, who lived through, such as Kido, Okubo, Ito, and Yamagata, became princes and viscounts for their services to the Imperial House. The leaders of the Kuomintang, led by Generalissimo Chiang Kai-shek, are mostly men suitable for the destruction of the old. The fact that few were men capable of construction was another big reason for the collapse of the Kuomintang. Only when a leader capable of both construction and destruction arises can revolution be truly successful.

The Chinese Revolution was a result of the explosion of long-smouldering resentment against the tyranny of the Ching Dynasty, set off by the example of Japan's Meiji Restoration. To evaluate highly this revolution, which overthrew the tottering Manchu Dynasty, as a great social and ideological revolution, based on the Three People's Principles, is to disregard the whole 4,000-year-old history of China. The Chinese Revolution was more of a dynastic change. Today's Chinese Communist Revolution, though having the superficial color of a social revolution, thanks to its attention to land questions, has essentially to a great measure the characteristics of a dynastic change, whose roots go deep into the tradition of China.

Running like a common thread through the successive revolutions in Chinese history is the trend for a new and rising influence to breathe a spirit of freshness, but once in power to tread the path of degeneration. The seriousness and sincerity of the Kuomintang lasted only until the abdication of the Ching Emperor. From that time, the Kuomintang had been coasting downhill. For a time, Japan injected into the Kuomintang a shot of preservatives. With the defeat of Japan the preservatives lost their efficacy.

If the Kuomintang really had been believers of the Three People's Principles, it would not have met with its present day defeat. Following my arrival in Chungking, I conversed often with the nuclear elements of the Kuomintang on the doctrine of the Three People's Principles. However, I found none who could converse intelligently with me on this subject. The political knowledge of the party officials was limited to that crammed into their brains in order to pass government examinations and then completely forgotten. Politics was being carried out in the same way as all things in China. An outstanding example of this characteristic tendency was the constant bickering and rivalry within the Kuomintang itself. Until the opposition party is overthrown and the reins of the government taken over, the various factions unite in a big whole. But once power is grasped, there always appears an opposition faction within the party.

Two years after the war, Generalissimo Chiang scolded the Third Session of the Central Committee of the Kuomintang:

"I do not fear the Chinese Communist Party, I rather fear the rivalry of factions within the Kuomintang."

While the CC Corps group, centering around Chen Li-fu, Chen Kuo fu and Chang Li-sheng, sought control of the Government through placing Sun Fo, son of Sun Yat-sen at the masthead, the Europe-America faction, composed of the Kung and Sung families, gained control of the main body of the Chekiang money clique and through China's economic world opposed the CC Corps group. In between these two big factions were number of others. The big first-rate political bosses, Chang Chun, Wu Ting-chang, Wu Tieh-cheng and Hsiung Shih-hui constituted the Political Science faction, making mischief through Su Chin and Chang I. There was also the Kiangsi Faction, with its Big Two, Li Tsung-jen and Pai Chung-hsi, joining hands together to constitute a huge enemy state in opposition to the Chiang Kai-shek clique. In addition, there was the pro-Communist faction, composed of such men as Shao Li-tzu and Chang Chih-chung, seeking only the safety of their own necks. In contrast, there was Chen Cheng, the hope of the young army officers, who, ambitious to become the Generalissimo's successor, had organized the Three People's Principle Youths Corps and stood in direct opposition to the Kuomintang.

It 'would be too tedious to list the other minor factions. But, placed on a pedestal above all these quarreling groups, was Generalissimo Chiang Kai-shek, the symbol of unity. The dictatorship of the Generalissimo was only maintained through the power of the Military Control Bureau (Blue Shirt Society). It was thus natural for the death of General Tai Li to be a mortal blow to the Generalissimo.

In the election of the Deputy Premier in the spring of 1947, Sun Fo, with the backing of the CC Corps, opposed Li Tsung-jen, only to be defeated. The position of the Generalissimo himself was threatened by this upset and big split in the unity of the Kuomintang Party was brought to the fore. Chinese politics is based neither on the opposition of friendly political parties nor of ideals or principles. It centers around the interests of individuals. Those who have common interests widened their field of influence through blood and home province connections and formed factions and parties. The sole objective was individual gain or protection and honor. The Generalissimo himself fully knew that there was no room for race or for state in the Chinese political structure. The same thing can be said to greater or less degree of the internal makeup of the Chinese Communist Party. In the largest sense, the party is split into the International Faction and the Racial Faction. From a regional point of view, the Hunan Province Faction stands opposed to other provincial factions. Then, the war leaders, Lin Piao, Chen I and Liu Pai-cheng are constantly struggling against each other to win war fame, with an eye to succeeding Commander-in-Chief Teh. Then, the political and military giants, who defected from the Kuomintang Party, L' Chi-chen, Chao Li-tzu, Chang Chih-chung, Sung Kei-ling (Madame Sun Yat Sen), Fu Tso-I and Cheng Chien constitute the malcontent group, constantly seeking to split up the Chinese Communist Party. Perhaps, it would be correct to view Mao Tze-tung's headaches the same as those of Generalissimo Chiang Kai-shek.

Japan has 80,000,000 people pushed into four small islands, and has an emperor for 2,000 years. Yet this small country has today various parties and various factions, with each individual engaged in doing what he so desires. In contract to this, China has a population of 350,000,000 people, almost six times that of Japan. This population is a conglomerate of some ten odd races, differing in language, religion, customs and history. In addition, this vast population is scattered over a vast area, with an inadequate communications system. When we think of these factors, it is more than clear to us that no matter how many great men arise in China; it will be difficult to completely unify the country. Thc age in China, when a number of influential individuals instituted politics in differing regions aimed at maintaining the security of borders and safeguarding the populace, was the period when the people sang loudest its praises of peace.

The efforts of General Marsha toll push through reconciliation between the Chinese Communists and the various parties and factions of the Kuomintang, without first having mediated the conflicts within the Kuomintang itself, met with pitiful result.

In the spring of 1948, when the Chinese Nationalists still held Manchuria and North China, Generalissimo Chiang Kai-shek addressed the Fourth Session of the Central Committee of the Kuomintang and indignantly denounced the corruption of the party:

"As regards organization, propaganda and discipline, we cannot but recognize the vast superiority of the Chinese Communist Party over the Kuomintang."

Organization, propaganda and discipline are the three basic essentials for carrying out a successful revolution. If the Kuomintang was inferior in these three respects to the Chinese Communists, what then did it have?

On the basis of the newspaper item reporting this speech by the Generalissimo, I engaged in frank conversation a friend who was a backbone member of the Kuomintang:

"Did you enter the Kuomintang, believing that the Three People's Principles constituted an ideal?"

"I studied the principles just to pass my examinations doctrine insuring me food."

"Do you think the Kuomintang can win over the Chinese Communists?"

"Of course not.....If we do lose, I'll pack up my luggage and return to my home town. I still have paddy-fields and gardens."

"Do you intend to apply yourself to a study of Communism?"

"Depending on time and circumstances....."

"How long will the fight between the Kuomintang and the Chinese Communists continue? Is there no way of affecting a compromise?"

"The problem of Chinese Communism will be solved by the Chinese themselves in a Chinese manner. If foreigners butt in they're sure to fail."

"If the Communists excel the Kuomintang in organization, propaganda and discipline, what has the Kuomintang left?"

He broke out in a loud laugh and then said:

"Western-style mansion and Number Two and Number Three mistresses."

In the manner in which the Chungking Government executed the members of the Nanking Government, without mercy and without pity, there was not to be seen a shred of magnanimity expectable between the people of the same Han Race. There was a cruelty involved that was not seen even when the soldiers of Chungking were filled with hatred and enmity towards Japan at the heights of the war. The people who were executed as traitors had many acquaintances and friends in the Chungking camp. When the tragedy of disregarding the ties of a common blood and a common race were being repeated in succession, I began to doubt the old Oriental tradition not even a hunter will kill a wounded bird that seeks refuge in his bosom.

What I watched with greater pain than this were the scenes in which the Chinese Communist and the Chinese Nationalist parties washed blood with blood with a cruelty and brutality several times greater than the vengeance wrecked on the Nanking Government.

Whenever the Chinese Communists captured a city or a farming village and arrested underground workers of the Chinese Nationalists, they would kill not only the worker but his entire family. When the Nationalists would retake the same place, they would discover the countless bodies of their comrades in sheds and closets, tied hand and foot with wire and hung upside down. They would find the eyes gouged out, the noses chopped off and ears torn loose. Covering eyes at this brutality, they would, none the less, repeat the same treatment to Communist Party members whom they caught.

Merchants would betray their rivals to both the Communist and the Kuomintang parties. These betrayed persons would be plastered with the label "Te Kung Wen Tzu" and their brutally murdered bodies would be found in cities and in hamlets. In this way hatred, several times greater than that engendered against foreign foe, was instilled in the hearts of the people, to be handed down unto the third and fourth generations.

Lieutenant-General Kuo-Peng-chu, who defected to the Nanking Government from the Kuomintang forces to become commander of the Hsuchow District, and concurrently provincial governor, was to me a familiar figure whom I recalled with a certain sense of nostalgia. Although he was a nuclear member of the Chungking camp, he realized the futility and meaninglessness of war against Japan and had crossed over to help Wang Ching-wei. Given authority as military administrator, he purged the politics of Hsuchow Province, established peace and order and through his ability won the admiration of even the Japanese.

The year after the defeat of Japan, when I first arrived in, Nanking, I watched first to see what attitude General Kuo would take. With the 50,000 troops placed under his command during his Nanking Government days he made, a friendly overture to the Chinese Communists and showed himself ready to resist with force his arrest as a traitor by the Chinese Nationalists.

The Hsuchow area lay astride the strategic supply route to North China and Manchuria and its possession by hostile hands was a grave disadvantage to Chiang Kai-shek. He mobilized crack Nationalist troops and encircled General Kuo in a far-flung envelopment movement. Mao tze-tung, however, stood by, watched General Kuo's fate with coldness. Mao would neither send supplies or reinforcements. At this attitude of complete disregard, General Kuo once again defected for the third time and entered the Kuomintang camp. There was probably no other way left to him but to do what he did.

"Kuo Army, 50,000 Strong Defects," screamed the newspaper headlines and the news threw Chungking into wild joy. Kuo's actions must have hurt the Chinese Communists sensibilities. Mao Tze-tung immediately ordered out the main force of his Third Field Army under General Chen I, in a fierce attack against General Kuo.

General Kuo naturally expected Chinese Nationalist reinforcements but all that he received were several telegrams of encouragement. Chief of Staff Chen Cheng also sent a cable ordering General Kuo to put up a dogged stand. General Chen and the Kuomintang's idea was to place General Kuo's forces at the vanguard of Nationalist forces fighting the Communists.

General Kuo, of course, could not contend against such odds and was slowly driven back by the Communists. Leading his last troops, he prepared for a last stand near his birthplace, the port town of Haichow. Completely surrounded on four sides by hostile forces, he quietly watched the sun go down as he racked his brains for some means of escape. Suddenly, there appeared in his room an officer of the Chinese Nationalist Army, who said, sincerity brimming in his voice:

"At the orders of Generalissimo Chiang Kai-shek, I have come to save Your Excellency. One crack battalion will escort you out of this trap. Please feel relieved and escape with me."

To the General, completely depressed by the turn of events in the past, this officer must have seemed like an angel from heaven. He breathed a sigh of relief and, led by this unknown battalion commander, slipped through the cordon of the Chinese Communists in the covering dark.

The next morning, the General woke from a deep and peaceful slumber to be greeted by the officer that had saved him the night before. But, to his surprise, the officer was today in the uniform of a Chinese Communist Army. The officer was commander of a commando unit of the Chinese Communist Army that had disguised themselves as Chinese Nationalists.

The General, caught so easily in the Communist trap, was then whipped, kicked and led away in handcuffs. After this, he was dragged from district to district and finally his nose cut off, his eyes gouged out and his tongue pierced through, he had been thrown to the wild dogs. This story was reported a few months after the incident in all Nanking papers.

The more queasy Japanese do not welcome this attitude of glib prostitution of principles or defections. However, the long history of China does not approve this action as natural tactics, but neither does it level censure against it.

The sincerity and the ability of General Kuo had won admiration and praise in the Wang Ching-wei Government. The fact that he changed color several times could be forgiven him in the light of the good characteristics that he possessed.

In the vortex of civil war, which a dispassionate heart would never have allowed, in particular in the internecine strife between a race of one blood, such tragedies as that of General Kuo were repeated in countless numbers.

Modern versions of the Hogen and Heiji riots... the picture of brother killing brother and father killing son... are being repeated today in the neighboring countries of China and Korea.

On the basis of my experience in the numerous battles I fought during the war in China, I would place the battle strength of the Japanese and Chinese armies at a ratio of 10 to one. This figure is the ratio of respective troops that could engage in an equal battle, assuming the same conditions of training and equipment prevalent during the Chinese War. In other words, it means that one division of Japanese troops could meet ten divisions of Chinese soldiers.

There were not a few cases of heroism among Chinese Nationalist troops at Taierhchwang, Shanghai, Changsha and the North Burma battlefields. If we were to assume that we could give the same equipment and the same training and assume that a fight is waged on the same battlefield, I would place, on the basis of actual combat experience in eight years of front-line fighting, the fighting strength of the soldiers of various nations that I fought against in the following order (Japanese soldiers excepted):

1. Chinese
2. Soviet
3. Indian-Ghurkas
4. American

5. Australians
6. Other Indian
7. English
8. Filipino
9. Burmese
10. Thai
11. Annamese
12. French

(Of course this is the subjective view of just myself).

The Chinese soldier, unparalleled in the world in his ability to fight on poor food and poor clothing and to bear fatigue and hardships, had completely lost the courage they possessed when they fought the Japanese and collapsed miserably in face of the Chinese Communists. This was impossible for the Japanese to understand, and a grave disappointment to the United States.

The reason for this surprise and disappointment, lay in the mistake of thinking of the civil war between the Chinese Communists and the Chinese Nationalists as the same as a war against a foreign army.

Both the Chinese Nationalists and the Chinese Communists are Chinese. Defections from the Kuomintang to the Chinese Communists are to be regarded in the same light as the election picture of every Tom, Dick and Harry running for a Japanese Diet seat under the Liberal Party banner, when the reputation of the Liberal Party is on the rise, It is the same as the splitting and joining, the establishment and dissolution of political parties in Japan. These defections differ essentially from surrendering to the Japanese Army.

The Chinese had expected life to become easier after the defeat of Japan and had suffered and borne patiently eight long years of war. War ended, but their hopes were dashed to the ground. The old landowners, who had been chased out by the Japanese Army, came back prouder and more overbearing than when they had gone away. In contrast to the districts under the control of the Kuomintang, where tenant farmers were being exploited to even a greater extent that before, the farmers in the Communist areas now owned their own land and tenant farmers had become small-holders. News of this spread faster than by radio throughout the whole of China. If the soldiers of the Chinese Nationalist Army had been the sons of capitalists, it would not have mattered. However, 100 per cent of the soldiers were the sons of poor farmers and laborers.

It was, thus, natural for even the most uneducated soldier to begin thinking:

"For what reason must we be fighting the soldiers of Mao Tze-tung, who gave us land?"

Generalissimo Ching lamented:

"The military spirit has deteriorated and the will to fight lost."

However, he was mistaken, from the very beginning there was no will to fight among the soldiers. Why then were the Chinese Nationalists able in the early stages of the war against the Chinese Communists to take North China, and Shantung and occupy Yenan?

This was because the Chinese Communists withdrew temporarily from these places for strategic reasons without putting up a fight, thereby opening up a path for the advancing Nationalists. It was the same as if the sea, without obstructing the passage of a vessel, would immediately flow back and cover up the place that the ship had just passed. The Chinese Nationalists only had the cities that they occupied. These cities were equivalent to the position at which the ship stopped. The control of the Nationalists was limited to within the ship itself, while the broad expanse of the ocean still remained under Communist control. The proof of this truth lay in the fact that the Chinese Nationalists captured in all of the places they took hardly any weapons and no prisoners. They had simply moved into empty nests vacated by the Communists.

While the Chinese Communist Army maintained no distinction in food, clothing or shelter between officers and men, and determined the rank of their men on their ability to command and lead troops, the Chinese Nationalist army made great distinction between officers and non-commissioned officers and between non-com's and privates. Ability did not count in the Chinese Nationalist Army. It was a man's scholastic background and pull. Many became commanders without any combat experience whatever and there were not a few lieutenant and major generals among the many general grade officers in Nanking who volunteered for the post of division commander. Once a division commander, a general would report to the Defense Department that he had his full quota of 10,000 men. In reality, the actual number would be more like 6,000 while the pay for the 4,000 ghost soldiers, registered under false, names, would be drawn out by bogus seals, and be placed into the pocket of the commander. After one year as division commander, a general could buy a beautiful European style house in Nanking or Shanghai and acquire a mistress or two.

The rank and file could not be kept in the dark forever as to what their commander was doing. They would say:

"If our commander is doing such a thing, we can also do the same."

Usually, they found vent for this feeling in looting, bringing added distress to the innocent masses.

In the winter of 1947, when a transport crashed near Fengtien (Mukden) Chinese Nationalist soldiers were rushed to the spot to help in the rescue of the many passengers who were critically wounded. However, when they arrived on the scene the soldiers completely neglected their rescue work and immediately began stealing watches, fountain pens and wallets from the wounded passengers. This story was reported in the Chinese papers.

When the Chinese Communist Army captured the steel works at Anshan, they found a group of fearful technicians all hiding in one room. The Communist officers and soldiers gave these, men food and did not loot anything from them. On top of this, the Communists allowed those who wished to stay to remain and help with the work of restoration and escorted those who did not desire to stay to Fengtien.

The battlefield instructions of Mao Tze-tung, differed from that issued by the Japanese Army, and demanded the most-common place things and strictly saw that these instructions were enforced:

"Return for sure anything that you borrow."

"Pay for sure for anything that you buy."

"Make your camp out in the open."

"Those looting or raping women will be shot."

The instructions contained no mention of Marxism or the Three People's Principles. Only the standards of conduct were simply and clearly stated.

While there were a number of army commanders and divisional commanders surrendering to the Chinese Communists together with their entire army or division, the Chinese Communists who were taken prisoners were practically all wounded.

When 16 middle and lower class officers were captured wounded by the Chinese Nationalist forces after the fall of Yenan, a reporter for the Takungpao interviewed each of the officers separately. Fifteen of the officers said that if they were released, they would again take up arms and fight the Kuomintang. Only one said that he would go back to his home and return to his previous business. This also was reported in the Takungpao. If the whole 16 had been questioned together in the presence of each other, they might have advocated continued resistance just to maintain their "face." However, the fact that 15 answered the same thing when interviewed individually indicates that a vast difference in morale and fighting spirit existed between the Chinese Communists and the Nationalists.

The heavy equipment and weapons given to the Kuomintang Army by the United States were reportedly sufficient to arm roughly 40 divisions. However, one third of this war materiel was sold to the Chinese Communists by Chinese Nationalist officers and men and not captured in battle. The remaining two thirds went over to the Chinese Communists intact, to improve their equipment, by being stripped from Chinese Nationalist forces either captured or surrendering in combat.

In the fall of 1947, when the two armies of Generals Chen Chi and Liu Pai-cheng crossed the Hwangho and pushed at one stroke to the Yangtze and there cut into two along the Tai Pieh Shan mountain range the plains between the two rivers, they brought with them no supply troops, requisitioning their own food on the spot and receiving reinforcements of weapons and munitions from the Chinese Nationalist forces. In extreme cases, the Chinese Communists even replenished their ranks with soldiers from the barracks of the Chinese Nationalist Army and equipped and trained in Kuomintang schools with Kuomintang funds.

On the eve of the Chinese Communists' all-out offensive in the fall of 1947, Mao Tze-tung stated in his instructions to the entire Communist forces:

"Your supply base for men, weapons, munitions, food and clothing lies in the front line."

The Chinese Nationalist forces that took over Manchuria were composed of troops from South China, who had fought in North Burma. They were not only unused to the cold Manchurian winters but were soldiers that could not fight unless fed on rice. On the other hand, the Chinese Communists had swallowed some 200,000 highly trained men of the Manchoukuo Army and police force, and had trained and organized their new armies with picked natives of Manchuria, who were not only used to the cold but could subsist on kaoliang, so plentiful in that country.

From the point of view of equipment, the Chinese Nationalists were armed with American heavy weapons. The ammunition, therefore, had also to be hauled all the way across the Pacific from the United States. In contrast to this, the Chinese Communists were equipped chiefly with Japanese light weapons or rifles, mortars and hand grenades made in small factories in the mountains. Ammunition for these weapons could be supplied by the Chinese Communists themselves. In addition, the shoulders of the men were the backs of pack-horses. They could walk freely over trackless plains and mountains. In contrast, the Chinese Nationalist forces could not move except by automobile. When railways were disrupted and bridges blown up, the Chinese Nationalists were absolutely bound down. The Chinese Communists, however, were not limited in their movements by topography. While the maximum mobility of the Chinese Nationalist forces was an average of 20 miles a day, the Chinese Communists were able to march day after day from 40 to 45 miles daily.

In a battlefield like the Chinese continent, characterized by undeveloped communications, it would be correct to view American-style heavy weapons as losing out to Japanese-type light weapons. The Chinese Nationalists greatly relied upon their small air force and whenever they met up with the slightest resistance, they called for aerial support and airborne supplies. On the other hand, the Communists skillfully utilized the cover of darkness, their legs and their hand grenades.

To compare the ability of commanders, we find among the Chinese Communists men that had studied in Japan, in Soviet Russia and in France. There were also many generals who had obtained formal military training as members of the first; second and third graduating classes of the Wangpoo Military Academy. In addition, the Chinese Communists had in Yenan an officer's school and an Army University, with the famous Lin Piao, a rising general endowed with youthful ardor, as principal. Here young officers were trained with all seriousness. Thus, in generalship, the Communists were superior but not inferior to the Chinese Nationalists, In particular, the old leaders, who had led the heroic year long march of the Chinese Communists, from Suichin west over 4,200 miles, were still active as front-line army commanders, sharing with their own bodies the hardships of battle with youthful soldiers. Among these leaders were Chu Teh, himself, and such men as Liu Pai-cheng, Hsu Hsiang-chien, Peng Teh-huai, Ho Lung, Lin Piao and Chen Chih.

In contract to the generals at Nanking who were absorbed in manipulating strings to obtain positions most lucrative, giving 70 percent of their time and energy to politics and only 30 per cent to military matters, the Chinese Communist Party, both young and old, were advancing often wearing only sandals woven from straw. When all these points are brought together, it was, natural for a considerable gap to appear between the attackers and the defenders.

The one side gained control of the agrarian masses and thereby consolidated through their hold on farming villages a self-sufficient supply of food. The other side, on the other hand, based itself in large consumer center cities and towns, maintaining only a thin line between these points. In this sense also, the, ratio of defeats to victory in clashes with the Chinese Communists was already established.

There was reason for victory on the side of the victorious and every effort was made to win the victory that did finally come. The triumph of the Chinese Communists was not in any way to be attributed to greater aid from the Soviet Union than that given the Chinese Nationalists by the United States.

The China policy of the United States could be characterized by the phrase:

"A spendthrift son of a rich man caught by an unprincipled hussy in the prime of her womanhood."

The China policy of the Soviet Union can be described as:

"A fortyish man trifling with a country maiden,"

During the eight years of China's war of resistance against Japan, the United States, transcending all considerations of gain or utility, aided China, she praised Chiang Kai shek as the greatest living man on earth and believed this to be true. After the war, the United States was shown the true character of China, with its covering removed, and could not hide its anger at being fooled. The Chinese, on the other hand, felt:

"The reason why the United States was able to win against Japan was because China, on her own strength, nailed down for eight long years on the Chinese continent several million Japanese soldiers."

Some Chinese went as far as to make the extreme accusation that it was natural for the United States to help China during this period and that what help did come was no more than a drop in the bucket.

General Joseph Stilwell, Chief of Staff of the Allied Army and Commander in Chief of the American forces in India, crossed over to China and even during the war decried the corruption of the Chinese Nationalist Government and praised the Chinese Communist forces because he denounced Chiang Kai-shek straight to, his face, General Stilwell was completely boycotted by the Chinese and was forced to leave China. His successor, Lieutenant-General Wedemeyer was superficially the faithful Chief of Staff of Chiang Kai-shek and won the confidence of both the Chinese and the rest of the world. General Marshall, who arrived after the war as a special envoy, planned from the bottom of his heart to bring about collaboration between the Chinese Communists and the Kuomintang and did not spare himself in his efforts to mediate the differences between the two parties. However, in the end he was forced to throw in the towel and return home a disappointed man. Clad in his five stars shining with distinction and honor, General Marshall travelled nine times up and down in a swinging sedan chair the steep Lusan Mountain were the Generalissimo had sought refuge from the heat and carried on his negotiations for peace in China. At other times, he flew on his own plane to negotiate on bended knees with Chou En-lai (the Chinese Communist representative) who had escaped to Shanghai.

Discriminating Chinese were moved to admiration for the General's perseverance and dogged determination. Although he did show a sincerity which moved even the Chinese, regrettable to say, several outrages, by an extremely small number of American soldiers and officers, were committed during this period of negotiations. These incidents, occurring in Peking, Nanking and Shanghai, were one of the causes of the failure of General Marshall's mediation efforts.

The most representative of these incidents took place on Christmas night in Peking. A drunken American Military Policeman pushed two Chinese Nationalist soldiers enjoying evening coolness off the Chunghochiao Bridge in the capital city. The two Chinese soldiers were killed in the fall.

"One dog barks and ten thousand dogs join in." This incident finally developed into a nation-wide anti-American student movement and Chinese feelings aggravated to such an extent as to completely ruin the friendly relations built up during China's eight years of war against Japan. Chinese street girls found walking with American soldiers were lynched by Chinese youths. A number of American soldiers who went hunting or for picnics out in the country were reported missing.

The United States poured into China a total of close to four billion dollars worth of loans. In addition, American aid to China, including gifts of planes and ships, reached gigantic proportions. No country outside of the United States could ever have afforded to give such help. Yet, the United States Government was forced root and all to evacuate from the Chinese mainland, with nothing to show for this vast aid and the enormous loans made to China. The basic reason for this tragedy must be characterized as the result of a policy arising out of America's lack of knowledge of China. Yet, the actions of a few unthinking American officers and soldiers cannot be ignored as one of the factors accelerating this tragedy.

The aid extended Communist China by the Soviet Union differed radically from that given Nationalist China by the United States. Assistance was accorded in intangible form through ideological comrades among the Chinese. Soviet Russia's material aid did not amount to one percent of that of the United States. Nay, the Soviets rather expropriated commodities from the Chinese. It is a fact that the Soviets gave the Chinese Communists a part of the munitions left by the Japanese in Manchuria. However, the Russians demanded in return compensation several times greater in value from the Chinese. These demands included logging rights to the forests of northern Manchuria and large quantities of soya beans.

In contrast to the United States, which stationed a military advisory group numbering over 1,000 men in Nanking, Shanghai and Tsingtao to help in the training of Chinese Nationalist soldiers, the Russians only sent a few advisors to supervise the operations of the Chinese Communists.

While the United States stood on the face value of its policy on nonintervention in the Chinese civil war, the Soviet method, in comparison, was indeed a crafty and cunning way of controlling Chinese Communist military operations. In addition, since the Russians did not station a single soldier in China (except Dairen and Port Arthur), they avoided stirring up the resentment of the Chinese through the misbehavior of their soldiers.

Though keeping completely aloof from the conflicting attitude of these two countries toward China, Great Britain revealed that it was the most seasoned diplomat of the three.

Historically viewed, Great Britain's policy toward China should have been the most bitterly criticized; symbolized as it was in its continued possession of Hong Kong, wrested from China in the Opium War. Despite this fact, Great Britain skillfully diverted the objectives of China's racial movement toward Japan, carefully guarded her already acquired rights and interests and, utilizing the gratitude of the Chinese for her aid in the reform of China's currency through Lease-Loss, continued to make a profit from her relations with China. Great Britain watched with cold detachment the return by the various Powers of leased territory in China, but could not be pried loose from Hong Kong even with a crow bar.

When the position of the United States became overwhelmingly great after the war, the British Ambassador isolated himself completely from the internal politics of China, shut himself in his study and thus avoided falling together with the dying Kuomintang Government. Today, following in the footsteps of the Soviet Union, she has recognized the newly arisen Chinese Communist Government and is monopolizing, in place of the United States, trade profits accruing through her possession of Hong Kong.

When the United States gave the Kuomintang Government ten odd small destroyers, Great Britain sent one old cruiser, a vessel much larger, after all, than a destroyer and let it lord over the American destroyers. In addition, Great Britain unveiled a bronze statue of Dr. Sun Yat-sen in London Park. In this manner, by expressing respect to China in a showy way without spending much money, Britain continued to gain the appreciation of the Chinese people.

There was, however, even in this seasoned diplomacy one blunder.

This took place at the end of 1947, when Great Britain used force in the compulsory evacuation of Chinese from the Kowloon Leased Territory. As a result, the Chinese populace became inflamed and mobs set fire to the British consulate at Shamien. Fears were expressed that the incident would become a big international issue. However, Great Britain skillfully settled the issue on a local level through negotiations with Sun Tzu-wen (T.V. Soong).

Again during the battle between the Communists and the Nationalists for Nanking, the British cruiser Amethyst was fired upon and damaged by Communist shore batteries on the banks of the Yangtze. However, the British Parliament did not lose its good judgment or composure. Britain's aim was told avoid as much as possible friction with the newly-risen forces in China and to monopolize in substance trade profits from China.

I wondered why there could be such a difference between Great Britain and Japan of the past, which furiously enraged over the Captain Oyama Incident, started war in Shanghai or maddened by the Captain Nakamura Incident, to set off the September 18 Affair.

We cannot but pay our respects to the veteran diplomacy of the Great British Empire.

Chapter 7 | A Handful of Earth

We found already aboard the ship, roughly 300 Japanese compatriots from Taiwan. They were a motley crowd–war crimes' suspects, professors from Taipei University, detained technicians, artisans and merchants. They were returning to Japan, after having lost all hope in face of the worsening plight of the Chinese Nationalists. The whole group was filled with pessimism over the future of the Kuomintang.

General Chen I, who had been appointed Governor of Taiwan immediately after the war to take over from the Japanese there, was unable to restrain or control his officials and soldiers from wrong-doing. The smouldering resentment of the native Taiwanese festered until it could no longer be held down and exploded in the February 28 Incident of 1948 [anti-government uprising, February 27 – May 16, 1947, in Taiwan, suppressed by the Nationalist Government of the Republic of China]. The blood of tens of thousands who were shot flowed into the hearts of the native population and stained their entire feelings towards the Kuomintang with lasting hatred. There even appeared among the population a desire to shake off the rule of the Chinese Nationalist Government and link hands and fate with Japan. On the departure of the group from Taiwan, Taiwanese police, hiding from the sharp eyes of Chinese Nationalist officers, afforded every convenience to the Japanese and showed toward them a protective attitude. After the end of the war, the Taiwanese had dreamed of the arrival of a magnificent army from their homeland to take the place of the Japanese forces. They experienced bitter disappointment on greeting a pack of starving wolves, carrying earthen pots and wearing straw sandals. Then came a wave of rapacious government officials who deprived them of their homes, looted their belongings and wealth and brazenly accepted bribes. "Fapi" unbacked by gold or silver bullion, flooded the country. At the beginning it was only resentment, which developed into opposition and eventually a bloody struggle. Here, too, the Kuomintang revealed the figure of a tottering and corrupt regime heading for oblivion.

Our ship left Shanghai on the evening of May 16 [1948].

I watched from the deck the fading outline of Shanghai, doomed once again to be wrapped in the flames of war. I watched this metropolis of Satan of the Orient, over which the red-flag was destined soon to fly. Those who stood on the pier waving good-bye, and those who waved back from the decks were all filled with indescribable feeling.

When the ship passed the entrance to Woosung Harbor, I recalled to mind the old battlefields of the First Shanghai Incident, 16 full years ago [January 28 to March 3, 1932]. The battlefield on which I had lost 16 subordinates retained only in its hills and streams the vestiges of the past.

My quarters were situated in the ship's hold near the stern of the vessel. We were packed in like so much freight, allowed only a half a mat of space per person. I could not even stretch out my legs. The person next to me was a young fellow by the name of Nagai. A student of the Tungwen University of Shanghai, he claimed to be the nephew of the novelist Sanjin (Kafu) Nagai [born December 3, 1879, in Tokyo, died April 30, 1959]. He himself was of the literary type. Yasuda of the *Domei News Agency* was also aboard. He was on his way home with a baby, still fed on milk, that had been born between himself and a Russian girl. The Russian wife, he had divorced. The child's name was Shepherd and soon became the darling of the ship. The hold, packed indiscriminately with men and women, was filled with nauseating odors, decadence and selfishness. I wondered if this ship's hold were an epitome of a defeated Japan.

The group of Japanese soldiers, who had been detained as war crimes suspects, were known as the Kiangwan Unit. Among these men was Major-General Fukuyama who had been a class-mate of mine in the Army University. He played go from morning to night, rubbing his bald head. He did not even guess that this pseudo university professor could be his classmate. Of course, my disguise was so designed as not to allow him to recognize me; although we met face to face on several occasions. After this experience I gained confidence and felt that at this rate I would not be discovered. Hiroshi Kadoya was the commander of the entire group of repatriates. I had known him too several years ago while at the Supreme Headquarters in Nanking. But he, too, failed to recognize me.

The only pleasure aboard the ship were the meals. Surprisingly, everybody seemed to hate being placed on kitchen duty I volunteered before anyone else. My work clothes were black from the coal tar painted over the ship's bottom. No matter how one could have looked at me they would not think me a university professor.

After two or three days of journey northward, the boat touched at Tsingtao. Here again, a customs inspection more strict than at Shanghai was held. The looting and stealing by the military police was worse than at Shanghai. The Tsingtao group of Japanese repatriates, mostly composed of women and girls, did not have the strength to repack their luggage which had been ripped mercilessly apart. Somebody suggested that we help the women. Roughly 20 of the strongest volunteered and in a cloud of dust blown up by a strong wind, we helped these pitiful compatriots. A mountain of luggage was immediately loaded aboard ship and tired though we were, we experienced the pleasure of having been able to a help our neighbors.

As our ship began leaving the pier, we heard a Japanese voice from among the Chinese police standing on the wharf.

"Please take good care of yourself. Do write, won't you? There isn't a single decent person among the Chinese dogs. Japanese are really good people...."

What a surprise this was!

The mystery soon cleared up when I heard that the voice had come from a Taiwanese police officer. This youth had been employed by the Japanese Navy and had crossed over to Tsingao. After the war he had become a Chinese Nationalist police officer. From the pier, packages of cigarettes were thrown aboard ship. One Chinese watched with sour face this exchange of friendly gestures between the Japanese and the Taiwanese. He seemed to be the officer in command, but he could do nothing.

After ten days of navigation, our vessel finally entered the harbor of Sasebo. Our quarantine ended; instructions concerning our processing after landing were shouted through megaphones by repatriation officials and former army officers. I was impatient with the innumerable delays was seized with the desire to step as soon as possible on the soil of Japan. Even a second seemed like a long time.

On the morning of May 26, [1948], we finally landed [at Sasebo, Nagasaki Prefecture, travelling under the name of Kiuchi Katauji, professor of Archeology].

As I placed my first step upon the soil of Japan, I quietly picked up a handful of earth, unnoticed by the others, and smelt its sweetness.

It was the first smell of my motherland in six years.

Could my longing for and love of the soil of my motherland be this strong?

Though the country was defeated, the hills and the streams were still left, together with the Emperor....

Impoverished as this soil may be, it was our earth, it was our land.

By no means was it Stalin's land.

Withered and dried though these hill and streams might be, they were our mountains and they were our rivers.

By no means were they Truman's.

We must make this land green once again, with our blood, our sweat and our love.

If fertilizer is needed, then shall I not hesitate to grind to powder these old bones.

THE END

The God of Strategy, A Biography of Masanobu Tsuji

by David G. Thomas

Masanobu Tsuji, November 28, 1931 photo, xinjiapo.news.

The God of Strategy, A Biography of Masanobu Tsuji

Masanobu Tsuji's life story is a tangled tapestry of self-reported truths and deliberate distortions. His primary voluntary account comes from *Underground Escape*, in which he details his three and a half years in hiding from Allied forces while evading charges as a suspected war criminal. Beyond this self-narrative, the most substantial external source of information about Tsuji is his CIA dossier, declassified and released to the public in 2006.

Tsuji's CIA dossier documents years of U.S. Intelligence efforts to track and interpret a figure repeatedly dubbed "Japan's mystery man" or "Japan's most dangerous man."

The dossier is heavily shaped by Cold War strategic priorities. Tsuji, broadly suspected of involvement in several of Japan's wartime atrocities, was viewed less through the lens of justice and more as a potential asset – or threat – in the shifting geopolitical landscape of postwar Asia. This focus obscures and minimizes his complicity in events such as Operation Sook Ching in Singapore or the Bataan Death March in the Philippines.

A key function of intelligence gathering is to evaluate an individual's personality, underlying motivations, and psychological vulnerabilities. In Tsuji's case, his actions were deeply rooted in ideological conviction. He maintained a wide network of private connections with former officers of the Imperial Japanese Army and was actively involved in several semi-secret ultranationalist organizations.

After 1952, Tsuji emerged as a prominent public figure, known both for his outspoken views and his notoriety as the "war criminal" who had evaded prosecution. He skillfully leveraged this controversial reputation to secure a seat in the Japanese Diet and to exert significant influence over public opinion.

Tsuji was not content with merely steering Japanese public opinion toward nationalism; he aimed to position himself as an independent diplomatic actor, operating beyond the bounds of the official government. In 1955, he embarked on a 40-day tour of China and the Soviet Union, where he held talks with military leaders and secured a private meeting with Russian Defense Minister Georgy Zhukov. Two years later, in 1957, he undertook a 60-day journey through 19 countries across the Middle East and Eastern Europe, meeting with prominent leaders including Egyptian President Gamal Abdel Nasser, Yugoslav President Josip Broz Tito, and Chinese Premier Zhou En-lai.

In 1961, Tsuji embarked on what would become his final journey – a so-called "private" inspection tour of Southeast Asia. Spanning 40 days, his itinerary included visits to Thailand, Laos, South Vietnam, Cambodia, Burma, and Singapore. For much of the trip, he planned to travel incognito, disguised as a Buddhist monk.

All available information the CIA was able to gather regarding Tsuji's disappearance during that final trip is presented in the following text. Despite extensive efforts to trace his movements, the details remain fragmentary and inconclusive. Intelligence reports, informant accounts, and speculative analysis were compiled in an attempt to piece together his fate, but no definitive conclusion was ever reached. All we have is what follows – an incomplete but revealing glimpse into the final chapter of one of Japan's most elusive postwar figures.

Information from sources other than Tsuji's CIA dossier are so noted in the text.

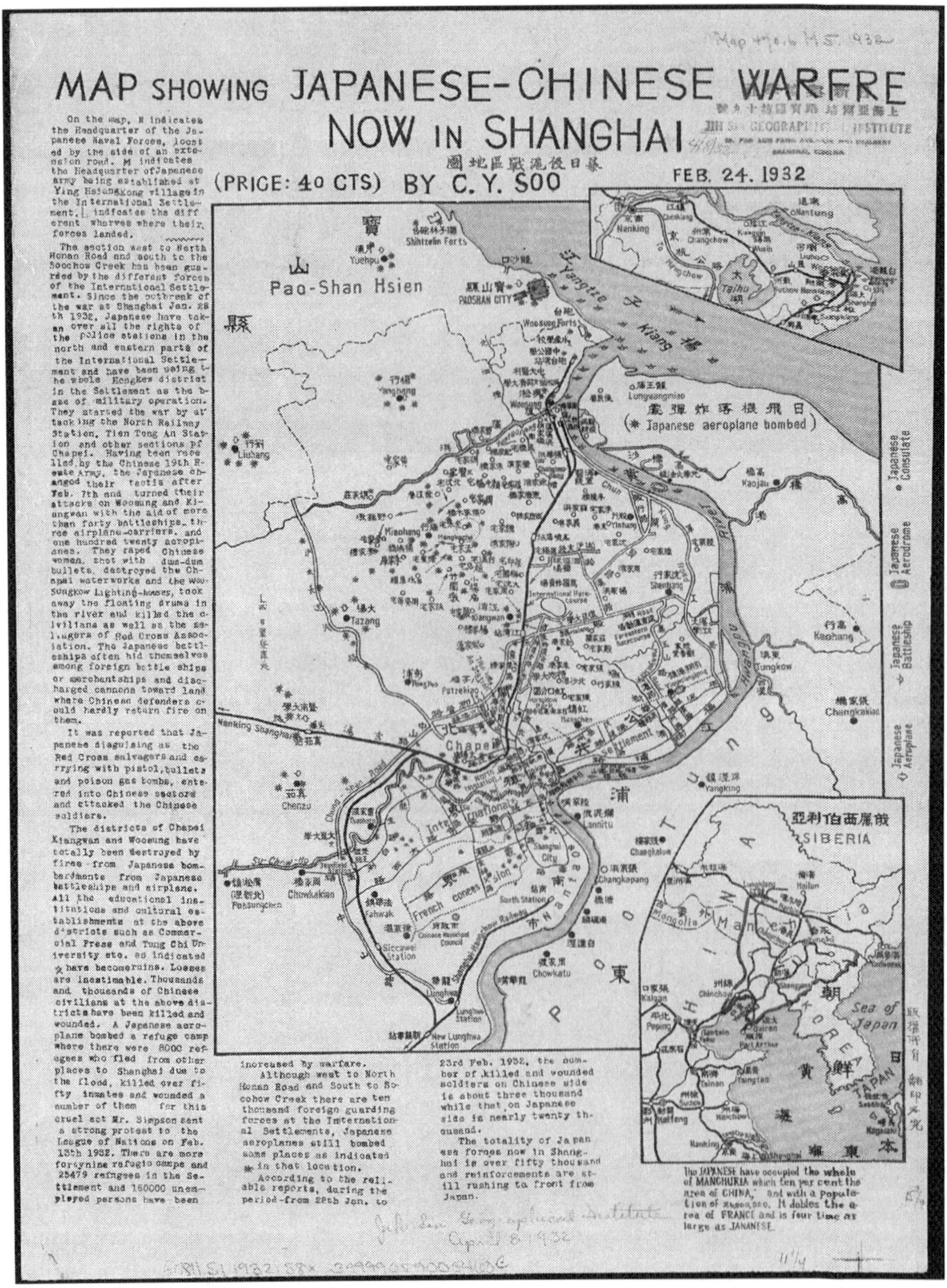

Map showing Japanese-Chinese warfare in Shanghai, China, February 24, 1932, John Curtin Prime Ministerial Library, Perth, Australia.

Oct 11, 1902 – Born in the village of Higashi Tanioka, Enuma District of Ishikawa Prefecture, Japan. His father was Kamekichi Tsuji, his mother Moto. He was the second son of six children, four boys and two girls.

1914 – Future wife Chitose born.

Sept 1, 1917 – Enrolled in Nagoya District Army Preparatory School.

July 18, 1924 – Graduated from Imperial Japanese Army Academy. Presented with a silver watch by Crown Prince and Future Emperor Hirohito for being top in his class.

Oct 25, 1924 – Appointed 2nd Lieutenant and assigned to the 7th Japanese Infantry Regiment. The 7th Infantry Regiment was formed in Sapporo, Hokkaido, on May 12, 1888.

1930 – Married Chitose.

Sept 18, 1931 – Japan's Kwantung Army invades the Manchurian city of Mukden. The pretext for the invasion, which led to the occupation of all of Manchuria, was a railroad bombing blamed on Chinese soldiers that was staged by Lieutenant Suemori Kawamoto of the 29th Japanese Infantry.[2]

Nov 28, 1931 – Graduated from Military Staff College, second in class. Joined a traditional Sword Group. Returned to the 7th Infantry Regiment after graduation.

Jan 28, 1932 – "January 28" or "Shanghai Incident" begins. The conflict erupted when several Japanese Buddhist priests were attacked by Chinese civilians in the Shanghai International Settlement. In response, Japanese soldiers launched retaliatory actions against Chinese merchants, quickly escalating tensions into full-scale combat between the two sides.

Feb 2, 1932 – Appointed Commander of the 3rd Company of the 7th Infantry Regiment.

Feb 25, 1932 – 7th Infantry Regiment dispatched to bolster Japanese forces in Shanghai. Tsuji sees active combat as Commander of the 3rd Company. Awarded the 5th Class Decoration of the Order of the Golden Kite.

March 3, 1932 – Chinese forces withdraw from Shanghai area, ending Chinese resistance to Japanese military occupation of Shanghai.

Sept 1, 1932 – Assigned to General Staff Office in Tokyo.

Early, 1933 – Appointed instructor at the Imperial Japanese Army Academy in Tokyo.

Aug 1, 1933 – Promoted to Captain.

Dec 20, 1933 – Reassigned to General Staff Office in Tokyo.

Aug 1, 1934 – Appointed Company Commander of the Cadet Unit at the Imperial Japanese Army Academy.

April 2, 1935 – Assigned to the 2nd Infantry Regiment.

Aug 4, 1937 – Assigned to the Japanese Occupation Army of China as a staff officer.

Aug 25, 1937 – Assigned to the North China Army Group as a staff officer.

Nov 1, 1937 – Assigned to the Kwantung Army in Manchuria as a staff officer.

March 1, 1938 – Appointed Major.

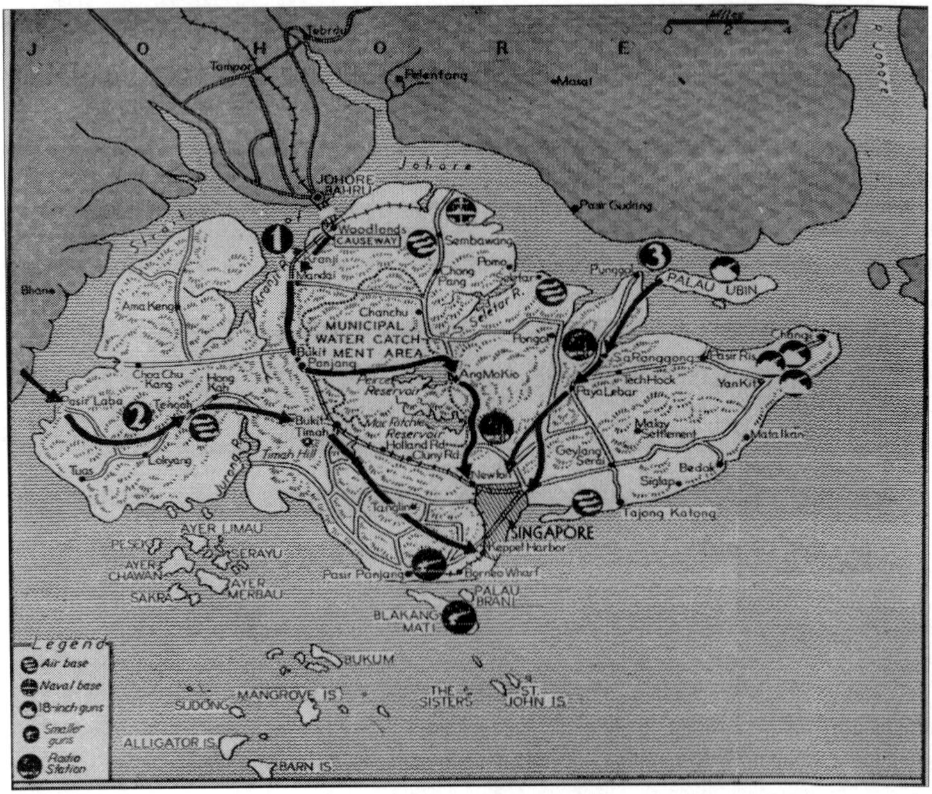

How Singapore Fell

1. With complete control of the air over the island, Japanese troops poured across the Johore Strait landing in the Kranji-Mandai area. Breaking "strong resistance" they pushed to Bukit Panjang, then struck southward toward the coast west of Singapore.

2. Japanese forces staged a second landing from Johore on southwest coast of the island in the Pasir Laba sector. They continued advance along highway to Bukit Timah, capturing the airdrome at Tengah, reached the western section of the city, taking the fortified island of Blakang Mati, south of Singapore.

3. The third force of invaders, hopping from the mainland to the island of Palau Ubin in Jahore Strait, then to the island of Singapre, attacking the city from the northeast.

New York Daily News, NY, February 16, 1942.

May 11, 1939 – "Nomonhan Incident" or "Battle of Khalkhin Gol" begins. The conflict, between the Kwantung Army and Soviet Russia, erupted over the border between Japanese occupied Manchuria and Russia-allied Mongolia. Tsuji sees active duty in the conflict.

Sept 7, 1939 – Assigned to duty at 11th Army Headquarters in Manchuria.

Sept 16, 1939 – Nomonhan Incident ends with a ceasefire agreement, after several Japanese battlefield defeats.

Feb 6, 1940 – Assigned to General Headquarters Expeditionary Force in China.

Aug 1, 1940 – Appointed Lieutenant Colonel.

Sept 22, 1940 – Japan invades Indochina.

Sept 26, 1940 – French forces in Indochina surrender to Japanese.

Nov 6, 1940 – Assigned to the Taiwan Army in Taiwan.

July 1, 1941 – Assigned to the 25th Army Headquarters as Chief of Staff, charged with planning the 25th Army's invasion of Malaya and Singapore. The 25th Army was under the command General Tomoyuki Yamashita.

Oct 22, 1941 – Reconnoitered the eastern coast of Malaya from an unmarked and unarmed twin-engine reconnaissance aircraft, taking photos of the beaches, towns, airstrips and military installations – *Singapore 1941-1942, The Japanese Version of the Malayan Campaign*, pp 41-57.

Dec 7, 1941 – Attack on Pearl Harbor.

Dec 8, 1941 – 25th Army invades Malaya from Indochina with about 30,000 troops.

Jan 31, 1942 – British, Australian, and Indian troops withdraw from Malaya.

Feb 8, 1942 – 25th Army invades Singapore from Malaya.

Feb 15, 1942 – Allied forces in Singapore surrender. Allied troop strength was about 85,000 men.

Feb 23, 1942 – Order given for what is known today as "Operation Sook Ching," a Chinese term meaning "purging through cleansing." The Japanese name for the operation was Dai Kensho, meaning "great inspection."

> *"The directive ordered Chinese males between the ages of 18 and 50 to report to designated screening centers.... The directive was [to arrest] those whose names appeared in lists of anti-Japanese suspects maintained and distributed by the Japanese."*
>
> *"Those unfortunate to be selected were loaded into lorries and taken to remote sites to be summarily executed. There was little attempt to conceal these killings. People living near the massacre sites and prisoners of war were ordered to dispose of the bodies. The Japanese estimated that there were between 5,000 and 6,000 executions, while the local Chinese put it between 40,000 and 50,000."* [3]

After the war, the Singapore War Crimes Court, in April, 1947, tried seven Japanese Officers for these killings: Lt. Gen. Takuma Nishimura, Lt. Gen. Saburo Kawamura, Lt. Col. Masayuki Oishi, Lt. Col. Yoshitaka Yokota, Major Tomotatsu Jyo, Major Satorou

Onishi, and Captain Haruji Hisamatsu. Lt. Gen. Saburo Kawamura and Lt. Col. Masayuki Oishi were sentenced to death. The other five men were sentenced to life in prison.[4]

General Tomoyuki Yamashita could not be tried for Operation Sook Ching because he had already been tried for war crimes in the Philippines, sentenced to death, and executed on February 23, 1946. [5]

During the trial of the seven officers, the prosecution presented witnesses—including defendant Major Onishi Satorou—who testified that the plan for mass killings following the occupation of Singapore had been part of the invasion strategy from the outset:

> *"Before we arrived in Singapore orders were issued from Gen. H.Q. which stated that due to the fact that the Army is advancing fast, and in order to preserve the peace behind us, it is essential to massacre many Chinese who appear in any way to have anti-Japanese feelings." – Major Onishi Satorou* [6]

Tsuji, as Chief of Staff, was the lead officer in planning the invasions of Malaya and Singapore. The overwhelming success of both campaigns earned him the nickname "the god of strategy" among his fellow officers. During the Singapore War Crimes Court's investigation into the Sook Ching massacres, his central role in orchestrating the campaign made him a prime war crimes suspect. A warrant was issued for his arrest–but he had vanished.

On June 22, 1959, former Major General Kiyotake Kawaguchi published a pamphlet in which he leveled the following accusations against Tsuji regarding his conduct during the Singapore campaign.[7]

> *"At the time of the commencement of the war, Mr. Tsuji was a senior operations staff officer of the 25th Army under the command of General Yamashita. Following the fall of Singapore, he drew up an order for the purge of two million Chinese merchants in the Malayan Peninsula and Singapore, claiming that it was necessary for maintaining security, and received the approval of the commanding officer. This order, however, was extremely abstract, and merely ordered the placing of the field military police corps under the command of Major General [Kiyotake] Kawaguchi, who was the Commanding General of the Singapore Defense Headquarters and under Lt. General [Takuma] Nishimura of the Konoe Regiment stationed near Singapore."*
>
> *"Mr. Tsuji, taking this order with him, personally called on Major General Kawaguchi and Lt General Nishimura and gave concrete instructions on the methods of liquidation. Concretely, he divided Singapore into six districts and assigned military police, auxiliary military police (infantry men wearing temporarily the armband of the military police) and pro-Japanese Chinese."*
>
> *"All Chinese were made to walk in a single file in front of these military police and pro-Japanese Chinese pointed out those who had joined the British Volunteer Army and were members or past members of anti-Japanese resistance organizations, such as the Communist Party or the Blue Robe Association. After a summary investigation, they were taken to the seashore and machine-gunned or were roped together, rowed out to sea, and with weights tied on them, thrown overboard."*

"The total of Chinese thus killed is said to be estimated at about 10,000 maximum, and even at a minimum estimate, is said to be 2,000. Staff Officer Tsuji made the rounds of these six military police stations, stepped up the liquidations, and witnessed the executions. If there were some who tried to escape, he admonished and egged on the military police and made them carry out the execution."

"After the end of the war, Major General Kawamura (had been promoted to Lt. General) and Lt. Colonel Onishi of the Military Police (had been promoted to Colonel) were given sentence of death by hanging by a British military tribunal and Lt. General Nishimura was given life imprisonment, which however, was changed to death by hanging by an Australian military tribunal. Besides these three, there were many other Military Police members who were sentenced to life imprisonment and who suffered many years of detention."

At the time these allegations were published, Kiyotake Kawaguchi was an outspoken political opponent of Tsuji. He was also a convicted war criminal. He was tried in 1948 for his involvement in atrocities committed by troops under his command in Borneo and sentenced to life imprisonment. He was released in 1954 after serving six years' imprisonment.

March, 1942 – Assigned to 14th Imperial Japanese Army Headquarters in Manila, Philippines, as a staff officer.

April 9, 1942 – Battle of Bataan in the Philippines ends with the surrender of U.S. Major General Edward P. King to Colonel Mootoo Nakayama of the 14th Imperial Japanese Army.

April 9, 1942 – Assigned to 16th Army Division Headquarters as a staff officer. The 16th Division was one of two divisions of the 14th Imperial Japanese Army.

Kiyotake Kawaguchi in his June 22, 1959, pamphlet makes the following accusations regarding Tsuji's conduct in the Philippines:

"Some time after the fall of Singapore, Mr. Tsuji was reassigned as Staff Officer of the Imperial General Headquarters (chief of operations group). Around the fall of Bataan, he arrived in the Philippines to direct operations. On April 9, he arrived at the 16th Army General Headquarters which was stationed near Mariveles on Bataan. His arrival coincided with the surrender of Major General King and many other officers of the U.S.-Philippine Army, who came down from the mountains to surrender. Mr. Tsuji demanded of the Senior Staff Officer of the 16th Army, Colonel Saburo Watanabe, to kill all the prisoners of war of the U.S.-Philippine Army surrendering at the time. Senior Staff Officer Watanabe naturally refused. When the Commanding General of the 16th Army, Lt. General Morioka, heard about this, he was extremely angered, and immediately sent Mr. Tsuji packing."

"In the May 22, 1950, issue of the weekly magazine ***Thriller*** [8]*, there appears a conversation between Mr. Tsuji and Mused Tokugawa. In this conversation, Mr. Tsuji implied that he, at the request of the Takasago Tribesmen, authorized the brutal chopping off of the heads of abut three hundred American soldiers with*

hunting knives near a mountain gorge (the Panchings River which separated the bivouacs of the 16th Army and Nara Corps). Mr. Tsuji confessed that this is what is called the Bataan death march."

"It should be self-evident, considering the situation which preceded the incident, whether it was the Takasago Tribesmen who request that they be permitted to brutally kill the white men, or whether Mr.Tsuji on his own initiative issued the order to the Takasago Tribesmen. After the end of the war, the U.S. military tribunal passed a sentence of death by hanging on Logistics Major General [Yoshitaka] Kawane and Col. [Kurataro] Hirano. At the detention camp in Manila, Philippines (this was before the military tribunal), I was in the same building with Colonel Hirano, with our bunks side by side, and Colonel Hirano repeated reproached the methods adopted by Staff Officer Tsuji. He, however, was after all executed."

Jose Abad Santos, 1941 photo, Wikipedia.org

The author was unable to locate the weekly magazine *Thriller*. The tale of Takasago tribesmen beheading prisoners appears to be a fabrication, likely blending fictional elements with accounts of the horrendous treatment endured by the soldiers who were taken prisoner in Bataan.

May 1, 1942 – Jose Abad Santos, Chief Justice of the Supreme Court of the Philippines, executed by Japanese troops in Malabang, Lanao del Sur, Philippines.

Abad Santos and his son José Jr. were captured April 11, 1942, by Japanese soldiers in Barangay Tubod in Barili, Cebum, while traveling by automobile.

"The Japanese high commander, Kiyotake Kawaguchi, took him and his son aboard a ship on April 26, 1942, thinking they were heading to Manila. Instead, they arrived on April 28 at Parang, Cotabato (now in Maguindanao del Norte). The next day they were brought to Malabang, Lanao, arriving on April 30. After two days' confinement at Japanese camps, Abad Santos was called in front of Kawaguchi and was informed about the order of his execution. Before he was shot to death, he was able to talk to his son Pepito. Among his last parting words to his son were, "Do not cry, Pepito, show to these people that you are brave. It is an honor to die for one's country. Not everybody has that chance." Abad Santos was executed at 2:00 pm, on May 1, 1942, under a tall coconut tree near a river bank. He refused to be blindfolded and refused the last cigarette offered to him." [8]

Kiyotake Kawaguchi in his June 22, 1959, pamphlet makes the following accusations against Tsuji regarding Abad Santos' execution:

> *"This incident involves me personally, and as it is a complicated and mysterious affair, I will leave out the details. It happened that Staff Officer Tsuji, after completing his tour of inspection of Bataan mentioned above, came to the Military Headquarters at Manila on April 10. I recommended by telegram that a high government official of the Philippines, Jose Abad Santo, whom my men had captured, be used for military administration. Staff Officer Tsuji's reaction to the cable I had dispatched was to issue a strong demand to the Military Headquarters in Manila to 'kill Santos immediately,' and forced them to issue the order. I tried two more times to save Santos and to take steps to get him to cooperate with our country, but I was unable to oppose the three successive orders from the Military Headquarters and was forced to carry out the execution.*
>
> *"This Santos was an extremely fine gentleman, and if he had been alive, he would very likely have become president. No words can express the loss the executing of such a man means. Because of this, I was sentenced to a six-year imprisonment, and spent seven years and three months in Sugamo Prison and the prison in Muntinlupa in the Philippines."* [9]

In spite of this statement, Kiyotake Kawaguchi was not convicted of war crimes for his actions in the Philippines. He was convicted of executing civilians and suspected resistance members during his command of Japanese forces in Borneo, where he was transferred after his service in the Philippines.

July 6, 1942 – Japanese forces of the 17th Army under the command of Major General Kiyotake Kawaguchi land at Lunga Point, Guadalcanal, to build an airfield.

Aug 7, 1942 – U.S. 1st Marine Division, commanded by Major General Alexander Vandegrift, launch an amphibious assault on Guadalcanal.

Aug 8, 1942 – U.S. forces capture the partially built Japanese airfield at Lunga Point and rename it Henderson Field.

Aug 19, 1942 – Kawaguchi Corps commanded by Kiyotake Kawaguchi arrive on Guadalcanal.

Late Aug, 1942 – Assigned to Guadalcanal as a staff officer.

Sept 12, 1942 – Battle of Edson's Ridge begins, a failed attempt to re-capture Henderson Field by the Kawaguchi Corps.

Kiyotake Kawaguchi in his June 22, 1959, pamphlet makes the following accusations against Tsuji regarding his actions in Guadalcanal:

> *"On Guadalcanal, Mr. Tsuji got himself appointed staff officer of the 17th Army (on the spot army) concurrently with staff officer of the Imperial General Headquarters. He came to Guadalcanal and sent the front line troops out to war to their death, and in this way, the greater part of the Second Regiment and Kawaguchi Corps died ignoble and useless deaths. Staff Officer Tsuji went out to the front lines personally and was very active. I admit his courage, but the duty of a Staff Officer of the Imperial General Headquarters is to direct the activities of the Japanese Army from a broader and higher standpoint."*

> *"On reflecting, the decision should have been made to abandon Guadalcanal and withdraw the forces around the time the Kawaguchi Corps failed in its offensive on September 12 and 13. Even so, Japan would probably have been defeated in the end, but even if Japan had to lose the war, there could have been other ways of losing the war. However, until the Imperial Council meeting held in the presence of the Emperor on December 31, 1942, when it was decided that the troops should be re-deployed, Staff Officer Tsuji insisted on carrying out a tough policy. In this way, Japan lost 20,000 lives and consumed a large amount of munitions, thus losing her war potential. From then on, Japan's eventual defeat became more and more evident."* [10]

Feb 1, 1943 –Appointed instructor at the Imperial Army War College at Tokyo.

Feb 8, 1943 – Last remaining Japanese soldier on Guadalcanal killed.

Aug 2, 1943 – Appointed Colonel.

Aug 20, 1943 – Assigned to the Expeditionary Army Headquarters in Nanjing, China, as a staff officer.

Nov 25. 1943 – Organized a memorial service for Wang Caiyu, the mother of Chiang Kai-shek.

July 3, 1944 – Assigned to the 33rd Army Headquarters in Burma as a staff officer.

March 9, 1945 – Japanese occupation forces stage a coup d'état in Indochina, rounding up and imprisoning all French residents.

May 24, 1945 – Assigned to the 39th Army Headquarters in Bangkok, Thailand, as a staff officer.

Aug 6, 1945 – Atomic bomb dropped on Hiroshima, Japan.

Aug 15, 1945 – Atomic bomb dropped on Nagasaki, Japan.

Aug 15, 1945 – Emperor Hirohito announces the surrender of the Empire of Japan.

Aug 16, 1945 – Left 39th Army Headquarters disguised as a Buddhist priest of the Shinshu Sect under the assumed name Norinobu Aoki and took residence in Ryab Temple, an ossuary temple containing the remains of many Japanese who had died in Thailand. The ossuary also contained the remains of Subhas Chandra Bose, founder of the Indian Independence Movement, who had died in an airplane accident in Taiwan.

Sept 2, 1945 – Japan's representatives sign unconditional surrender papers.

Sept 9-15, 1945 – British and Indian forces occupy Bangkok.

Oct 28, 1945 – The Blue Shirt Society, a secret Chinese organization formed March 1, 1932, in response to the Japanese invasion of Manchuria by members and graduates of the Whampoa Military Academy, agrees to smuggle Tsuji out of Thailand.

Nov 1, 1945 – Left Bangkok by train disguised as a Chinese merchant named Shih Chenghsin, escorted by two members of the Blue Shirt Society.

Nov 2, 1945 – Arrived Korat City, Nakhon Ratchasima Province, Thailand.

Nov 3, 1945 – Arrived at Thai border town of Ubon Ratchathani.

Nov 4, 1945 – Crossed Mekong River by boat and arrived at Vientiane, capital of Laos.

Nov 15, 1945 – Arrived at Savannakhet, Laos.

Nov 29, 1945 – Arrived at Hanoi, Vietnam, disguised as a Chinese merchant named Shih Kung-yu.

Nov 29, 1945 to Mar 9, 1946 – In Hanoi.

Mar 9, 1946 to Mar 19, 1946 – In Kunming, Yunnan Province, China.

Mar 19, 1946 – Arrived at Chungking, Sechwan Province, China.

April 18, 1946 – Sent letter to Chiang Kai-shek written in blood.

May 22, 1946 – U.S. Army Intelligence Report:

> *"The Japanese Army had a lot of gold. It is rumored that they had 3 tons which had been taken from the French in F.I.C [French Indo-China]. They intended to use this to finance the government in Thailand. However, the war ended and in order to get rid of it they distributed it to various Japanese companies. The idea was that it should be hidden or somehow saved from confiscation by the Allies so that in about ten years or so the Japanese would have goods here to start the merry game all over again. Nearly everybody in camp is convinced that it is only a matter of time before Japan makes a come-back."*
>
> *"It was divided into two parts:"*
>
> *"Part One was taken by Col. Tsuji of the General Staff and distributed among his junior officers whom he told to desert and go to their friends in Bangkok. Tsuji himself was reported as having committed suicide but everyone here believes he is hiding in Bangkok with a Chinese girl he knew."*

July 1, 1946 – Arrived at Nanking, Jiangsu Province, China. Assumed the alias Wut Chieh-nan.

Sept 11, 1946 – Warrant issued for Tsuji's arrest as a suspected war criminal by the Supreme Commander for the Allied Powers.

Dec 10, 1947 – U.S. Military Intelligence files the following report:

> *"It is well established that high-ranking Japanese officers who surrendered in China are being utilize there behind the scenes in intelligence training and military advisory capacity. One source claims that about 20 Japanese general and field grade officers are employed in the Second Department of the Ministry of National Defense, under the personal supervision of Lt. Gen. Cheng Chie-min, successor to Tai Li as head of Chinese Intelligence...."*
>
> *"Colonel Masanobu Tsuji, who, though the Japanese Demobilization Bureau reports him dead by hara-kiri, is still in Nanking according to CIC [Counter Intelligence Corps] reports. A China source says that Tsuji was at one time head of the Fourth Department of General Headquarters in Nanking. This department is elsewhere said charged with over-all direction of counter-intelligence activity. CIC reports further say that Chinese intelligence-collecting in Japan is directed from Nanking under the guidance of Colonel Tsuji and that Japanese repatriation from China who intended to work for the First Section of the Chinese Mission generally bear letters of introduction from him."*
>
> *"Once a young officer of the fire-eating, militarist clique, Tsuji became noted as a brilliant staff officer of the Japanese Expeditionary Forces who served in*

the war in China, and later in Singapore, Guadalcanal and again in China, where he reportedly disappears in August, 1945."

"A fanatic, high-principled disciplinarian, he is said to have worked his utmost to bring about peace negotiations with the Chungking government and blamed then Premier [Hideki] Tojo for incompetence that Japan failed to settle the China Incident before undertaking a Greater East Asia war. Repudiated by Tojo and his home government, it is said that Tsuji's efforts won over the admiration and sympathy of many Chinese. Later he was said largely responsible for drawing up the plans that resulted in the capture of Singapore and for ordering the 'liquidation' there of alleged Chinese Communists. Tsuji is wanted as a suspected War Criminal on SCAP [Supreme Commander for the Allied Powers] order to the Japanese Government dated 11 September 1946."

May 15, 1948 – Departed Nanking, China, by train, for Shanghai.

May 16, 1948 – Departed Shanghai, China, by ship, for Japan.

May 26, 1948 – Landed at Sasebo, Nagasaki Prefecture, Japan, after 10 days of voyaging, under the assumed name Dr. Kiuchi Katauji, Professor of Archeology.

Oct 31, 1949 – U.S. Army Intelligence Report:

"The plan to send Japanese pilots to Formosa was discussed at the Supreme Council of the Nationalist Government in Hangchow on April 20 [1949]."

"The plan was put into effect from the 21st, but at 11 pm of that day, a secret directive from Shanghai was sent to a Japanese man, who was at one time in custody in Sugamo [Prison] as a war criminal suspect. On the 23rd, Chu Shih-ming arrived as a representative of the Chinese Mission to Japan to carry out this plan."

"General Okamura in China and Colonel Masanobu Tsuji, Chief Staff of Japanese Forces in Burma, also took part in the planning. The volunteer corps was to be called the 'Japan Volunteer Corp' and was to be composed of air, ground and sea forces."

Dec 7, 1949 – The Nationalist government of the Republic of China, accompanied by many tens of thousands of refugees, retreats from mainland China to the island of Taiwan.

Aug 16, 1949 – *The Newcastle Sun*, Newcastle, NSW:

"Generalissimo Chiang Kai-shek is secretly recruiting a force of 500 Japanese veteran fliers to fight Chinese Reds.... to spend 12,000,000 dollars on enrolling this unofficial expeditionary Japanese air force. He has made his contacts with disbanded Japanese air force officers through former Japanese war crimes suspects released after 12 months' detention in Sugamo Prison in Tokio."

"It is hinted broadly here that U.S. Occupation authorities, who are aware of this unofficial traffic, will discreetly keep a blind eye to the telescope...."

"More than 300 former Japanese fliers – accustomed to both bomber and fighter operations – have already been enrolled...."

"Their conditions of enlistment are: a bonus of 100,000 yen for enrolment; 100,000 yen for uniform and clothing; 15,000 yen for travel expenses; 60,000 yen a month for pay."

"A mysterious figure in the background of these secret negotiations is Li Ching Guen, who is at present in Formosa. Li's real name is believed to be Masanobu Tsuji."

"A Japanese staff officer in Burma when the war ended, Li fled to French Indo-China disguised as a Buddhist priest. Then he flew to Chunking and presented a plan for using Japanese troops against Chinese Communists."

Jan 7, 1950 – Finished manuscript of *Underground Escape* and gives it to a friend, Toshio Kodama. The Japanese title of the book is *Senko Sanze-ri* (Traveling in Disguise for 3000 Ri). One ri equals 8 kilometers.

April 6, 1950 – *St. Louis Post-Dispatch*, St Louis, MS, by Keyes Beech:

"Japan's No. 1 mystery man is a scarred, shadowy ex-colonel in the Japanese Imperial Army, a hard, brilliant, resourceful man who for four and one-half years eluded allied sleuths who wanted him as a major war criminal."

"His name is Masanobu Tsuji...."

"His whereabouts is unknown, except possibly to Maj. Gen. Charles A. Willoughby, Gen. MacArthur's tall, multi-lingual, Prussian-like intelligence chief...."

"Japanese authorities believe that Tsuji is on G2's payroll, which may account for the fact that they have pursued their search for Tsuji with conspicuous lack of success. They had little difficulty in rounding up other war crimes suspects...."

"There are many theories about Tsuji, but few facts."

"But it is a fact that he was high on the list of Japanese wanted by the British as war crimes suspects. It also is a fact that last December, after more than four years of futile search, the British dropped him and packed up their war crimes machinery."

"Officially, the reason was that it was uneconomic to maintain a war crimes staff for just one suspect. Besides, everybody was sick of war crimes anyway."

"Today the only charge that can legally be made against Tsuji is that he failed to register with the Japanese government as a 'purgee.' However, normally industrious Japanese investigators seem curiously apathetic toward Tsuji."

"The British wanted to try Tsuji for the execution of three British prisoners of war. In Singapore, he reportedly visited a P.O.W. camp, asked if anyone had any complaints, the personally beheaded three who did complain...."

"When Allied forces marched into Bangkok after the surrender, Tsuji was disguised as a monk and living in a Buddhist monastery...."

"Tsuji made his way to Shanghai, where he smuggled himself aboard a repatriation ship bound for Japan."

Masanobu Tsuji, 1950 photo. Wikipedia.org.

"Every repatriate was individually screened, but Tsuji hid his identity behind a crust of dirt and a huge beard. He said he was a student of ancient Chinese art."

"Back in Japan, Tsuji lived a hunted life.... He had no regular supply of food. Food was scarce and rationed and Tsuji had no ration book. He lived in shrines and often slept under the moon."

"Shortly after his name was dropped from the war crimes list, Tsuji appeared at the home of a friend and left with him a 2,000 page diary which detailed his activities over the past four years."

April 9, 1950 – *The Sunday Sun and Guardian*, Sydney, NSW:

"Mystery man of Asia, Colonel Masanobu Tsuji, who has been hunted for four years as a war crimes suspect, is arrogantly publishing his memoirs in Tokyo."

"Despite official amnesty for war criminals by the United States and Britain and the recent release of many suspects held up to three years by the Australians, he still fears arrest and interrogation by the Australians."

"A leader of the fanatical 'young colonel' clique in the Imperial Japanese Army, Tsuji planned high strategy in the Malayan, Burma and Guadalcanal campaigns...."

"As his appearance is well known – he is a tall, gaunt, ascetic Japanese, with piercing eyes and straggly chin-whiskers – there is little doubt that he enjoys protection from the Japanese police allegedly hunting him...."

"Tsuji, whose personal probity at least has never been questioned, claims that in China he once burned down publicly a pleasure-house frequented by Japanese officers whose corruption and immoral living had affronted him."

"He also declares that he sent a cable to General Tojo several months before the Prime Minister's resignation reading: 'Nincompoop Minister! You are ruining the nation! Get out!'"

"His Japanese publishers say that the wanted colonel delivered his 2,000-page manuscript to them by proxy."

"The Japanese police claim, of course, that he is in Formosa."

*"**The Asahi**, leading Tokio daily, front-paged news of the publication of the Tsuji memoirs this week and referred to him by his old army nickname of 'God of Strategy.'"*

"A US Occupation intelligence officer said dryly: 'Wherever he is and whatever he is doing, I don't think even General MacArthur would claim that Colonel Tsuji is yet a democratic Japanese.'"

June 25, 1950 – Korean War begins.

Mid-January, 1951 – Received a letter asking for his help with the repatriation of former Japanese soldiers in French Indo-China. He took no action.

March 3, 1951 – Delivers a speech to the Jiyujin Kurabu (Free People's Club) located in the Ginichi Building, Chuo-ku, Ginza, Tokyo. Though given in secret, the speech was covertly transcribed in shorthand and leaked to the Japanese Communist Party (JCP), which published it in its banned, underground newspaper, *Inside Information*.

The subject of the speech was the comparative fighting strength of the United States and the Soviet Union. Several quotes:

> *"The documents concerning the fighting power of the Soviet Union, which the Kwangtung Army surveyed from various angles, were burned at the termination of the war except for a few documents. I precisely examined those for 1 1/2 years. In view of what I learned from them, and, in addition, various circumstances since then, I judge as follows."*
>
> *"The Soviet Union can mobilize 20,040,00 men at the minimum, among which 16 or 18 million can be dispatched anywhere. The United States will probably be able to mobilize 16 million, among which she will be able to send 8 million to the front. The Soviet Union has military power more than double that of the United States...."*
>
> *"[The Soviet Union's yearly production of planes] is 75,000, almost the same as the United States. Russian small planes are superior to American small planes in quality and quantity, but Russian strategic bombers are inferior."*
>
> *"The Soviet Union possesses 50,000 tanks, while the United States has 5,000...."*
>
> *"The expected strategy of the United States and the Soviet Union."*
>
> *"On the opening of the war, the Soviet Union would take a large-scale offensive in Europe and Asia with its large land and air force. It would take only 2 or 3 months for it to run from Suez to Normandy. According to newspapers, the United States has 500 atom bombs, while the Soviet Union possesses only 50. Judging by common sense, it would be more effective to drop 50 bombs in the United States than to drop 500 over the scattered industries in the Soviet Union...."*
>
> *"Thus, it seems that the victor of the next war will not be the United States. Both the Soviet Union and the United States will devote their heart and soul to it, and the result will be a long-term war of at least for 5 or 10 years. All civilization on the earth involved in the war will be completely destroyed."*

June, 1951 – Organized the Patriotism and Devotion Society.

July 27, 1951 – Visited hospital complaining of a painful stomach ulcer.

Aug 1, 1951 – Indicted by the Japanese Special Investigation Bureau (SIB) under Article 15 of the Purge Ordinance banning political activities by "purgees" for his speech to the Free People's Club. The Purge Ordinance had it origins in an order issued by the Supreme Commander for the Allied Powers (SCAP) in 1946 that banned political activities by militarists, ultranationalists, and wartime collaborators from public office, education, and media. In 1951, more than 200,000 individuals were on the official purgee list.

Aug 2, 1951 – Treaty of San Francisco officially ends the Allied Occupation of Japan.

Aug 3, 1951 – U.S. Intelligence report on July 1, 1951, article in *Tokyo Shimbun* with headline:

> *"The Misguided Speeches of Tsuji"*

Whatever purpose they are intended to serve, the mouthings of ex-Japanese army officer Masanobu Tsuji definitely do not serve the interests of Japan. *"It is because of idolizing such misguided mentality as Tsuji's that Japan today finds herself confined to four small islands."* A repetition of such idolatry and *"Japan will probably find herself completely eliminated."* Tsuji makes the mistake of the former Imperial Army headquarters of *"exaggerating the Soviet strength and underrating the American."* It may not be an unconscious mistake. His *"fascistic neutral"* theories are evidently suiting Communist Party purposes. Communist cells distribute copies of his speeches throughout the country. And though he professes *"equal aversion"* for both the United States and the Soviet Union, *"sometimes, in the heat of his speech making, he lets slip the phrase, Our USSR.'"*

Aug 6, 1951 – U.S. Intelligence Report:

> *"Tsuji met Joseph Fromm [regional editor of* ***U.S. News & World Report****] on the afternoon of 6 August. Fromm, scheduled to return to America shortly, had asked Tsuji to put his views on current affairs in writing, which Tsuji anxiously had completed before his operation, not knowing the outcome. The views were written on 150 pages, had been translated into English, and were given to Fromm on 6 August under the title of 'Message to Those of the Ages of Twelve to Forty Five' (the number 45 may be an error)."*
>
> *"Fromm was quite worried about the recent indictment case against Tsuji and offered to appear in court himself as a witness for Tsuji's defense. Tsuji merely answered that he would have to visit him the next time in jail and refused Fromm's offer with gratitude. Fromm again offered to help in any way possible. A Canadian newsman was present during all of that interview."*

Aug 9, 1951 – Entered the Red Cross Hospital in Tokyo for an operation on his stomach.

Aug 12, 1951 – Appendix and caecum removed in a double operation. (The caecum is the beginning of the large intestine.)

Aug 21, 1951 – An unnamed source reported to U.S. Military Intelligence that he *"visited [Tsuji] in his room at the hospital for the first time since he had entered. He was very weak, though protesting his good health, and could eat nothing and had to be fed intravenously. The doctor stated that it may be a full year before they find out for certain if Tsuji will recover or not. Tsuji requested me to make arrangements for a place for him to stay at Asami at a rest hotel beginning [in a week]."*

Sept 7, 1951 – Left hospital for home.

Sept 22, 1951 – *Newcastle Morning Herald and Miners Advocate*, Newcastle, NSW:

> *"Colonel Masanobu Tsuji, a notorious Japanese war criminal suspect, who successfully evaded British Army man hunt for five years, is again wanted, this time for 'Libel.'"*
>
> *"According to a placard displayed at a Communist Party headquarters raided last week at Hichioji, west of Tokyo, Tsuji made defamatory remarks against the*

occupation, 'constituting criminal libel of the Americans and British and of American and British policy' in a private lecture to members of the exclusive Industrial Club in Tokyo."

"The members of that stately white marble club are bland amiable Japanese industrialists, temporarily 'purged,' but patiently awaiting the end of occupation."

Sept 27, 1951 – U.S. Intelligence Report:

Source had a lengthy conversation with Tsuji who divulged the following information to him concerning an interview he had a short while before with an American newspaper correspondent [Joseph Fromm].

"A North American Newspaper Alliance correspondent visited my home the other day with an interpreter. He wanted an interview with me. I did not ask his name, nor did he offer it either. I could see that he was still a young and pure-minded individual, not a trickster. I declined his request for an interview. Since he hung around the gate of my house for some time afterwards, and did not show any signs of going home, however, I finally called him in out of pity and met with him on the condition that the period would not be more than ten minutes. Actually, however, my talk with him exceeded ten minutes and lasted about an hour."

"At first he asked for my opinions on the arming of Japan. My answer was, 'A variety of opinions are reported not only here but throughout the world on the rearmament of Japan. Such debates are already a waste of time. What steps did Stalin take for Eastern Germany? Did he hesitate over petty arguments? How is the rearmament of Japan treated in the draft of the Japanese Peace Treaty? Even Stalin has tacitly recognized the necessity of rearmament for Japan. The only question remaining is, what means should be taken for the implementation of prompt and efficient rearmament.'"

"'The step-by-step reinforcement of the existing Police Reserve is one way. To cope with an emergency, however, this course is all but worthless. In view of the slowness of the training program for the Police Reserve, little help is held for the future of it. Approximately six million men of the Japanese Army were mobilized in the late Pacific War. Only 10% of them, it is true, could be promptly made available for defense even if equipped with arms. Nevertheless, I believe it to be far more advisable to raise approximately 600,000 of these old soldiers of strong health and excellent mental ability, organize a "Homeland Defense Volunteers Crops" in each local district, provide them with U.S. arms and equipment, give them a quick re-training, and turn them into a supplementary or auxiliary force to the Police Reserve.'"

"However, we must take cognizance especially of the fact that America has never given military aid to foreign countries since the end of the war without dispatching a military mission there also, which interfered with even the most petty details of training.'"

Nov 15, 1951 – U.S. Intelligence Report:

"Around the 10th of November, Tsuji moved into a new house at #883 Narimune Cho, 2-chome, Suginami Ku, Tokyo, which he had bought from the proceeds of the sale of his previous house at Wakaboyashi Cho, Setagaya Ku."

"After he had undergone a cancer operation, Tsuji's health proved much more robust than was expected. Now he does not differ at all, at least in appearance, from a normal, healthy man. Of course, he had not fully gained his physical vigor in full, since for example he eats only a little of the food he used to eat before. In appearance, however, he has recovered to a point where somebody could hardly believe him if he said he were a sick man."

March 4, 1952 – *Chillicothe Gazette*, Chillicothe, OH:

"Japan's soldiers and sailors are fighting the Pacific war all over again – with best sellers."

"Memoirs of the former high command and a flood of 'I Was There' thrillers are heaped high on the counters of every Japanese book store...."

"The boom started last year. Former Navy Captain Mitsuo Fuchida, who led the 360 Japanese planes in the attack of Pearl Harbor, made a hit by writing the full details of the Midway battle – a disastrous defeat for Japan [Hawai Sakusen Nikki, 'Hawaii Operation Diary']."

"Another best seller was former Col. Masanobu Tsuji's 'Hiding Away 6,000 Miles.' It recounted his underground adventures while he was hiding for nearly three years in Thailand, Indo-China and China...."

April 1, 1952 – *Junkao Yomiuri* newspaper reports on statements that Tsuji supposedly made to the SIB (Special Investigation Bureau) critical of the United States, calling General MacArthur *"an obsolete battleship,"* and saying he supports the Viet Minh. In response, Tsuji issued the following statement:

"Some of the material in the article comes fairly close to facts which I have disclosed to no one but an official of the SIB who was interrogating me. I believe I laid bare all my iniquities before that SIB official. I cannot but conclude that the SIB has sold its confidential documents to some cheap publication, or else that the documents were stolen. Although I do not intend to sue the SIB, I think it is an indisputable fact that the government office is rotten, and its authority is completely lost if such a thing is allowed to happen."

April 2, 1952 – U.S. Intelligence Report:

Tsuji does not consider that his article in the 1 April 1952 issue of the *Junkao Yomiuri* was anti-American and therefore a second violation of the Purge Ordinance. In case he were indicted on the basis of the article, he told Kawaguchi of the Tokyo District Prosecutors Office, who took his purge violation case, that he would use three Americans – Lt. Col. O'Neill, head of GHQ Legal Section, Lt. Col. Breakie, FEAF intelligence officer, and Joseph Fromm, correspondent for *U.S. News and World Report* – to act as character witnesses to testify that he was not anti-American.

April 3, 1952 – U.S. Intelligence Report, extract from April 1, 1952, *Junkao Yomiuri* newspaper biographical article:

"No one takes a neutral position regarding Tsuji. His acquaintances make the following comments on him."

"1. Is it true that Tsuji is a genius in operations?"

"Negative: Despite his position as a staff officer, he consistently went to the front, ostensibly encouraging a partial success, and thus deprived himself of the necessary objective viewpoint. Most of the operations he directed failed, except in Malaya."

"Affirmative: There is a climax to every operation. When the time arrives, it is the duty of every staff officer to go to the front to see the results first-hand. He has a genius for grasping essentials. His operations failed only because the situation was impossible."

"2. His character as a soldier."

"Negative: Although he may be a flamboyant actor, he is otherwise a very shrewd and calculating person. For instance, he left his subordinates behind and escaped from Bangkok alone."

"Affirmative: He is a rough, passionate man of action. He is apt to be misunderstood by others because he does not hesitate to clash with them in the line of duty. He acts on the basis of principles derived from the whole of history. Time will show that he has been misunderstood."

"3. Is he insubordinate?"

"Negative: Because he was born the son of a poor farmer, he hates power. He sympathetically listens to his subordinates' complaints. In his way, he is a boisterous, ambitious man."

"Affirmative: He will fight with anyone, superior, equal, or subordinate whose acts are unjust. He is a simple, naïve man."

"4. He is called a man who can be trusted. Is that true?"

"Negative: He holds to the theme of the "Naniwabushi [Rokyoku]," stressing personal obligations."

"Affirmative: There is no perfect man, and it is not fair to mention a man's defects only, especially only generally. Tsuji never compromises with the sophisticated world, has violent feelings of love and hate, and goes his own way in the closest possible union with justice."

"Tsuji's income from his books in 1951 is estimated at ¥ 3,500,000 [about $98,000 at the time]. His family consists of two sons and three daughters. Then, how and where does he spend his money?"

"He claims that he gives half of his income to the families of war victims. He sends money regularly to about eight families, including those of Colonel S.M.K. and Captain K.I., and occasionally to about fourteen other families.

These people have asked for a larger lump sum rather than small frequent amounts, but Tsuji replies, 'It is better to sprinkle water on parched soil.'"

May 14, 1952 – A source reports to U.S. Intelligence:

"[Tsuji] visited the Soviet Mission... to ask that a letter he had written to his old friend, Col. Mishina Takayuki be delivered. Mishina is a graduate of the class of 1924 at the Military Academy, a graduate of the War College, and a one-time staff officer in political affairs with the Kwantung Army. He was a classmate of Tsuji and the two worked together in the Kwantung Army Headquarters later. Mishina was captured by the Soviets in 1945 and is still a prisoner of the USSR."

Masanobu Tsuji, 1952 photo, *Asahi Graph,* September 3, 1952.

May 19, 1952 – *Battle Creek Enquirer,* Battle Creek, MI:

"Former Col. Masanobu Tsuji, famous wartime strategist, has called General MacArthur an 'obsolete battleship' and demanded withdrawal of American garrison troops and the rearming of Japan to meet the Communist danger in its own way."

"Colonel Tsuji further states that the Viet Minh opponents of the French in Indo-China are 'nationalists, not communists,' whom he would join, if he could, to help 'kick' the French out."

July 30, 1952 – U.S. Intelligence Report:

"The ***Nagasaki Nichi Nichi*** *newspaper sponsored a public meeting at the Hitsubishi Kaikan, Nagasaki City, with Masanobu Tsuji as principal speaker. Tsuji spoke on 'Asia for the Asiatics.' His speech stressed complete neutrality for Japan, national 'self-defense,' and included remarks 'bordering on' anti-American and anti-Soviet propaganda. The meeting reportedly drew a large crowd which was 'wildly enthusiastic' over Tsuji's lecture."*

Aug 1, 1952 – U.S. Intelligence Report:

"Police have learned of a rightist plot to assassinate Masanobu Tsuji, war-time operations colonel, from a former air force major arrested for robbery. The National Rural Police and Special Investigation Bureau have assigned agents to check on the secret society that allegedly ordered Tsuji's death. Tsuji, after reading about the plot in a newspaper yesterday, left his home in Satagaya telling his wife, 'I won't be gone long.' According to his wife, Tsuji knew shortly after

he began publishing his war memoirs that his life was in danger. 'Whenever he leaves the house,' she said, 'he tries to put on a disguise. But he as never asked for police protection.'"

Aug, 1952 – Begins campaign as an Independent for the House of Representatives of the Diet.

Aug 16, 1952 – An unidentified source reports on a Tsuji campaign speech delivered to 30,000 people at Kenroku Park in Kanasawa City in which he said: *"Truman is the No. 1 war criminal and Stalin the No. 2 war criminal."*

"Tsuji supports the Emperor system, belongs to the Far East League, and advocates the Greater Asia Principle. His financial backer is Tomokich Nakamura, manager of the Yoshida Paper Store...."

"Through his election campaign speeches, Tsuji clarified the following points: He is for the instigation of world war and against Japan's entry into the war; for an anti-American and anti-Soviet attitude; and for Japan's self-defense and neutrality; against Japanese becoming a mercenary army and for a militia system; and against converting Japan into a colony and for the revival of the Emperor system."

"On the premise that World War III is inevitable, Tsuji predicted that an American-Soviet war would break out by the end of this year or the fall of next year...."

"His remarks created the illusion that an American-Soviet war would offer a good chance for Japan to achieve complete independence."

Oct 1, 1952 – Elected by a landslide to the House of Representatives of the Diet as the representative of the First Electoral District of Ishikawa Prefecture. He won 64,912 votes out of 251,411 cast, a plurality of 23,408 votes over his closest opponent.

"An important source of Tsuji's strength was a large group of voters, who as young men called up for conscription, were trained under him in the late 1020s and early 1930s. When Tsuji was commissioned second lieutenant about 1925, he was assigned as an instructor for the district's conscripts to the Seventh Regiment at its permanent station in Kanazawa-shi, Ishikawa-ken, his native prefecture.... It is estimated that Tsuji trained about 16,000 young men in the eight years he held the assignment."

"The Liberal Party emerged from the Lower House elections with a bare majority of nine seats.... The precarious Liberal majority consequently is constantly threatened by death, illness, abstention, or defection from the majority party's ranks. The Progressive, right wing Socialist, left wing Socialist, and Labor-farmer parties all have publically announced that they will act as opposition parties in the Diet. This leaves the Independent Club, as the only remaining organized Lower House body, in a strong bargaining position. The only reason for the 11 members banding together in a club was to improve their bargaining position."

Nov 30, 1952 – *Palladium Item*, Richmond, IN:

"Japan's wartime 'god of strategy' told the parliament Saturday, 'In my opinion the war between the Russia and United States already has begun and nobody knows who will win.'"

"Masanobu Tsuji... threw the Diet into an uproar."

"Amid cries of 'You are criminal!' and 'Shameless fellow, stop, stop, stop!' Tsuji blared:"

"'It is clearly erroneous for the Japanese to rely on the U.S. to preserve Japan's security. Strategically, America is going to sacrifice the Japanese for her own safety.'"

"Leftwing Socialists and the radical Labor Farmer minority supported him with 'Yes! Yes!' But angry government benchers shouted 'You aren't in the Imperial headquarters now!'"

Masanobu Tsuji, 1953 photo, *Times-Tribune* (Scranton, PA), December 22, 1953..

Feb 8, 1953 – *Oakland Tribune*, Oakland, CA, interview by Ernie Hill:

"Japan's old-line militarists are much stronger today and more likely to come to power in the land of the rising sun than are the communists."

"The ultra-Nationalist military crowd is on the ascendency while the communists are going downhill. In an exclusive interview with the most volatile of the super-militarists, Col. Masanobu Tsuji, this reporter was given a full review of what we might expect from Japan should the military crowd bounce back into power."

"Japan would never help us in an ideological war unless it saw some advantage or the possibility of territorial expansion."

"While the militarists like the United States slightly better than they like the Russians, they would try to remain neutral in any war between the two."

"Col. Tsuji [is] the country's out-spoken champion of a return to the emperor system, the citizen-soldier system and the abandonment of American democracy introduced during the occupation."

"'Japanese,' Tsuji told this reporter, 'are tired of democracy. The Diet (Parliament) is a mess. If every citizen could watch the Diet in action, they would be unanimous in their desire to return to the old emperor system.'"

"Tsuji's movement to return to the old way of life claims some 100,000 followers...."

"Since most of Japan's big military names were hanged or jailed after the war, Tsuji is one of a very few operating full steam on the public scene today."

"By some, he is considered the most dangerous man in Japan today."

April 19, 1953 – Reelected to the House of Representatives of the Diet as an Independent.

May 30, 1953 – U.S. Intelligence Report:

"On 17 May 53, approximately 20 members of the Kansai and Shikoku Districts of the ultranationalistic East Asia League Comrades Society held a meeting in Kobe and voted unanimously to demand the expulsion of ex-Col. Masanobu Tsuji from the Society. The group then forwarded this recommendation to the National Headquarters and other districts of the Society for consideration. The main reason for opposition to Tsuji is ostensibly that since he was elected to the House of Representative in Oct 52 he has been furthering his own interests rather than those of the Society."

July 27, 1953 – Korean War ends with signing of the Korean Armistice Agreement.

Dec 22, 1953 – *Times-Tribune*, Scranton, PA, interview by Vic Suzuki:

"Masanobu Tsuji, one of Japan's most brilliant military strategists in World War 2, today warned the United States to pull its troops out of Japan as soon as possible or the day will come when she will regret it."

"He also warned, 'a strong anti-American feeling prevails and is increasing in this country....'"

"He said... '[the United States should] admit the war crimes trials were a mistake.'"

"'If the United States withdraws her troops from Japan this country will at once buckle down to prepare for any aggressive acts by the Communists.'"

May 1, 1954 – Sent a letter to President Richard Milhous Nixon. Some quotes:

"An Open Letter to American President Nixon"

"Paying my respect for your generous attitude, I, as a member of the Diet would like to ask you the following questions which I think are important for the future of Japan-U.S. relations...."

"You recognized the mistake of American policy toward Japan in 1946 and boldly urged Japan to rearm. But I wonder why with all your courage and conscience, you do not further admit that the war crimes trial in Tokyo was a mistake."

"Is it ever rational for the victors to try the vanquished? In my opinion, former President Truman, the supreme responsible person of your country,

who deliberately killed and injured more than 300 thousand non-combatants in Hiroshima and Nagasaki with atom bombs and hundreds of thousands of innocent citizens in other cities by indiscriminate bombing, is No. 1 war criminal, and the late Stalin of Russia, who committed atrocities unprecedented in history of mankind in Manchuria and Korea and killed hundreds of thousands of Japanese and German civilians and prisoners of war through starvation and maltreatment, No. 2 war criminal."

"If you wish for permanent unity between Japan and America, you should frankly admit the mistake of the Tokyo trial and promptly free war criminals related to your country in Sugamo Prison...."

"Anti-Americanism in Japan has become intensified as a result of the long occupation by American forces. This is especially true of the Japanese who are employed by the American forces in the neighborhood of their bases. If America desires that Japanese to defend their country, I believe it is most advisable for America to withdraw all her armed forces in Japan as soon as possible...."

"The Japanese are charged with a mission to advise leaders of America and Russia to avert an atomic war in the capacity of the only victims of the atom bomb in the world. Should the two countries ignore this advice, they should wage the war at the North Pole.... "

"It is a well known fact that America made Japan conclude a peace treaty with the Chiang Government on Formosa. America should fully reconsider that, as a result, the Peking Government has become antagonistic toward America only to please Russia. If you try to establish normal relations between Japan and China by letting Japan to promptly start economic and diplomatic negotiations with the Peking Government, and if America gives through Japan more support to China's construction than Russia, America will be able to grasp a chance to win over China from behind the Iron Curtain to a neutral camp friendly to America."

"Masanobu Tsuji"

May 6, 1954 – *Daily News*, New York, NY:

"A Japanese colonel named Masanobu Tsuji told the United Press yesterday in Tokyo that 'it is too late to save Indo-China unless the United States sends five or six Army divisions there.'"

"That would make U.S. intervention in Indo-China about as big a project as the Korean War...."

"Maybe we'd do well to attend carefully to what Tsuji says about possible U.S. intervention in Indo-China."

May 7, 1954 – Battle of Dien Bien Phu ends with total Viet Minh victory over the French defending forces, ending French military presence in Indochina.

June 2, 1954 – An unnamed source reported to U.S. Intelligence that *"Tsuji submitted a request to the Philippine Mission in Tokyo for travel to the Philippines at an early date. His aim, if permission is granted, will be to round up Imperial Japanese Army stragglers still at large in the Philippine countryside and escort them to Japan."* Request denied.

Dec 8, 1954 – *Alton Evening Telegraph*, Alton, IL:

"Rightwing members of the Japanese Diet today verbally attacked former U.S, Presidents Franklin D. Roosevelt and Harry Truman and told a new veterans group 'both the United States and Japan are to blame for the war on a 50-50-basis.'"

"Ex-Col. Masanobu Tsuji... told an audience of more than 1000:"

"'If God were to give a fair trial to war criminals, Truman would be the No. 1 war criminal for killing 300,000 civilians at Hiroshima and Nagasaki, because he gave the order to atom bomb those cities.'"

"'Stalin would be the No. 2 war criminal for starving to death hundreds of thousands of Japanese in Siberia.'"

Feb 27, 1955 – Reelected to the House of Representatives of the Diet as a member of the Japanese Democratic Party.

Aug 15, 1955 – U.S. Intelligence reports that as of this date, Tsuji has published the following books:

Senko Sansen-Ri (Underground Escape)

Jugo Tai Ichi (Fifteen Against One)

Guadal-Canal (Guadacanal)

Singapore (Singapore)

Asia Ho Kyokan (Common Feeling of the Asian Nations)

Nomonhan (Battle of Nomonhan at the Manchurian-Russian Frontier)

Kono Nipon Wo (What About This Japan)

Tsugino Sakai Taisen (Next World War)

Aug 23, 1955 – Left Japan for a 40-day trip to China and the Soviet Union as one of several fact-finding Diet members of the Democratic Party.

"Most of the time [was] spent in the USSR. The tours were strictly controlled and guided, but Subject evaded the Soviet schedule, and was able to get about a bit by himself. Subject reportedly had a good knowledge of both Russian and Chinese, and was able to talk with over 400 of the common people in an attempt to find out the true feelings of the people."

"Subject's real intention in making the tour was to ascertain the truth of the 'peace offensive' program of the USSR. Subject's military background assisted him immeasurably in talking to military leaders in the USSR and Communist China, and also made him an outstanding target for constant surveillance."

"During the tour, Subject has several meetings with representatives of [Defense Minister] Marshal [Georgy Konstantinovich] Zhukov, one of them Zhukov's chief of staff (a Maj. Gen.) and on the night before leaving Moscow, Subject had a 3-hour interview with Zhukov at the latter's request."

"A special messenger instructed Subject to follow a man wearing a red tie and carrying a newspaper who would be walking in front of the hotel. Subject,

fearing liquidation, told the leader of his group that he was being taken to meet Zhukov, and then met the contact man, who led him several blocks, motioned him into a car which took him to Zhukov. Zhukov showed great interest in US forces in Japan, asking questions on the books Subject had written, and questioned Subject about the attack on Pearl Harbor, being under the impression that Subject was one of the planners of that attack."

"Prior to Subject's meeting with Zhukov, Subject received a message from Zhukov asking if there was anything he could do for Subject. Subject replied that he would be very grateful to receive Zhukov's permission to meet former Col. Mishina Takayuki, who was imprisoned on war crimes charges at Ivanovo prison. Mishina had been a classmate of Subject's at the Military Academy and a very close friend. The following day Mishina was brought to Subject's hotel room with permission to remain there overnight."

"Upon noticing that Mishina's teeth were in very bad condition, Subject called the fact to Zhukov's attention through one of the messengers, and the very next day Mishina received dental treatment."

It was reported that three members of the Japanese Diet delegation which visited the USSR were recruited by the Soviets. The Japanese intelligence services suspected that Subject might have been recruited. These suspicions were based on the fact that Subject left the group on two occasions, at Kiev and on the Volga, and on one occasion was separated for two days from the rest of the group.

Oct 2, 1955 – Returned to Japan from USSR-China trip.

Oct 20, 1955 – Spoke to a group of U.S. Army officers about his trip to the USSR:

"This interview was sought by Tsuji through an intermediary.... A meeting was held with Tsuji at the Army Foreign Area Specialist Training School in Tokyo and was attended by the senior Army language students, representatives of this office, and Security Group G2."

"At the conclusion of Tsuji's remarks, R.O. asked Tsuji whether as a result of his trip he had concluded that his former belief that neutrality was the only policy for Japan had been changed. In a very impassioned and lengthy statement, Tsuji stated, in brief, that he was only the more convinced that there was no alternative course for Japan and that there was no reason for Japan being drawn into the struggle in the event of a third war because Japan would not serve as a significantly important base for the United States. Japan is too far distant from the heart of Russia, he says, to be a strategically important area. Therefore, his sole desire was to spare Japan from the horrors of a nuclear war and, if possible, see the entire world spared from such a conflict...."

"Tsuji further commented that United States policy with regard to the encirclement of the Communist bloc and increasing pressure was wrong, because it only served to build up counter-pressure within the Communist bloc. He stated that he felt the better policy would be to penetrate or pierce the Communist bloc at its most vulnerable point, i.e., Communist China...."

> *"During the approximately one and one-half hours that he was talking, Tsuji's attitude underwent a small but perceptible change. It appeared at first that he was somewhat on the defensive, not knowing what sort of reception his remarks would have. As the meeting developed and it appeared that we at least were willing to hear him out, his attitude became somewhat warmer and more friendly. He is a very convincing and powerful speaker and seemed to be full of self-confidence and self-assurance. The nature of his remarks indicates that he is a very keen and detailed observer regardless of whether or not the conclusions that he draws from his observations are reliable."*

Oct 29, 1956 – Suez Canal Crisis begins with the invasion of Egypt and the capturing of the canal by Israeli forces. France and the UK land troops in Egypt two days later.

Nov 20, 1956 – Resigned from the Diet to take responsibility for the conduct of his 25-year-old son, Tohru Tsuji. He told the press, *"One who cannot manage a household well is unfit for the task of ruling the country. My conscience does not allow me to remain in the Diet post."* Tohru had visited his former wife, whom he divorced on November 5, and asked her to return to his home for the happiness of their child. When she refused, he inflicted injuries with a broken beer bottle that required one month of medial treatment to heal. Consequently, Tohru was arrested.

Nov 26, 1956 – Withdrew his resignation from Diet:

> *"A report stated that, according to reliable information emanating from newspaper circles, the real reason behind Subject's attempt at resignation from the Diet was that he had been invited by South Vietnam's Govt to become the commander-in-chief of its army. It was also stated in this report that Subject had revealed this fact to Lower House Speaker Shuji Masutani, who discouraged him from accepting the job."*

Dec 24, 1956 – French and British troops withdraw from Egypt.

Jan 1, 1957 – Left Japan for a tour of the Middle East and Eastern Europe, as a member of the Foreign and Military Affairs Committee of the Diet.

Jan 21, 1957 – *"Subject had a talk with the US Army Attaché in Damascus, Syria. During the interview, Subject stated that he was on a mission for the Japanese Prime Minister to survey Middle East problems, particularly the Suez Canal situation and the magnitude of the Communist threat in that part of the world. Subject also stated that the results of his survey would have a bearing on Japanese relations with the Afro-Asian Bloc in the United Nations."*

Jan 22, 1957 – Left Damascus, Syria, and arrived in Cairo, Egypt. On his return, he wrote a newspaper article about his trip. Some quotes:

> *"I arrived in Cairo January 22 after a plane ride from Syria that took about two hours. I found no trace of bombing in the city...."*
>
> *"The shops lining the streets were well supplied with goods – even luxury items. The only exception was that practically no medicines were obtainable at pharmacies."*
>
> *"Contrary to what I expected, I found that prices were fairly stable in Cairo. Yet the Suez Canal was blocked; the Egyptian international balance of payments was in the red and defense expenditures must have gone up. So credit must be*

paid to [President Gamal Abdel] Nasser's policy of storing enough foodstuffs to last half a year and controlling exports and imports...."

"Late in the afternoon of the last day of my stay in Cairo, I received a brief notification to the effect that the president would receive me in his residence at 7 pm. There were only three guards outside his house – rather light in view of the fact that he is the top leader of a nation under marital law."

"It was a very modest house – just like that of the managing director of a second-class firm in Japan. I was led into quite a spacious living room. On the mantelpiece facing the door were displayed photos of Chou En-lai, Nehru, Sukarno, and Tito...."

"Tsuji presented Nasser with 'a dagger I have kept with me ever since the Malayan operation started to protect myself....'"

"Tsuji asked Nasser, 'The other day I visited the Suez Canal. According to UN technical experts, ships under 10,000 tons will be able to pass through by March. In May they said, the entire canal would be open. Will there be any political issues concerning passage through it?'"

"Nasser replied, 'We are waiting for a UN decision. The UN decided upon three conditions: Ceasefire, the clearing of the canal and the withdrawal of troops. Of these, the first and second have been agreed upon. But the withdrawal of troops is still a problem. Technically speaking, passage through the canal will be possible by the beginning of March, but the passage of ships is not a technical, but a political problem.'"

"Tsuji: 'I am leaving for Yugoslavia tomorrow. Would you send a telegram to Tito making him to see me?'"

"Nasser: 'With pleasure. I will send it tonight.'"

"When I told him that I wanted to take his picture, he willingly gave me permission. I took two pictures – one standing and one sitting down. I shall never forget the way he looked as he saw me off at the door. I was so deeply moved by the interview that I could not sleep until daybreak."

Jan 29, 1957 – Left Egypt for Yugoslavia. Interviewed President Josip Broz Tito. Tito *"informed [him] that the Yugoslav form of communism is nationalist and that this represents the only kind of communism that can last."*

Feb 3, 1957 – Left Yugoslavia for Turkey.

Feb 27, 1957 – Arrived in Peking, China.

"When Tsuji arrived in Peiping, he immediately requested an interview with [Premier] Chou En-lai, but he was informed that Chou was ill. Tsuji was entertained by Liao Cheng-chih at a welcoming dinner attended by about 20 persons. During the dinner, Liao said to Tsuji, 'We did not wish to invite you to our country because last summer you denounced us in an article you published in the ***Hokkoku Shimbun****.' Tsuji replied, 'I published that article because it was based on reliable evidence, and I have no intention of retracting my statements*

in that article.' Liao was so infuriated with Tsuji's reply that he left the table, and the party broke up automatically."

"Following this quarrel with his host, which amounted to a breach of etiquette, Tsuji remained in his hotel room, expecting that he might be placed under 'house arrest.' Four days later, Liao paid a sudden visit to Tsuji at his hotel and informed him that Chou En-lai would give him an interview. Liao escorted Tsuji to Chou's bedroom where the interview took place. Throughout the interview, Tsuji took precise notes of the conversation, and before leaving, asked Chou if he could make public the contents of the notes. Chou replied that he too had taken notes, and wanted the notes compared before they were made public. Tsuji has not yet disclosed the contents of his notes."

March 2, 1957 – Returned to Japan.

"Regarding his journey, Tsuji said that he had visited 19 countries, including Egypt, Saudi Arabia, Iraq, and Yugoslavia (where he had conversed with President Tito). From Karachi he had sent a telegram to Liao Cheng-chih, the Vice-Chairman of the Communist Party Overseas Chinese Affairs Commission, whom he had met about a year earlier when visiting Peking. Liao had responded by inviting Tsuji to visit him. Tsuji went to Rangoon whence he took an aeroplane to Peking. After talks with Liao, he had left Peking to resume his pre-arranged journey. The subjects of his discussions with Liao are not known."

March 14, 1957 – U.S. Intelligence report:

"The following views and statements were presented by former Colonel Tsuji to a group of former Japanese naval officers shortly after Tsuji's return from his recent trip to the Middle East."

"The United States' insistence upon a quid pro quo in extending aid to foreign nations is hurting the United States, particularly since the USSR has been handling the same problem much more skillfully."

"Tsuji was surprised, in his interview with Chou En-lai, to hear Chou state laughingly the Communism will not succeed in Japan because the Japanese are too nationalistic and passionately patriotic."

"Trade with Communist China is 'a snare and delusion,' because with Japanese individuals on one end of a transaction and the Chinese Communist Government on the other the Government will get the better of the individual every time."

March 19, 1957 – Israeli troops withdraw from Egypt, ending Suez canal military occupation.

June 2, 1959 – Elected to the House of Councillors of the Diet with third highest number of votes cast in the election.

June 22, 1959 – Pamphlet published by former Major General Seiken Kamaguchi denouncing Tsuji and accusing him of war crimes. Kamaguchi's accusations against Tsuji for his actions in Singapore and the Philippines, and the execution of Abad Santos, are detailed above. Here are addition accusations made in the pamphlet:

"It is well-known and witnessed by many that Staff Officer Tsuji often killed prisoners-of-war, took out their liver and made his own medicine out of it, and at times encouraged the use of this medicine to others. Around July 23, 1942, I heard with my own ears Staff Officer Tsuji boastfully telling this story at the South Pacific Hotel in Palau and it is something which I shall never forget."

"There is another story told about him. At one time, he waved to a Philippine officer, as prisoner-of-war, with his left hand to step up, and when the officer came up to several steps from him, he suddenly shot the man with the pistol he held in his right hand and killed him on the spot. He is reported to have boasted that: 'This is the way a prisoner-of-war should be treated. I have shown you the correct example.'"

"In short, Mr. Tsuji's acts of atrocity rival those of the Jewish manhunt by the Nazis and the Kastyn Forest[12] *by the Soviet Union (where a large number of Polish people were killed). German and Soviet atrocities were carried out by an organization, but in the case of Mr. Tsuji, he did it on his own initiative...."*

"It is true that in the last war there were men other than Mr. Tsuji in our Japanese Army who committed atrocities. The major responsibility for the bringing down on the Japanese Army the infamous reputation for atrocities and for destroying the good name and honor of the Japanese Army, however, rests with Mr. Masanobu Tsuji...."

"It should not be permitted that a man like him should be a member of the National Diet. My conscience and my sense of justice will not permit me to tolerate his being elected a member of the House of Councillors through his deception of 680,000 good people with his insufferable words. In particular, I strongly advice Mr.Tsuji to resign as a member of the House of Councillors in the name of my past friend Lt. General Kawamura who lost his life because of Mr. Tsuji, in the name of the souls of those people who were executed, in the name of the ghosts of my former men who gave up their lives on the southern islands, and in the name of their bereaved families." [13]

Seiken Kamaguchi was tried in 1948 for his involvement in atrocities committed by troops under his command in Borneo and sentenced to life imprisonment. He was released in 1954.

June 27, 1960 – *The Age*, Melbourne, VIC:

"An Australian 'digger' yesterday began to relive some of the Malayan battles of the second world war with the 'diggers' of the former Japanese Imperial Army."

"He is Lt. Gen. H. Gordon Bennett, former commander of Australian forces in Malaya, who arrived here by air today with his wife for a three weeks' holiday."

"He was met at the airport by some of the commanding officers of the Japanese 25th Army, which drove the Australian and British troops out of Malaya and Singapore."

Book advertisement, *Sydney Morning Herald*, Sydney, NSW, December 10, 1960.

"Among the welcoming party at Tokio's International Airport were Colonel Masanobu Tsuji, now a member of the Upper House of the Japanese Diet, who planned the Japanese Malayan offensive...."

"He and his wife were presented with a large crystal vase by Colonel Tsuji."

"In return, General Bennett presented the former Japanese officer with a bark drawing made by aborigines in Arnhem Land...."

"General Bennett said a book written by Colonel Tsuji on the Malayan campaign would be published within a month in Sydney. In the book the general was reported as killed in action."

*"General Bennett said the book, '***Hinge of Fate,***' told the Japanese side of the Malayan campaign."*

Oct 12, 1960 – *The Age*, Melbourne,VIC

"A British general was angry today with a Japanese colonel over a book about to be published in Australia on the war in Malaya."

"The general was Arthur Ernest Percival, the British commander who surrendered Singapore to the Japanese in 1942."

"The colonel was Masanobu Tsuji, Chief of the Operations and Planning Staff of the Japanese 25th Army."

"Colonel Tsuji... claims in his book that the Japanese did not, as maintained by General Percival, outnumber the British in men and equipment during the Malaya campaign."

"'The British regarded the Indian divisions as "goods for consumption," he writes.'"

"He accuses the British of abandoning the Indians when the Perak bridges were demolished."[14]

"Today, 72-year-old General Percival said, 'There are hundreds of my former men who know these claims are absolute nonsense.'"

"'At the start of the Malaya campaign, 128,000 British, Australian and Indian troops faced 95,000 Japanese.'"

"'But few of our chaps were sufficiently trained and, within no time at all, there were 135,000 Japanese, without us being reinforced....'"

"'So far as the allegation that we abandoned Indian troops is concerned, I was personally right up at the northern front, close to the Indians at the Perak River bridges,. I know they were not abandoned.'"

"'When we became prisoners, the Japanese were forever trying to impress us with the greatness of the Nippon army.'"

"'It seems they never stop forgetting,' he said."

Dec 31, 1960 – *The Age*, Melbourn, VIC (Book Review):

"Japanese Version of the Fall of Singapore"

"Some day a military historian will write a book showing not only how Malaya fell in 1942, but why. No account of the campaign yet published throws much light on this mystery. Colonel Tsuji's book does nothing to elucidate it. On the contrary, it adds something to the complexity of the riddle."

"This is not to say that Colonel Tsuji has not, in writing ***Singapore: The Japanese Version****, made an important contribution to the recorded history of the Malaya campaign; his book is essential reading for anyone wishing to understand what happened in Malaya between December 8, 1941, when the Japanese attack began, and February 15, 1942, when Singapore surrendered...."*

"His book conveys the impression that he is also an honest man, scrupulous with facts and telling the truth as he sees it. The absence from his writing of any apology for Japan's action in launching the Pacific war is convincing evidence

Masanobu Tsuji dressed as a Buddhist monk with Katsumi Akasaka. Last known photograph of Tsuji, April 21, 1961, xinjiapo.news.

that he is no misty-eyed humbug, seeking to ingratiate himself with Japan's conquerors...."

"There appears to have been one basic psychological difference between the Japanese victors and the British vanquished in Malaya: the Japanese, or many of them, at any rate, were dedicated men, while the British were not."

"Thus we find Tsuji, who in general betrays none of the fanatic's talent for strange ecstasies, describing his thoughts on the eve of the campaign:"

"'Sitting on a rush mat, I vowed to the gods that, day and night, I would abstain from wine and tobacco. I forgot instinctive desires and worldly passions, to say nothing of lust and appetite – and even life and death. My whole mind was concentrated on gaining the victory.'"

"'We will win. We will win....'"

"One looks in vain, however, for any mention of the atrocities committed by the Japanese military machine on defenseless prisoners, until at last, on page 264, one finds this:"

"'The numerous and disgusting later breaches of military discipline must be considered in comparison with the far more numerous fine and noble actions on the battlefield. Beside them any discreditable actions will in time be swept into oblivion.'"

April 4, 1961 – Requested a leave of absence for 40 days from the secretariat of the Diet.

Left Tokyo by Air France flight for a "private" inspection tour of Southeast Asia. Planned to visit Thailand, Laos, South Vietnam, Cambodia, Burma, and Singapore. His life was insured for $10,000 by Overseas Travel Insurance, a US company. The expiration date for the life insurance coverage was June 3, 1961.

Arrived at Ton Son Nhut airport in Bangkok and visited Colonel Chikashi Ito, an old acquaintance, a Self-Defense Force officer attached to the Japanese Embassy in Bangkok.

April 14, 1961 – Appeared in Vientiane, Laos, with Colonel Ito. Visited Tokyo Bank in Vientiane and converted American dollars to Laos currency. (Converted enough money to cover two or three days' stay.) Tsuji and Ito parted with each other. Before parting, Tsuji asked Ito to send the briefcase he was carrying with him back to Japan.

April 20, 1961 – Former Lt. Gen. Yasuyuki Miyoshi, a resident of Odawara, Japan, received an airmail letter from Tsuji informing him, *"I am shortly leaving Bangkok for home."*

April 21, 1961 – Photographed in the dress of a Buddhist monk with Katsumi Akasaka. Last known photograph of Tsuji.

April 22, 1961 – A reservation made in the name of Masanobu Tsuji on Air Thailand leaving Vientiane for Bangkok was canceled.

April 27, 1961 – Tsuji's family received a postcard dated April 21, 1961. The card read:

> *"I shall come home about 10 May. I am going to teach at Military Academy. I met both Mr. Astsu and Rabaion."*

Both Atsu and Rabaion, full names unknown, were officers in the Thailand Defense Force. The report says the card was mailed from Bangkok, but that is likely a mistake and the card was actually mailed from Vientiane.

April 29, 1961 – The suitcase which Tsuji asked Colonel Ito in Vientiane to send back to Japan arrived in Haneda by Japan Air Lines. Right beside the space for addresses was written in red pencil: *"I am to return about 10 May. Please hold it until then at the JAL office in Tokyo."*

May 10, 1961 – Japanese Foreign Ministry received a classified communication from the Japanese Embassy in Phnom Penh, Cambodia, saying that Tsuji's personal effects, mostly clothing, were being shipped to either Hong Kong or Tokyo.

May 13, 1961 – Leave of absence from Diet officially ends.

May 16, 1961 – U.S. Intelligence Report:

> *"Sometime in April 1961, the Finance Section of the Japanese Foreign Ministry submitted to the Chief of the Foreign Ministry Secretariat a written request, bearing a high security classification, for approval of payment of $1,300 in U.S. currency to Masanobu Tsuji, an Independent member of the House of Coucillors, as 'remuneration.'"*
>
> *"A few days later, the Foreign Ministry received from the Japanese Ambassador in Saigon a classified communication which stated that Tsuji had been in Saigon and from there had proceeded under clandestine conditions to North Vietnam, disguised as a Buddhist monk. Tsuji had told the Ambassador*

that he also intended to enter Communist China and return to Japan by way of Hong Kong."

"The Ambassador reported that he tried to dissuade Tsuji from making this trip due to the risk involved; however, Tsuji countered by stating that his trip had the sanction of Foreign Minister Zentaro Kosaka and Prime Minister Hayato Maeda, and that he had made arrangements for appointments with senior government officials in North Vietnam and Communist China as well as for safe passage in those countries. The communication from Saigon requested confirmation of Tsuji's statements from the Foreign Ministry."

May 26, 1961 – *"According to articles in the Japanese press on 26 and 27 May 1961, Tsuji's family had not heard from him since 27 April.* ***The Tokyo Shimbun*** *said the Japanese Embassy in Bangkok had reported that Tsuji had been there until 10 April; it added that the Japanese Foreign Ministry was worried over the possibility that Tsuji 'might have infiltrated into war-torn Laos disguised as a monk.'"*

May 27, 1961 – *"Japanese Embassy in Bangkok investigating to determine Tsuji's whereabouts since his departure from Bangkok on an undetermined mission. The dispatch cited reports that he had entered Laos disguised as a Buddhist monk and carrying, 'as a sort of passport,' a photograph of himself shaking hands with President Nasser of Egypt."*

June 12, 1961 – Kokichi Nakachi gives the CIA a letter he received from Tsuji.

"On June 12, 1961, the undersigned forwarded for information a personal letter to CIA headquarters of the United States from Masanobu Tsuji, a member of the Japanese House of Concillors and a former staff officer of the Japanese Army in the higher echelon. However, there has been no reply from CIA. Has it been pigeonholed? It is indeed regrettable that no word has been received from CIA. The letter would have provided sufficient information on the whereabouts of Tsuji and what he has been doing since he has been missing in Southeast Asia."

"Fortunately the envelope of said letter which had been retained by the undersigned is still kept intact and is enclosed herein. If you will have the signature checked through the Untied States Intelligence Agency in Japan, you may be able to find out about Tsuji. The undersigned is confident that the signature is Tsuji's although he has used a fictitious name and address. – Kokichi Nakachi."

June 24, 1961 – Hiroyasu Tani, managing director of Maruichi Trading Co, reported that:

"When I met Mrs. Tsuji on 13 June, she said she heard from her husband around 11 May but had not heard from him since then. She said she heard from some source that her husband asked for a guide at the Japanese Embassy in Bangkok and the last known about him was that he left with this guide, but this story was not true when she had the Foreign Ministry check with the embassy in Bangkok."

July 31, 1961 – Japanese Foreign Affairs Section received telephone call:

"About 2220 on 31 July, a person who called himself a relative of Tsuji telephoned from Tsuji's home:"

"'I am a relative of Upper House member Masanobu Tsuji. About 30 minutes ago a Chinese called and said, "I am sorry for what happened to Tsuji (rest redacted). I am calling because he said the (redacted) knows all about it and because I want to know the truth."'"

"The police on night duty answered, 'I don't know a thing about it.'"

"On the following morning, 1 August, Inspector Machida contacted the Foreign Ministry about the telephone call. The reply from the Foreign Ministry was that the Chinese must have read the article carried by a Hong Kong newspaper reporting that Tsuji was shot to death in Communist China."

Aug 28, 1961 – U.S. Intelligence Report:

"The Japanese Embassy in Saigon, hearing rumors that Masanobu Tsuji was in Vang Vieng, Laos, and had received permission to visit Pathet Lao leaders, sent an Embassy staff member named Yoshikawa to investigate."

"In a place call Phon Hong, Yoshikawa met a Chinese who identified a photograph of Tsuji and revealed that he had met and assisted Tsuji. The Chinese said that Tsuji had arrived in Vang Vieng around 15 May 1961. In early June, Tsuji received permission to visit leaders of the Pathet Lao, and on 7 June 1961 departed Vang Vieng for Xiangkhouang, Laos. Tsuji arrived in Phon Hong on 10 June 1961."

April 4, 1962 – *Daily Mail*, Hagerstown, MD:

"One of Japan's most intriguing mysteries is a year old. It concerns what has happened to Masanobu Tsuji...."

"Tsuji, then 59, left Tokyo a year ago today for what ostensibly was a 40-day tour of Southeast Asia. A month later his wife received a package from Bangkok, Thailand, containing his business suit, other clothing and personal; effects."

"There have been nothing but rumors since."

"One unconfirmed report said Tsuji reached Vientiane, Laos, April 14, and left for Communist-held northern Laos five days later in route to Red China."

*"**The Tokyo Times** recently published a report, which could not be confirmed, that Tsuji is teaching guerrilla warfare at a Communist military academy in Kirin, northeastern China."*

"It has also been rumored that Tsuji is under arrest in Red China. The Japanese Red Cross has approached Peiping officials in an attempt to confirm this, but a Red Cross official said, 'there has been absolutely no information.'"

"'I've had no word at all,' says Tsuji's wife, 'and can't hazard a guess.'"

April 13, 1962 – The fate of Upper House Masanobu Tsuji has been unknown since he disappeared during his trip to Southeast Asia last year. According to Foreign Minister Kobaka, the Chinese Red Cross Society was requested in December last year to investigate the matter on the basis of the information that he had entered China from Hanoi, but the Society's answer showed that there was no evidence of his having entered that country.

There is, however, the rumor that he was seen somewhere in the northeast district (Manchuria). According to the Chief of the Upper House General Affairs Department, Makoto Chiba asked Hsieh Nan-kuang of the Peoples' Institute of Foreign Affairs to trace Tsuji's whereabouts when Chiba visited China last January as a member of the Socialist Diet members' mission. Hsieh is said to have replied that Tsuji actually entered China and stayed there for ten days or so. Hsieh gave the unconfirmed information that he left China and was shot to death by American troops when he was trying to enter the Pathet Lao areas of Laos.

Some attribute his stay in China for only ten days to the fact that he was treated coldly there; the old China, toward which he used to entertain friendly sentiments, has already changed greatly. The new China would not have welcomed an 'adventurer' having no background. It is said that the Foreign Minister intends to ask the U.S. Government about the rumor of his having been shot to death, but it will be difficult to ascertain the truth in that area.

April 28, 1962 – *Daily News*, New York, NY:

> *"U.S. Ambassador Edwin Reischauer today categorically denied a rumor that ex-Col. Masanobu Tsuji, an Independent member of Japan's Parliament who vanished last spring, was shot to death by U.S. forces in Laos. Tsuji masterminded the conquest of Malaya and Singapore in World War II. He disappeared last April on a tour of Southeast Asia."*

April 30, 1962 – U.S. Intelligence Report, extract of article from *The Japan Times*, Tokyo, Japan:

> *"Toru Tsuji, the eldest son of Masanobu Tsuji, a member of the House of Councillors who has been missing since his tour of Southeast Asia last April, will leave here Friday for Saigon in search of his father."*
>
> *"He will begin his search in Saigon and then visit Bangkok and Vientiane in an attempt to locate his missing father. He will be accompanied by Shoei Omote, 27, president of the Meiji Pearl Co., who was formerly Tsuji's secretary."*
>
> *"Young Tsuji, 31, said he would stay in South Vietnam for about three weeks to meet people who have been in contact with his father."*

May 10, 1962 – U.S. Intelligence Report:

> ***"Estimate of Present Status of Tsuji's Whereabouts"***
>
> *"(1) One of the letters written by Tsuji to his family and friends is dated 10 April 1961, which contained, inter alia, as follows:"*
>
> *"On 20 April, I will go to the interior region. Fortunately, no one up to now knows my identity. It appears that the rainy season will arrive a little earlier this year."*

"(2) Based on the above-mentioned letter, it is believed that Subject commenced his full scale inspection tour from 20 April 1961."

"(3) The route which Subject studied before his departure from Japan was one leading from Vientiane to Xieng Khoua via Vang Vieng and Phou Khao Kham. Based thereon, it is estimated that Subject left Vientiane on 20 April for this route."

"(4) There is no reliable information concerning the whereabouts of Tsuji after his departure from Vientiane. Hence, it is not known whether he is captured or dead."

"(5) The area of contact between the pro-Communist main strength and that of the rightist on 20 April 1962 was the northern region of Vang Vieng, approximately 130 kilometers north of Vientiane. At that time, a rightist army headquarters was located in Vang Vieng."

"(6) It is estimated that Subject first proceeded to Vang Vieng where he arrived on/or about 22 April, at which time a fierce battle was being fought in that area. The pro-Communist forces occupied the town of Vang Vieng on 24 April, at which time he could have made himself available for capture as he anticipated."

"(7) During all his previous travels, Subject never neglected to correspond with his family, even while traveling incognito during the early days of the end of World War II. As his family had not heard from him since 19 April 1961, it is estimated that Subject has been captured."

"(8) In November 1961, there were rumors to the effect that Subject had been captured and taken to Peking. Based thereon, the Japanese Red Cross on 2 December 1961 requested the Communist Red Cross to investigate whether or not Tsuji was being held in Communist China. In January 1962, the Communist China Red Cross notified the Japanese Red Cross that Tsuji was not in Communist China and that the rumor that he was in the country was utterly groundless."

"(9) According to unconfirmed reports, Subject was disguised as a Buddhist monk when he left Vientiane. If such is the case and Subject fell into native hands, it is possible that he could have met with foul play because of such deception."

"(10) According to members of Tsuji's family, Tsuji sent a preliminary report on his trip to Prime Minister Ikeda through the Japanese Embassy in Thailand."

"(11) Available information also discloses that when Subject reached Bangkok, he requested Colonel Chikashi Ito, Japanese Military Attaché and former student at the Military Staff College of the defunct Japanese Imperial Army when Subject was an instructor thereat, to cooperate with him. As Ito concurrently serves in the Japanese Embassy in Laos, he proceeded to Vientiane with Subject. However, Ito has not been able to shed any present light on the whereabouts of Subject."

"(12) Extracts from a letter sent from Minoru Kadowaki, serving in the Japanese Embassy in Laos, in response to queries from a friend concerning news of Tsuji follow:"

"Although the diplomatic staffs of the various Embassies in Laos took refuge in Thailand when street fighting broke out in December 1960, the Japanese staff remained behind. As a result, they won the confidence of the Laotian people. Japanese are very popular with the Laotian people and only the Japanese are permitted to go as far as 70 kilometers from the suburbs of Vientiane. It is dangerous for a white man to approach that area."

"We, too, are anxious about Professor Tsuji. However, I believe that the reports in the newspapers relating to his disappearance are approximately 70 per cent correct. I think that the story in the Bunshun is incredible. Concerning Tsuji's activities in Laos – when I have an opportunity, I will tell you about them."

"I am deeply impressed with the motive of Tsuji's trip. With reference to an article that appeared in a certain weekly stating that when Tsuji was in Vang Vieng, he had contact with me – I am sorry that I cannot comment on it at this time. When the opportunity presents itself, I think I can give you some information on Professor Tsuji's activities in Laos. Off hand, I can say that we have no information beyond the statement of the Foreign Minister that Tsuji disappeared in Kieno Khouang."

"I believe that the Japanese Embassy on the spot is probably keeping certain facts secret."

"Motivation for Trip to Southeast Asia in April 1961"

"(1) From the early part of 1961, the Southeast Asian issues, particularly the Laotian problem, drew the Subject's attention. Subject believed that the Laotian domestic situation might touch off a Third World War because of American aid to Laos, which in turn would force the Soviet Union to increase aid for the pro-Communist forces."

"(2) In order for Subject to obtain the necessary intelligence for him to formulate his own policy on the elimination of a Third World War which might be caused by the Laotian problem, he decided to proceed to the area of conflict to obtain the necessary data."

"(3) Subject first planned to inspect the route connecting Vientiane, Thakhek, Savannakhet and Tchibo. However, subsequently, Subject realized that this route was of slight political and military value and therefore planned to inspect the route connecting Vientiane, Xieng Khouang, Nong Het, Vinh and Hanoi. Subject was particularly desirous of inspecting the Communist aid supply depots, which he estimated to be located along the Vietnamese-Laotian border and within North Vietnamese territory."

"(4) Prior to his departure, political operations were not on Subject's agenda. However, he stated he might engage in political operations during his trip if he

found that such would decrease tension in the area between the East and West blocs."

"(5) Subject had in mind prior to his departure that he would like to meet with Captain Kong Le[15] *in order to negotiate the release of US military and civilian personnel held captive by the pro-Communist forces."*

"(6) Prior to his departure, Subject was of the opinion that it would be very difficult to prevent Laos from going Communist. He was of the opinion that Laos could not be saved from Communism, but felt that all necessary measures should be taken to keep Thailand from the same fate."

"(7) Subject firmly believed that a military and ideological strong bridgehead should be established in Vietiane, Thakhek, and Savannakhet in order to protect Thailand from the threat of Communism."

"(8) Prior to starting on his trip, Tsuji stated that he was very pleased with the attitude of the Ministry of Foreign Affairs in connection with his obtaining the necessary documentation for his trip. Prior to his application to travel, Subject advised the Foreign Ministry and Prime Minister Ikeda about the nature of his trip."

May 11, 1962 – Results of Laboratory Tests Conducted on an envelope allegedly bearing Tsuji handwriting:

"Some time in the latter part of May 1961, an envelope allegedly bearing the handwriting of Masanobu Tsuji was handed over to [redacted] for analysis at the [redacted] laboratory. On 24 August, [redacted] reported that the extensive laboratory tests resulted in the conclusion that the handwriting was indeed Tsuji's own."

May 12, 1962 – *Tokyo Sankei*, Tokyo, Japan, by Correspondent Mamoru Noda:

"As to the whereabouts of Mr. Masanobu Tsuji, a member of the House of Councillors who vanished while on a tour of southeast Asia late in April last year, the North Vietnamese trade mission in Phnom Penh disclosed that there are no traces of Mr. Tsuji having crossed the North Vietnamese border."

"According to informed sources here, during his stay in Phnom Penh from 8 to 10 April last year, he was introduced to the mission by a visiting member of the Mitsubishi Shoji Company. He asked the mission for a letter of introduction to the DRV (Democratic Government of Vietnam) Government authorities. The mission, however, rejected his request on the grounds that it was not an authorized diplomatic office, and informed the DRV Foreign Ministry by telegram that he would visit Hanoi. His trail ended there and there have been no traces of him after that. The Japanese Embassy here asked the North Vietnamese trade mission through the Mitsubishi representative to trace his whereabouts. The embassy on 22 July last year received an informal reply from the DRV Foreign Ministry that they had not found any Mr. Tsuji among those who have crossed the border so far."

"A certain North Vietnamese source disclosed recently, 'There has been no trace of Mr. Tsuji in North Vietnam. We were expecting his visit to our country.

It would have been impossible for him to visit China without crossing the North Vietnamese boundary.'"

"Mr. Tsuji visited the border areas both in Cambodia and South Vietnam and inspected the Viet Cong activities. He reportedly met in Saigon a former lieutenant colonel of the Japanese army intelligence service who hid himself in Vietnam at the end of the war. It is almost certain that he tried to investigate the situation in Vietnam both in the North and South. However, since neither the DRV Foreign Ministry nor the Japan-DRV Association delegates visiting Hanoi last year had heard about Mr. Tsuji at all, he could not have reached North Vietnam. Informed sources here speculate that he may have died in an accident or from illness somewhere near Xieng Khouang."

May 16, 1962 – *Tokyo Kyodo*, Tokyo, Japan:

"Masanobu Tsuji... missing since last June while on a mission to Laos is possibly being 'informally confined' in North Vietnam. This was disclosed today by Tsuji's son Toru, who returned 16 May from a visit to Southeast Asia to probe the whereabouts of his father. He was speaking to newsmen at a press conference held shortly after he submitted a report on his missing father to the secretariat of the Upper House. He said he confirmed in his survey that the former Japanese Army staff officer entered the northern part of Laos from Vientiane on 21 April last year, clad in the robes of a Buddhist priest. He said he found that rumors that his father was in Communist China were without foundation. He said that according to investigation circles, Tsuji went to Hanoi from northern Laos."

Oct 10, 1962 – *Sankei Shimbun* newspaper reports that Tsuji *"is held incommunicado by the Communist Chinese in Yunan Province pending opportunity to exploit [him] to the maximum, and that there is a possibility of getting [him] out from the hands of the Chinese in exchange for ransom money."*

April 5, 1963 – *Anaheim Bulletin*, Anaheim, CA, report by Kip Cooper:

"A top-ranking Japanese soldier-diplomat may be in Cuba helping Fidel Castro bolster his Communist regime."

"This is the latest Asian rumor as to the whereabouts of Masanobu Tsuji, a senior parliamentarian who disappear mysteriously after departing Japan two years ago on a 40-day tour of Southeast Asia...."

"Some stories say Tsuji is in a Pathet Lao prison. Others say he is in Red China. A few say he is dead, shot by American troops in Vietnam."

"No one knows for certain where he is nor what he may be doing. Tsuji, who has been called an Oriental Quixote, reportedly has a flair for grandstand plays and heroic stances."

"The story he now is in Cuba first was reported in the Hong Kong newspaper ***Hsing Tao Jih Pao****.... The Cuban Embassy in Tokyo brushed aside a news query. 'We know nothing about that matter,' an embassy spokesman said."*

"The Hong Kong newspaper said a Chinese Communist named Tao Hsiao-chieh helped Tsuji enter Red China and then assisted him onto a plane for

Prague. From Prague, the story goes, Tsuji flew direct to Havana on the personal invitation of Prime Minister Fidel Castro. The newspaper said Hsiao-chieh and Tsuji had been friends since the Wang Ching-wei days (Wang was a Chinese who collaborated with Japanese invaders in 1938). Tsuji stayed with Hsiao-chieh in Kirin a few days before leaving for Prague, it is claimed...."

"Is Tsuji in Cuba? The one person who should know will not talk."

"Mrs. Tsuji bluntly turns down all requests for interviews."

April 18, 1963 – *Nottingham Evening News*, Nottingham, UK:

"Masanobu Tsuji... has been missing since April, 1961."

"One Japanese MP who travelled through India said he heard that Tsuji was teaching military strategy at a Chinese Army officers' school in Kirin province."

"A Japanese businessman claimed he had read an item in the ***Peking People's Daily****, official newspaper of the Chinese Communist Party, saying Tsuji had been captured by the Pathet Lao in Laos and had been sent under escort to Peking."*

"The businessman said the newspaper reported that Tsuji had married a 27-year-old Chinese woman on June 25, 1961."

"Another rumor said that Tsuji had had an accident near Xieng Khouang, Laos, and had died."

"In the meantime, as there is not law providing for the cancellation of his seat in Parliament, his wife is continuing to draw his pay."

Feb 19, 1967 – *Miami Herald*, Miami, FL:

"They seek him here, they seek him there – Japan's 'man in the saffron robe' who grew from fame to infamy by spying for five major countries during World War II."

"Regarded as the most colorful figure in the old Imperial Army, Col. Masanobu Tsuji first acquired legendary fame by mapping a strategy that helped Gen. Tomoyuki Yamashita smash through the Malayan Peninsula to Singapore in the early days of the Pacific war."

"Little more was heard of him, except that he was soon trading intelligence information with the United States, Great Britain, the Soviet Union, Communist China, and Nationalist China."

"Later reports, confirmed by a Japanese security official, said he established contacts with intelligence sources of those five countries in Yenan during the war...."

"Tsuji's political career was stormy even as he moved from the House of Representatives to the House of Councillors."

"In 1959 he was ousted from the Liberal-Democratic Party for accusing Prime Minister Nobuske Kishi of being a 'warmonger.' The theory advanced by some informed Japanese was that he was working in line with Communist

Chinese policies to divide the conservative party over the revised U.S.-Japan Security Treaty...."

"Foreign Office officials here believe he crossed the North Vietnam border into Communist China. They have an intelligence report that he was executed by the Reds because the Communists, who had thought Tsuji's contact with other powers was to collect intelligence information for them, had suddenly reversed their opinion."

"All of this, however, is pooh-poohed by friends of Tsuji. Shigeharu Asaeda, long-time associate of the spy in the saffron robe, said Tsuji was not rated high enough by the Chinese Communists to be killed."

"Tsuji's wife, Chitose, also never believed he was a master spy. She insists he is 'a simple-minded man. I still wonder how he could even have been a staff officer.'"

June 15, 1968 – *Sydney Morning Herald*, Sydney, NSW, by a Special Correspondent:

"What has become of Colonel Masanobu Tsuji, who disappeared four years ago and has not been heard from since...."

"Immediately after the Japanese surrender Tsuji was ordered by Tokyo to 'evade capture by the Allies and return to Japan only when you can again serve the Emperor.'"

"Listed as a war criminal, with Allied Intelligence agents hot on his trail, he assumed the role of a Buddhist monk, and for the next 12 months begged his way through Siam and Vietnam."

"Arrested in Hanoi by Chinese Nationalist troops, he escaped and made contact with Ho Chi Minh to whom he revealed his identity and who offered him a command in the Viet Minh forces then being organized for the campaign against the French."

"Tsuji declined on the ground that, as a Japanese officer, he could serve in no other army without the permission of the Emperor, but for three months he continued to live with Ho and General Giap on very friendly terms, until, hard pressed by the French, they passed him over the border to Mao Tse-tung. Mao received him cordially but found him and embarrassing guest, in constant danger of assassinating by his communist rank and file."

"Having once been of considerable service to Chiang Kai-shek, Tsuji made his way to Peking and secured employment with the Chinese Government, for which he worked for more than two years – held incommunicado except for contact with a small group of staff officers. Released on the order of the Generalissimo, he then made his way home and landed in Japan with 10 yen in his pocket to learn he was sill wanted as a war criminal and that the American authorities were offering a reward of one million yen for information leading to his arrest."

"He immediately went into hiding and, with the help of former colleagues, collected evidence to refute the charges against him. At the end of 12 months he surrendered, stood trial, and was acquitted on all counts."

"Adopting the slogan 'no meat in the sandwich,' he stood for Parliament as an Independent, vigorously opposing Japanese participating in the cold war – which at that time threatened to become a hot one. He was returned to the Diet with the largest majority ever gained by a candidate at a Japanese election, and later moved to the House of Councillors – still retaining his Independent status."

"Lone Journey"

"Early in 1964 he was sent on a lone mission to Bangkok and Saigon to negotiate for the Japanese Government on matters which have not been disclosed...."

"On April 25 the manager of his hotel in Saigon rang the Japanese consulate there to report that Colonel Tsuji and his baggage had vanished from his room during the previous night...."

"A month later, when no information was forthcoming, the Japanese Government sent a delegation consisting of a very senior Public servant, several intelligence officers, Tsuji's son, and his parliamentary secretary to carry out an exhaustive investigation into the Colonel's disappearance."

"After three months they had learned nothing, and returned to Japan believing that Tsuji had been kidnapped and possibly murdered – either by the Vietcong, by South Vietnamese Government agents, or by bandits operating independently in the hope of profit from loot or ransom."

"In August, 1964, a broadcast by Radio Peking announced that Colonel Tsuji had been shot in Laos by American Intelligence agents – an allegation promptly denied by U.S. authorities...."

"Early in 1964 a note handed to the Japanese Consul in Hong Kong demanded half a million yen ($50,000) ransom for Colonel Tsuji, but all efforts to make contact with the authors failed. A year later a similar note containing Tsuji's signature on a sheet of paper was handed to the consul, and, on this occasion, brief contact was made with the writer who, however, disappeared before serious negotiations could be commenced."

"This week a letter from Saigon threw some light on the mystery. It was written by a close friend of Tsuji's to another friend in Sydney."

"The letter says that after completing his work for the Japanese Government in 1964, Colonel Tsuji walked across the border from Vietnam into Laos with two Buddhist monks as guides."

"The monks took him to two senior officers of the dissident Pathet Lao forces who escorted him to Vientiane, the capital."

Memorial statue of Masanobu Tsuji at Aratani machi, Yamanaka Onsen, Kaga, Ishikawa, August 17, 2008, Wikipedia.org.

"There he conferred with members of the Laotian Government, then caught a Russian plane to Hanoi. He was cordially received by Ho Chi Minh, and announced that he intended to travel to Peking."

"'What happened after that is unknown,' the letter said, 'but most of his friends believe he is still alive.'"

July, 1969 – Officially declared dead at wife's request.

July 26, 1970 – *The Miami Herald*, Miami, FL:

"Nine years ago, Japanese parliamentarian Masanobu Tsuji packed his bags and left on a fact-finding tour of Indochina."

"He never returned."

"His disappearance, coupled with his background as a top strategist in World War II, has developed into a mystery."

"A younger brother, Masayoshi, believes Tsuji is alive in China, 'a prisoner accused of espionage which he never committed.' Masayoshi bases his belief on 'instinct....'"

"North Vietnam denies that Tsuji ever stepped into its territory. Communist China called the report unfounded. Radio Peking reported – and then retracted – that the U.S. military killed Tsuji in 1962 when he tried to re-enter Laos...."

"The missing man, who would be 67, was officially declared dead by his wife last July. But Masayoshi says the wife 'still has a firm belief of her husband's survival.'"

Feb 25, 2007 – *Rapid City Journal*, Rapid City, SD, by Joseph Coleman:

"Col. Masanobu Tsuji was a fanatical Japanese militarist and brutal warrior, hunted after World War II for massacres of Chinese civilians and complicity in the Bataan Death March."

"And then he became a U.S. spy."

"Newly declassified CIA records, released by the U.S. National Archives and examined by The Associated Press, document more fully than ever how Tsuji and other suspected Japanese war criminals were recruited by U.S. intelligence in the early days of the Cold War."

"The documents also show how ineffective the effort was, in the CIA's view."

"The records, declassified in 2005 and 2006 under an act of Congress in tandem with Nazi war crime-related files, fill in many of the blanks in the previously spotty documentation of the occupation's authority's intelligence arm and its involvement with Japanese ultra-nationalists and war criminals, historians say."

"In addition to Tsuji, who escaped Allied prosecution and was elected to parliament in the 1950s, conspicuous figures in U.S.-funded operations included mob boss and war profiteer Yoshio Kodama and Takushiro Hattori,

former private secretary to Hideko Tojo, the wartime prime minister hanged as a war criminal in 1948...."

"Historians long ago concluded that the Allies turned a blind eye to many Japanese war crimes, particularly those committed against other Asians, as fighting communism became the West's priority...."

"Tsuji, for instance, was wanted for involvement in the Bataan Death March of early 1942, in which thousands of Americans and Filipinos perished, and for allegedly cosigning an order to massacre anti-Japanese Chinese merchants in Malaya."

"Yet none of that seemed to matter much to American intelligence."

Footnotes

1. Tsuji is referring to the August 9, 1945, invasion of Manchuria, northern Korea, and Mongolia by the Soviet Red Army. In just over two weeks, the Soviets crushed Japanese resistance. On August 20, the Kwantung Army surrendered to the Soviets. Over 600,000 Japanese soldiers were taken prisoner and sent to labor camps in Siberia. The surrender lead to Soviet forces committing widespread looting, rape, and indiscriminate killings of civilians. These atrocities paralleled what had occurred in Germany and Eastern Europe earlier in 1945. As heinous as the Red Army's actions were, they were far less severe than the atrocities committed by the Imperial Japanese Army in Nanking in December 1937 and January 1938.

The "Rape of Nanking" was a calculated massacre by the Imperial Japanese Army, marked by mass rape, public executions, and sadistic violence fueled by racial hatred and military ideology. An estimated 200,000 were killed and tens of thousands of women raped in just six weeks. In contrast, the Soviet occupation of Manchuria it was less organized but still devastating. Soviet troops committed widespread rape, looting, and killings, driven more by revenge and breakdown of discipline than by policy.

2. G. William Whitehurst, *The China Incident Igniting the Second Sino-Japanese War*, (McFarland & Company, Inc., 2021), pp77-78.

3. "Operation Sook Ching Is Carried Out," Singapore National Library Board (NLB), https://www.nlb.gov.sg/main/article-detail?cmsuuid=d0523464-3a43-4520-a864-f195a8aef418, accessed Jan 15, 2025.

4. Wai Keng Kwok, Justice Done? Criminal and Moral Responsibility Issues in the Chinese Massacres Trial Singapore 1947, Genocide Studies Program, Working Paper No. 18, 2001 (Yale University, 2001), p12.

5. Ibid, p12.

6. Ibid, p4.

7. Quoted in Masanobu Tsuji's CIA dossier, Vol. 1 No. 0033.

8. *Weekly Thriller*, May 22, 1959 (Volume 1 Issue 4), Edited by Hiramoto Kazuki, (Moriwaki Bunko, Singapore, 1959)

9. https://en.wikipedia.org/wiki/Jos%C3%A9_Abad_Santos, accessed Jan 15, 2025.

10. Quoted in Masanobu Tsuji's CIA dossier, Vol. 1 No. 0033.

11. Quoted in Masanobu Tsuji's CIA dossier, Vol. 1 No. 0033.

12. The Katyn massacre was a series of mass executions of nearly 22,000 Polish prisoners of war carried out by the Soviet Union in April and May 1940; https://en.wikipedia.org/wiki/Katyn_massacre, accessed January 15, 2025.

13. Quoted in Masanobu Tsuji's CIA dossier, VOL. 1 No. 0033.

14. The Battle of Grik Road, December 16-23, 1941; https://cove.army.gov.au/article/smart-soldier-battle-grik-road-ww2, accessed January 15, 2025.

15. Captain Kong Le was a Laotian military officer and prominent military figure in Laos during the 1960s.

Doc45 Publications

Incident at Ple Tonan, An Imperial Japanese War Crime and the Fate of U.S. Navy Airmen in French Indochina

This book began 33 years ago, in early 1990, with a question posed to the author by his father. The author's father was just completing his memoir of his World War II military experiences and he was greatly troubled by not knowing the fate of two of his fellow fliers in his USS ESSEX based air squadron, Torpedo Four (VT-4). The two men whose fate still haunted the author's father 45 years after the fact were shot down in an air raid on Japanese-occupied French Indochina.

The answer to this question would have remained hidden but for an Imperial Japanese war crimes trial. In this book, you will find the events, investigations, statements, related documents, and stories of the men that led to that trial – and the riveting testimony of the trial itself.

Not surprisingly, uncovering the stories of the two missing squadron members' fates also uncovered the stories of other men. Many of these others, as disclosed in this book, paid the ultimate price – some for heroism, and some for acts later judged war crimes.

A courtroom criminal trial is unlike any other human institution. It is combat without physical weapons. And as in combat, the stakes can be as high as death.

Here you will find the statements and interrogations that led to the belief that multiple war crimes were committed at Ple Tonan in French Indochina on April 27, 1945. You will find the pre-trial statements of witnesses. You will find the charges presented against the defendants. You will find the daily transcripts of the 13 day trial. You will find the closing arguments and sentencing. You will find the decisions of the appellate authority.

You will find the pleas for clemency submitted for each defendant. You will find their last letters home to their families. You will find the details of their punishments.

A courtroom criminal trial can reveal a defendant's inner being in a way no other human institution can. A defendant that appears on the stand must try to explain and justify the actions for which the person is being tried. And that person must do so in an adversarial setting in which every assertion is open to challenge and refutation.

The men who appeared in the trial presented in this book all appeared on the stand in their own defense. This book is no cold reporting. With their trial testimony, each defendant got to explain and justify his actions. This book, based on the way in which it is presented, puts you in the position of, first, an investigator trying to determine whether a war crime was committed, and, second, a jury member who must decide the guilt or innocence of a defendant after hearing the evidence against the defendant.

59 images, many never published before.

Paperback, 320 Pages, ISBN 978-1-952580-14-7

Doc45 Publications

Carrier Warriors:The Men of Air Group 4 in World War II

This In the long and brutal sweep of World War II, few forces proved as versatile, fearsome, and transformative as the United States Navy's aircraft carriers. They brought the fight to enemy shores, one launch at a time - projecting American power across thousands of miles of open sea with precision, daring, and deadly effectiveness. This book tells the stories of the men who flew from those decks - the pilots, crewmen, and air group leaders whose actions shaped the outcome of the war in the Pacific and beyond.

Their journey began in places like the North Atlantic, where in October 1943, OPERATION LEADER marked the first American carrier strike against German shipping off the coast of Norway. Though often overshadowed by later Pacific operations, it was a milestone in naval air power - a swift, unexpected blow that signaled the reach and potential of carrier-based warfare. From that point forward, American carriers were no longer defensive tools - they were strategic weapons, capable of hitting fast, far, and hard. Carrier Warriors: The Men of Air Group 4 in World War II uniquely includes two remarkable accounts written by German soldiers who survived that attack, offering rare insight into what it was like to be on the receiving end of a U.S. Navy air assault.

The scale and ferocity of the war in the Pacific soon dwarfed that early raid. At Iwo Jima, carrier pilots played a pivotal role in softening the island's defenses, flying multiple sorties a day in support of Marines locked in some of the fiercest combat of the war. Day after day, in skies thick with flak and clouded by smoke, these aviators provided close air support - all to secure an airfield whose location would bring Japan within bomber range.

Then came the daring raids on Tokyo, where carrier task forces struck the Japanese homeland itself. These were not symbolic gestures; they were coordinated attacks against aircraft factories, shipyards, and military installations at the very heart of the enemy's war-making capacity. The risk was enormous - ships operating deep inside enemy waters with no margin for error - but the impact was undeniable. These raids broke the illusion of Japanese invulnerability and signaled that the end was approaching.

Behind these legendary operations were the men who made them happen - young, often barely out of training, flying complex machines under extraordinary pressure. ***Carrier Warriors*** brings together firsthand accounts from the men of Air Group 4, a tightly knit unit of Navy aviators and aircrew who flew from the decks of the USS Ranger, USS Bunker Hill, and USS Essex. Collected by the author over more than 35 years, these narratives - written by the men themselves: pilots, gunners, radiomen, and mechanics - offer an unfiltered, personal perspective on what it meant to fight as part of one of the Navy's most experienced and battle-tested air groups.

211 photos, most never published before.

Paperback, 304 Pages, ISBN 978-1-952580-18-5

Doc45 Publications

La Posta – From the Founding of Mesilla, to Corn Exchange Hotel, to Billy the Kid Museum, to Famous Landmark, David G. Thomas, paperback, 118 pages, 59 photos, e-book available.

"For someone who grew up in the area of Mesilla, it's nice to have a well-researched book about the area – and the giant photographs don't hurt either.... And the thing I was most excited to see is a photo of the hotel registry where the name of "William Bonney" is scrawled on the page.... There is some debate as to whether or not Billy the Kid really signed the book, which the author goes into, but what would Billy the Kid history be without a little controversy?" –Billy the Kid Outlaw Gang Newsletter, Winter, 2013.

Giovanni Maria de Agostini, Wonder of The Century – The Astonishing World Traveler Who Was A Hermit, David G. Thomas, paperback, 208 pages, 59 photos, 19 maps, e-book available.

"David G. Thomas has finally pulled back the veil of obscurity that long shrouded one of the most enduring mysteries in New Mexico's long history to reveal the true story of the Hermit, Giovanni Maria de Agostini. ...Thomas has once again proven himself a master history detective. Of particular interest is the information about the Hermit's life in Brazil, which closely parallels his remarkable experience in New Mexico, and required extensive research in Portuguese sources. Thomas's efforts make it possible to understand this deeply religious man." – Rick Hendricks, New Mexico State Historian

Screen With A Voice - A History of Moving Pictures in Las Cruces, New Mexico, David G. Thomas, paperback, 194 pages, 102 photos, e-book available.

The first projected moving pictures were shown in Las Cruces 110 years ago. Who exhibited those movies? What movies were shown? Since projected moving pictures were invented in 1896, why did it take ten years for the first movie exhibition to reach Las Cruces? Who opened the first theater in town? Where was it located? These questions began the history of moving pictures in Las Cruces, and they are answered in this book. But so are the events and stories that follow.

There have been 21 movie theaters in Las Cruces – all but three or four are forgotten. They are unremembered no longer. And one, especially, the Airdome Theater which opened in 1914, deserves to be known by all movie historians – it was an automobile drive-in theater, the invention of the concept, two decades before movie history declares the drive-in was invented.

Billy the Kid's Grave – A History of the Wild West's Most Famous Death Marker, David G. Thomas, paperback, 154 pages, 65 photos.

"Quien es?"

The answer to this incautious question – "Who is it?" – was a bullet to the heart.

That bullet – fired by Lincoln County Sheriff Patrick F. Garrett from a .40-44 caliber single action Colt pistol – ended the life of Billy the Kid, real name William Henry McCarty.

But death – ordinarily so final – only fueled the public's fascination with Billy the Kid. What events led to Billy's killing? Was it inevitable? Was a woman involved? If so, who was she? Why has Billy's gravestone become the most famous – and most visited – Western death marker? Is Billy really buried in his grave? Is the grave in the right location?

These questions – and many others – are answered in this book.

The Stolen Pinkerton Reports of the Colonel Albert J. Fountain Murder Investigation, David G. Thomas, editor, paperback, 194 pages, 28 photos.

The abduction and apparent murder of Colonel Albert J. and Henry Fountain on February 1, 1896, shocked and outraged the citizens of New Mexico. It was not the killing of Colonel Fountain, a Union Civil War veteran and a prominent New Mexico attorney, which roused the physical disgust of the citizenry - after all, it was not unknown for distinguished men to be killed. It was the cold-blooded murder of his eight-year-old son which provoked the public outcry and revulsion.

The evidence indicated that although Colonel Albert J. Fountain was killed during the ambush, his son was taken alive, and only killed the next day.

The public was left without answers to the questions:

- Who ambushed and killed Colonel Fountain?
- Who was willing to kill his young son in cold-blood after holding him captive for 24 hours?

The case was never solved. Two men were eventually tried for and acquitted of the crime.

The case file for the crime contains almost no information. There are no trial transcripts or witness testimonies. The only reports that exist today of the investigation of the case are these Pinkerton Reports, which were commissioned by the Territorial Governor, and then stolen from his office four months after the murders. These Reports, now recovered, are published here.

These Reports are important historical documents, not only for what they reveal about the Fountain murders, but also as a fascinating window into how the most famous professional detective agency in the United States in the 1890s - the Pinkerton Detective Agency - went about investigating a murder, at a time when scientific forensic evidence was virtually non-existent.

Paperback, 196 Pages, ISBN 978-0-9828709-6-9

Torpedo Squadron Four – A Cockpit View of World War II, Gerald W. Thomas, paperback, 280 pages, 209 photos, e-book available.

"This book contains more first-person accounts than I have seen in several years. ...we can feel the emotion... tempered by the daily losses that characterized this final stage of the war in the Pacific. All in all, one of the best books on the Pacific War I have seen lately." – Naval Aviation News, Fall 2011.

When New Mexico Was Young, Harry H. Bailey, paperback, 186 pages, 10 photos.

The autobiography of Harry H. Bailey (1868-1954) Mr. Bailey was a pioneer New Mexican who took a major role in the development of the Mesilla Valley. In 1900, he built the "Natatorium," the first public swimming pool in El Paso, Texas. Three years later, he built the Angelus Hotel. In 1906, part of the Angelus Hotel building became the Crawford Theatre. After leaving El Paso and returning to New Mexico, he helped develop Radium Springs as a health resort and built the hotel and baths there.

His autobiography contains many stories about the early-day Mesilla Valley settlers who were his companions. Among the individuals he knew were Sheriff Pat Garrett, Colonel Albert J. Fountain, Attorney Albert B. Fall, Oliver Milton Lee, Sheriff Mariano Barela, Demetrio Chavez, Humbolt Casad, and George Griggs. He was a close friend of Western author Eugene Manlove Rhodes. For almost a year, he lived in the courthouse building in Mesilla where Billy the Kid was sentenced to hang.

Paperback, 186 Pages, ISBN 978-1-952580-0-17

Doc45 Publications

The Frank W. Angel Report on the Death of John H. Tunstall

"In the matter of the cause and circumstances of the death of John H. Tunstall...."

So begins the single most important contemporary document recounting the origins of the Lincoln County War. That document is the "Report of Special Agent Frank Warner Angel on the Death of John Henry Tunstall," known today to historians as the "Angel Report."

The 395-page, hand-written Report that Angel submitted on October 3, 1878, on Tunstall's unprovoked, sadistic murder is published for the first time in this book.

The Report documents the events leading to Tunstall's murder – the testimony of the men present at the brutal killing – including Billy the Kid's eye-witness account – and the violent consequences that followed.

It includes sworn accounts by William "Frank" Baker, Robert W. Beckwith, Henry N. Brown, James J. Dolan, William Dowlin, Pantaleón Gallegos, Godfrey Gauss, Florencio Gonzales, John Hurley, Jacob B. Mathews, Alexander A. McSween, John Middleton, Lawrence G. Murphy, John Wallace Olinger, Juan B. Patron, George W. Peppin, David P. Shield, Robert A. Widenmann, and 18 others.

Supplementing the Report are an extensive introduction, notes, contemporary documents, associated letters, biographical details, and a timeline.

The book also reveals the brazen attempt by two powerful politicians – Thomas Catron and Stephen Elkins – to destroy the Report, depriving history of its priceless contribution.

Forty three images, most never published before.

Available in both paperback and hardcover.

Paperback, 254 Pages, ISBN 978-1-952580079
Hardcover, 254 pages, ISBN 978-1-952580055

Doc45 Publications

Water in a Thirsty Land

by Ruth R. Ealy
David G. Thomas, Editor

"Water in a Thirsty Land" is a chronicle of Dr. Taylor Filmore Ealy's 1874 to 1881 sojourn as a medical missionary in Indian Territory (Oklahoma) and New Mexico Territory, compiled by his daughter Ruth R. Ealy.

Dr. Ealy's first assignment was Fort Arbuckle, Chickasaw Reservation, Oklahoma Territory. His second was Lincoln, New Mexico Territory. His final assignment was Zuni Pueblo, New Mexico.

Dr. Ealy's faithful accounts of his struggles and challenges at these — at the time — exotic locations make for fascinating reading. His daily records of eye-witnessed events in Lincoln are of exceptional historical value. He arrived in Lincoln on February 19, 1878, the day after John Henry Tunstall was murdered. The unprovoked, sadistic murder of Tunstall kicked off the bloody Lincoln County War. Dr. Ealy was present at Tunstall's funeral, the killing of Lincoln County Sheriff Brady and Deputy Hindman, and the five-day shootout that ended with the firing of Alexander McSween's home, and the heinous slaughtering of McSween and four others as they frantically fled the blazing conflagration.

There are many details about the Lincoln County War in Dr. Ealy's account not recorded in other sources. Here are examples:

- Tunstall's funeral was held at 3 pm. His bullet-holed, bloody clothes were lying on the dirty ground in McSween's back yard during the service.
- The book provides many details about Tunstall's store: *"The floors were good ones and the windows were large."* One room was *"12 feet high, 18 feet long, and 18 feet wide, with a huge window and a door with a large glass in it."* That room was *"large enough to hold three hundred people."* The store lot was five acres in size and fully fenced.
- When the McSween house was fired during that 5-day shootout, one of Elizabeth Shield's children stepped in the coal oil used to ignite the fire.
- Among the items in McSween's house destroyed by the fire were an elegant piano, a Brussels carpet, costly furniture, rich curtains, and fine paintings.
- After Taylor testified at the Dudley Court of Inquiry, he was warned by anonymous note that he would be killed before he got back to his home in Zuni (a "coffin note").

From Lincoln, Dr. Ealy went to Zuni Pueblo. His keen observations are one of the primary, early sources of halcyon life in Zuni in 1878.

The Editor has added an extensive introduction, contextual notes, footnotes, appendices, and an index to the text of this extremely rare book.

Paperback, 208 Pages, ISBN 978-1952580-10-9
Hardcover, 208 Pages, ISBN 978-1952580-11-6

Doc45 Publications

Killing Pat Garrett, The Wild West's Most Famous Lawman - Murder or Self-Defense?

Pat Garrett, the Wild West's most famous lawman – the man who killed Billy the Kid – was himself killed on leap day, February 29, 1908, on a barren stretch of road between his Home Ranch and Las Cruces, New Mexico.

- Who killed him?
- Was it murder?
- Was it self-defense?

No biographer of Garrett has been able to answer these questions. All have expressed opinions. None have presented evidence that would stand up in a court of law. Here, for the first time, drawing on newly discovered information, is the definitive answer to the Wild West's most famous unsolved killing.

Supplementing the text are 102 images, including six of Garrett and his family which have never been published before. It has been 50 years since a new photo of Garrett was published, and no photos of his children have ever been published.

Garrett's life has been extensively researched. Yet, the author was able to uncover an enormous amount of new information. He had access to over 80 letters that Garrett wrote to his wife. He discovered a multitude of new documents and details concerning Garrett's killing, the events surrounding it, and the personal life of the man who was placed on trial for killing Garrett.

- The true actions of "Deacon Jim" Miller, a professional killer, who was in Las Cruces the day Garrett was killed.
- The place on the now abandoned old road to Las Cruces where Garrett was killed.
- The coroner's jury report on Garrett's death, lost for over 100 years.
- Garrett's original burial location.
- The sworn courtroom testimony of the only witness to Garrett's killing.
- The policeman who provided the decisive evidence in the trial of the man accused of murdering Garrett.
- The location of Garrett's Rock House and Home Ranches.
- New family details: Garrett had a four-month-old daughter the day he killed Billy the Kid. She died tragically at 15. Another daughter was blinded by a well-intended eye treatment; a son was paralyzed by childhood polio; and Pat Garrett, Jr., named after his father, lost his right leg to amputation at age 12.

Garrett's life was a remarkable adventure. He met two United States presidents: President William McKinley, Jr. and President Theodore Roosevelt. President Roosevelt he met five times, three times in the White House. He brought the law to hardened gunmen. He oversaw hangings. His national fame was so extensive the day he died that newspapers from the East to the West Coast only had to write "Pat Garrett" for readers to know to whom they were referring.

2020 Will Rogers Medallion Award Finalist for Excellence in Western Media
2020 Independent Press Award Distinguished Favorite, Historical Biography
2019 Best Book Awards Finalist, United States History
2019 Best Indie Book Notable 100 Award Winner.

The Trial of Billy the Kid

This book is about Billy the Kid's trial for murder, and the events leading to that trial. The result of Billy's trial sealed his fate. And yet Billy's trial is the least written about, and until this book, the least known event of Billy's adult life.

Prior biographies have provided extensive — and fascinating — details on Billy's life, but they supply only a few paragraphs on Billy's trial. Just the bare facts: time, place, names, result.

Billy's trial the most important event in Billy's life. You may respond that his death is more important — it is in anyone's life! That is true, in an existential sense, but the events that lead to one's death at a particular place and time, the cause of one's death, override the importance of one's actual death. Those events are determinative. Without those events, one does not die then and there. If Billy had escaped death on July 14, 1881, and went on to live out more of his life, that escape and not his trial would probably be the most important event of Billy's life.

The information presented here has been unknown until now. This book makes it possible to answer these previously unanswerable questions:

- What were the governing Territorial laws?
- What were the charges against Billy?
- Was there a trial transcript and what happened to it?
- What kind of defense did Billy present?
- Did Billy testify in his own defense?
- Did Billy have witnesses standing for him?
- Who testified against him for the prosecution?
- What was the jury like?
- What action by the trial judge virtually guaranteed his conviction?
- What legal grounds did he have to appeal his verdict?
- Was the trial fair?

Supplementing the text are 132 photos, including many photos never published before.

Available in both paperback and hardcover.

Paperback, 254 Pages, ISBN 978-1-952580024
Hardcover, 254 Pages, ISBN 978-1-952580048

2022 Will Rogers Medallion Award Finalist for Excellence in Western Media
2022 Pasajero Del Camino Real Award Winner.

"Dirty Dave" Rudabaugh, Billy the Kid's Most Feared Companion

This book is about David Rudabaugh, a man whose life is both obscure and wildly mythologized.

Rudabaugh's obscurity begins with the spelling of his surname. In the only U.S. census in which he appears, his name is spelled Radenbaugh. His father, in his Civil War service record, spelled his name Rodenbaugh. Rodebaugh is the spelling David uses in his confession to an attempted train robbery. Yet, throughout his life, he answered to Rudabaugh, and was usually referred to as such in newspaper accounts and legal documents.

An oft-repeated myth about Rudabaugh is that he was a *"nasty, treacherous bully"* who *"stole and killed and brutalized people... Dirty Dave would try anything, as long as it was crooked."* Not true. Another fictitious accusation is that Rudabaugh shot a jailer in cold blood. The true account of jailer Antonio Lino Valdez's fatal shooting is presented for the first time in this book, based on the never-before-published trial transcript. The unquestionable trial evidence shows that it was another man who shot the ill-fated jailer, not Rudabaugh.

Following the jailer's killing, Rudabaugh fled. Now a wanted man, Rudabaugh teamed up with Billy the Kid and participated prominently in Billy's final gun battles with authorities. Famously, Rudabaugh was captured along with Billy at Stinking Springs by Deputy Sheriff Pat Garrett and his posse. After his capture, Rudabaugh was tried for Valdez's killing and sentenced to death by hanging. He escaped jail and went to Mexico.

On February 18, 1886, Rudabaugh was killed by a Winchester rifle shot to the chest in Parral, Mexico, by a grocery man named José. Following his killing, Rudabaugh was decapitated by José. His head was placed on a pole and paraded around the Parral plaza. Present at Rudabaugh's beheading was Albert W. Lohn, a nineteen-year-old photographer.

Lohn took four photographs of Rudabaugh's decapitated head. The two negatives he printed were confiscated by Mexican authorities. The other two negatives remained in Lohn's files for 57 years, entirely forgotten by him. The story of how these two negatives were acquired by an avid collector of Western memorabilia is given in the book.

Paperback, 194 Pages, ISBN 978-1-952580-20-8
Hardcover, 194 pages, ISBN 978-1-952580-21-5

Made in United States
Orlando, FL
04 November 2026

90204433R00155